Dynamics, Challenges, and Perspectives for Sustainable Forest Management

Said Moukrim
https://orcid.org/0000-0002-3172-3776
Faculty of Sciences, Mohammed V University in Rabat, Morocco

Said Lahssini
https://orcid.org/0000-0003-1986-8240
National School of Forest Engineers, Morocco

IGI Global
Scientific Publishing
Publishing Tomorrow's Research Today

Vice President of Editorial	Melissa Wagner
Director of Acquisitions	Mikaela Felty
Director of Book Development	Jocelynn Hessler
Production Manager	Mike Brehm
Cover Design	Jose Rosado

Published in the United States of America by
IGI Global Scientific Publishing
701 East Chocolate Avenue
Hershey, PA, 17033, USA
Tel: 717-533-8845 | Fax: 717-533-7115
Website: https://www.igi-global.com E-mail: cust@igi-global.com

Library of Congress Cataloging-in-Publication Data

Names: Moukrim, Said, 1977- editor | Lahssini, Said, 1975- editor
Title: Dynamics, challenges, and perspectives for sustainable forest management / edited by Said Moukrim, Said Lahssini.
Description: Hershey, PA : IGI Global Scientific Publishing, [2026] | Includes bibliographical references and index. | Summary: "This book offers a multidisciplinary examination of forest ecosystems, detailing their biodiversity, pressures, and management needs. It integrates climate impacts, participatory approaches, and emerging technologies, providing a much-needed global reference for researchers, policymakers, and forestry professionals committed to sustainable forest management"-- Provided by publisher.
Identifiers: LCCN 2026002816 (print) | LCCN 2026002817 (ebook) | ISBN 9798337362397 h/c | ISBN 9798337362403 s/c | ISBN 9798337362410 ebook
Subjects: LCSH: Forest management--Mediterranean Region | Forest conservation--Mediterranean Region | Forest ecology--Mediterranean Region | Forests and forestry--Mediterranean Region | Sustainable forestry--Mediterranean Region | Forest microclimatology--Mediterranean Region | Nature--Effect of human beings on--Mediterranean Region
Classification: LCC SD217.M44 D96 2026 (print) | LCC SD217.M44 (ebook) | DDC 634.9/2091822--dc23/eng/20260309
LC record available at https://lccn.loc.gov/2026002816
LC ebook record available at https://lccn.loc.gov/2026002817
British Cataloguing in Publication Data
A Cataloguing in Publication record for this book is available from the British Library.

Editorial Advisory Board

Table of Contents

Detailed Table of Contents

Section 1
Sub-Mediterranean Forests: Characteristics and Dynamics

Establishes the ecological and institutional foundations of the topic

The forest ecosystems of Morocco present a great diversity shaped by the interactions between biotic and abiotic factors. Their management has moved from a traditional community system to a productivist model established during the French Protectorate and today ensured by the Moroccan Water and Forests administration. This model has generalized the regular high forest and simple coppice, whose prolonged application has reduced the ecological resilience of stands in the face of growing climatic and biotic disturbances. Moreover, since the 1990s, a paradigm shift has been occurring in Europe, favoring other models such as irregular high forest and continuous cover forestry, which present increased resilience and support superior structural biodiversity. In Morocco, and in the context of climate change, their application requires the integration of three dimensions: control of grazing pressure to secure natural regeneration, human development at the level of forest territories, and adoption of adaptive management with scientific monitoring.

Mediterranean forest issues are increasingly framed through the lens of governance. International reports often portray forest degradation as a structural feature of so-called developing regions, particularly in Africa and South America, promoting participatory governance as a normative response to environmental crises. This paper critically examines forest governance in Mediterranean contexts by questioning the modern representation of nature underlying dominant policy frameworks. Governance principles- participation, transparency, sustainability, equity, and accountability- are analyzed through the notion of a loss of the object, understood as the shift from an appropriated to an administered nature. This procedural normativity reveals conceptual and political limits that weaken the analytical relevance of governance models, especially in the southern Mediterranean. Participation is further examined as a dispositif marked by hybridized social solidarities and expanding power relations that obscure the socio-political nature of forest conflicts.

In North African countries (Morocco, Algeria, Tunisia), silvopastoral ecosystems display high diversity shaped by biotic and abiotic interactions, yet face critical challenges from overgrazing pressure by millions of small ruminants on declining forest areas, resulting in forage deficits and widespread desertification. Silvopastoral management has evolved from traditional community-based systems, dominant until the colonial period, through a productivist model introduced during the French Protectorate, toward contemporary integrated participatory frameworks. This multifunctional approach combines technical prescriptions, rotational grazing, controlled stocking rates, fodder shrub plantations, and assisted natural regeneration, with organizational measures involving local communities through pastoral associations. This approach integrates climate adaptation, ecosystem services valuation, and emerging technologies for real-time pasture monitoring, offering promising pathways toward sustainable silvopastoral management facing global change challenges in North Africa.

Section 2

Ecosystem Services, Biodiversity, and Climate Mitigation

Focuses on the ecological functions of forest ecosystems and their contribution to ecosystem services, biodiversity conservation, and climate change mitigation

Mediterranean temporary ponds in Morocco are essential components of forest ecosystems, characterized by a seasonal alternation between flooded and dry phases driven by Mediterranean climate. Their hydrological functioning strongly influences the structure, composition, and dynamics of biological communities, including numerous rare, endemic, and threatened species. These ponds maintain close functional relationships with surrounding forest formations, jointly contributing to hydrological regulation, carbon storage, soil protection, and biodiversity maintenance. Their distribution covers a wide bioclimatic gradient, from the Rif to the High Atlas, with high concentrations in coastal forest areas. However, temporary ponds and

Climate change increasingly affects forest ecosystems, but forests remain key components of climate mitigation through carbon sequestration. More than half of terrestrial carbon is stored in forest soils and biomass, making forests strategic carbon sinks. This review presents a national-scale synthesis of carbon stocks in Moroccan forest ecosystems, integrating biomass with soil organic carbon. It is based on a structured analysis of published studies. The results highlight strong variability among ecosystems, with soil organic carbon representing the largest carbon pool, particularly in Mediterranean and semi-arid forests. The highest average carbon stocks were observed in zeen oak and cedar forests (197.88 and 187.16 tC ha^{-1}), followed by holm oak (134.06 tC ha^{-1}), while cork oak ecosystems showed lower values (76.08 tC ha^{-1}). These differences mainly reflect species composition and soil properties. Overall, Moroccan forests show a high potential for carbon sequestration, supporting their role in climate change mitigation and sustainable forest management.

Eucalyptus plantations are an important component of Morocco's forest landscapes. They provide numerous socioeconomic benefits and alleviate pressure on natural forests. However, their contribution to carbon sequestration remains poorly documented. This chapter assessed the carbon sequestration capacity of five Eucalyptus clones of different ages in the Gharb region in Morocco. Destructive sampling of 20 trees per clone–age combination was conducted to quantify carbon stocks in tree biomass, litter, and soil through laboratory analyses, and to develop clone-specific allometric models for aboveground carbon estimation. Mean biomass carbon fractions ranged from 53.60 ± 1.27% to 54.96 ± 1.80%, exceeding commonly applied default values. Aboveground carbon stocks varied from 3.86 to 32.87 t C ha^{-1}, while soil represented the largest carbon pool (37.67–74.70 t C ha^{-1}). These findings highlight strong clonal variability and emphasize the need for informed clonal selection and management to enhance climate mitigation strategies through carbon sequestration.

Explores technological and methodological innovations that are transforming forest monitoring and governance

Remote sensing and artificial intelligence are transforming forest monitoring and governance in Morocco's climate-vulnerable Mediterranean ecosystems. This literature review synthesizes advances in machine learning, satellite imagery, and geospatial analysis for monitoring deforestation, carbon stocks, wildfire risk, and land degradation across Morocco's 9 million hectares of forest. Through systematic

Foreword

The forests of the Mediterranean basin have fascinated naturalists, ecologists, and geographers. Their biodiversity, shaped by millennia of climatic oscillation, geological upheaval, and intimate coevolution with human activities, makes them among the most scientifically rich and historically significant terrestrial ecosystems on the planet. The forests of the southern shore of the Mediterranean, stretching across Morocco, Algeria, Tunisia, and Libya, that have remained most persistently underrepresented in the international scientific literature. These sub-Mediterranean forests, dominated by cedar, cork oak, holm oak, argan, and Aleppo pine, prevails within ecological and socioeconomic conditions that differ profoundly from those on the northern counterparts. They are facing challenges which, in the context of climate change and human pressure, call for a science that is as distinctive as the ecosystems themselves.

For several decades, researchers and foresters working in North Africa have accumulated a body of knowledge and good practices related to the peculiar dynamics of Atlas cedar under drought stress, to the delicate balance that sustains cork oak systems in the face of pastoral pressure, to the hydrological role of forest cover in the watersheds that supply water to millions of downstream inhabitants. Much of this knowledge has circulated in national reports, in French-language journals, and through the informal networks of practitioner communities, but it has rarely found its way into the international scientific conversation with the prominence the field deserves. The volume edited by Said Lahssini and Said Moukrim represents a step toward bridging that gap. It brings together rigorous and practice-oriented contributions on the dynamics, challenges, and management of sub-Mediterranean forests.

What distinguishes this volume from previous efforts is its architecture -not merely a multi-authored compilation of chapters juxtaposed without regard for thematic coherence. The four sections of the book follow a logic that will be immediately recognizable to any forest practitioner: from understanding, through diagnosis, to management response, and finally to the technological tools that are redefining what management science can deliver. This structure reflects not only editorial rigor but a deep familiarity with the realities of forest management in the field, an awareness that knowledge without application is incomplete, and that the most sophisticated modelling framework is of limited value if it cannot be translated into a decision that a forester can make on a field.

The integration of traditional ecological knowledge with the most advanced computational methods (remote sensing, machine learning, spatial big data), reflects a broader shift that is reshaping forest science globally. The chapters on artificial intelligence and explainable machine learning illustrate an approach to technological innovation that keeps the practitioner, not the algorithm, at the center of the analysis, an orientation the field urgently needs as the proliferation of remote sensing datasets and algorithmic tools risks outpacing the capacity of institutions to absorb and apply them wisely. This volume is the first of two companion works. While the present volume addresses forest management, governance, ecosystem services, and technological innovation, the second, Forest Vulnerability and Climate Risk Assessment: Dynamics, Disturbances, and Resilience, focuses on forest vulnerability assessment, disturbance regimes, and climate risk dynamics. Together, they are designed to offer a comprehensive scientific reference for sub-Mediterranean forest ecosystems, spanning from the biophysical understanding of their functioning to the institutional and technical dimensions of their stewardship.

I have had the privilege of working alongside Mediterranean forest scientists for much of my career, and I am convinced that this book stands as a fitting synthesis of their collective effort. It will advance our understanding of the dynamics, vulnerabilities, and management options of sub-Mediterranean forest ecosystems, and I hope it will serve as a foundational reference for the next generation of forest scientists, practitioners, and policymakers engaged in the consequential work of sustaining these ecosystems through the challenges ahead. The forests of the Southern Mediterranean have endured a great deal over the millennia. With the scientific knowledge assembled in this volume, and the management wisdom it is designed to inform, there is good reason to believe they can endure what is coming.

Riccardo Valentini is a Full Professor of Forest Ecology at the University of Tuscia. He was lead author of the third and fifth reports of the IPCC (Intergovernmental Panel on Climate Change) on climate impacts and adaptation and author of the IPCC special report "Land and Climate". Member of the European Academy of Sciences and Medal of the Academy of Sciences of the "XL" for Physics and Natural Sciences. He dedicated his research to understand forest role on global biogeochemical cycles with a special focus on Mediterranean forests.

Riccardo Valentini

University of Tuscia, Italy

Preface

INTRODUCTION

The forests of the Southern Mediterranean occupy a unique and precarious position in the global landscape of terrestrial ecosystems. Shaped over millennia by the interplay of climate, geology, fire, and human activity, sub-Mediterranean forests sustain a wide range of ecological functions. They regulate water cycles, stabilize soils, sequester carbon, shelter remarkable biodiversity, and support the livelihoods of millions of rural households. Yet these forests are also among the most threatened on the planet. Accelerating climate change, mounting anthropogenic pressures, and the chronic underfunding of conservation institutions are driving degradation processes that in many areas have already crossed critical thresholds. For scientists, land managers, and policymakers working in the region, the challenge is both urgent and shared: to understand these dynamics more rigorously, to develop management responses commensurate with the scale of the threat, and to equip practitioners with the most advanced tools available.

The idea for this volume emerged from years of collaborative research, field observation, and academic exchange among forest scientists and practitioners working across Morocco and the broader Mediterranean basin. Through this work, a persistent gap became evident: the absence of a comprehensive English-language synthesis devoted specifically to sub-Mediterranean forest systems, their ecology, their vulnerabilities, the pressures they face, and the scientific and institutional resources available for their management. Much of the foundational scholarship on these forests has historically been produced in French or contained within national technical reports with limited international circulation. As a result, a substantial body of knowledge of both scientific depth and practical relevance has remained fragmented and relatively inaccessible to the wider research community. At the same time, practitioners in the region have often faced difficulties in engaging fully with methodological and technological advances emerging from the global forest science community. This book is conceived as a bridge across that divide, appearing at a moment when international attention to forest governance, climate adaptation, and biodiversity conservation in dryland and semi-arid environments has never been more pronounced. It is the first of two companion volumes: while the present volume addresses forest management, governance, ecosystem services, and technological innovation, the second volume focuses on forest vulnerability assessment, disturbance regimes, and climate risk dynamics.

The chapters gathered here reflect the state of the art in forest science as applied to sub-Mediterranean and North African contexts. They are the work of researchers from multiple institutions and disciplinary traditions, encompassing ecology, climatology, soil science, remote sensing, artificial intelligence, silviculture, ecological economics, and environmental governance. The structure of the volume follows a deliberate logical progression. The opening chapters establish the ecological foundations of the forest systems themselves. Subsequent contributions examine the threats and degradation processes currently shaping them. Later sections explore management responses and restoration strategies, while the final chapters focus on the digital and computational tools that are redefining contemporary forest science and governance. This structure reflects not only the organization of the book but also the logic of evidence-based forest management: sound governance must underpin ecological management; ecological management must generate measurable ecosystem services; and the monitoring of those services increasingly depends on advanced analytical capabilities.

Sub-Mediterranean forests today face an interlocking set of challenges whose combined effect is more severe than any single factor considered in isolation. Climate change is altering the environmental envelope within which these ecosystems have evolved over thousands of years. Rising temperatures, shifting precipitation regimes, and more frequent and intense droughts are already reducing the vitality of key species, disrupting natural regeneration, and increasing disturbance risks across large portions of the Mediterranean basin (Giorgi & Lionello, 2008; IPCC, 2021). Scientific projections consistently identify the region as one of the most climate-vulnerable areas on Earth, with substantial increases in mean annual temperature and significant declines in precipitation expected over the course of the century. The consequences are increasingly visible in the field: widespread forest decline linked to accelerating aridity (Miranda et al., 2023), emerging patterns of drought-induced tree mortality, and growing uncertainty regarding the long-term stability of some emblematic forest species.

These climatic pressures are compounded by long-standing anthropogenic stresses. In Morocco and throughout much of North Africa, rural communities have historically depended on forests for fuelwood, grazing, and a variety of non-timber forest products. While national policies have made important progress in regulating access and reducing open-access exploitation, the interface between forest ecosystems, agriculture, and pastoralism remains complex and often fragile. Overgrazing continues to compromise regeneration in many forests stands, while agricultural expansion fragments landscapes and reduces the ecological connectivity required for species migration and climate-driven range shifts. In many instances, institutional frameworks governing forest management, some of them inherited from earlier administrative systems, struggle to reconcile ecological sustainability with contemporary socioeconomic realities (Agrawal et al., 2008). These pressures make the governance, management, and monitoring dimensions addressed in this volume not merely academically interesting but operationally urgent.

Land degradation represents one of the most insidious aspects of this dynamic. In the steep terrain of the Rif, the Middle Atlas, and the High Atlas, soil erosion in degraded forest catchments can reach levels that threaten not only forest productivity but also the hydrological services upon which downstream agricultural systems depend. The loss of soil is in many respects irreversible: it is the loss of the biophysical foundation upon which forest regeneration and long-term ecosystem stability ultimately depend.

Wildfire constitutes another increasingly significant challenge. Although fire has historically played a natural ecological role in many Mediterranean ecosystems, current conditions, marked by climate-driven fuel desiccation, landscape homogenization associated with land abandonment in some regions and overgrazing in others, and insufficient early detection infrastructure, have contributed to a wildfire regime of growing intensity and frequency. This evolving pattern poses serious risks not only to ecosystems and rural communities but also to the substantial investments made in afforestation and reforestation across the region.

Responding effectively to these challenges requires a genuinely interdisciplinary approach. Ecological understanding, management innovation, institutional reform, and technological advancement must be integrated into a coherent framework capable of supporting adaptive forest governance. The contributors to this volume approach these issues from diverse perspectives, yet they share a common conviction: that the sustainable management of sub-Mediterranean forests remains achievable if science, policy, and practice can be more effectively aligned.

Recent advances in ecological modelling, species distribution analysis, and forest dynamics simulation have significantly improved our ability to anticipate how forest systems may respond to climate change and land-use pressures (Moukrim et al., 2019; 2022). At the same time, innovative management approaches, such as integrated silvopastoral systems, offer promising pathways for reconciling productive land use with conservation objectives in North African landscapes (FAO, 2022). Economic valuation frameworks are increasingly capable of capturing the full spectrum of forest ecosystem services, including carbon sequestration, water regulation, biodiversity conservation, and cultural value (Costanza et al., 1997). Such frameworks provide powerful arguments for investing in forest protection and restoration.

Meanwhile, rapid technological progress is transforming the tools available to forest scientists and practitioners. Remote sensing platforms, drone-based surveys, LiDAR, and high-resolution satellite imagery, combined with machine learning and artificial intelligence, are enhancing our capacity to monitor forest conditions, detect disturbances, estimate biomass, and assess wildfire risk. One of the central questions addressed in this book is how these capabilities can be translated into operational tools that are accessible to managers and embedded within governance systems capable of converting scientific insight into practical decision-making.

Ultimately, safeguarding the future of sub-Mediterranean forests will require sustained investment in research, institutional capacity, and the training of a new generation of forest professionals. It will also require governance frameworks capable of aligning local livelihoods, national policies, and global environmental commitments. This volume does not claim to resolve all of these challenges. Its ambition is more modest, but no less important: to assemble a scientifically rigorous, practically oriented, and accessible body of knowledge for those engaged in the study and stewardship of these forests.

We hope that this book will serve as a resource for researchers, practitioners, students, and policymakers alike. More importantly, we hope it contributes we hope it contributes to the collective effort to ensure that the forests of the Southern Mediterranean continue to sustain both ecological integrity and human communities in the decades to come.

ORGANIZATION OF THE BOOK

Section I

From Paradigms to Practice: Governance and Integrated Management

The opening section of the volume establishes the conceptual, ecological, and institutional foundations necessary for understanding the dynamics of sub-Mediterranean forest systems. The chapters gathered here examine the historical evolution of forest management paradigms, the governance structures that shape forest use and conservation in the Southern Mediterranean, and the role of integrated land-use systems, particularly silvopastoral practices, in reconciling ecological sustainability with rural livelihoods.

In **Chapter 1**, Aissa Chkhichekh traces the evolution of forest ecosystem management paradigms from traditional production-oriented approaches toward contemporary models grounded in sustainability, ecosystem resilience, and adaptive governance. The chapter situates forest conservation within a broader historical trajectory of policy and scientific thought, highlighting how shifting environmental pressures, particularly climate change, require new frameworks capable of integrating ecological processes, social demands, and long-term resource stewardship. By examining both theoretical developments and practical management perspectives, the chapter provides a conceptual foundation for the sustainable management strategies explored throughout the remainder of the volume.

Chapter 2, by Kamal Menzou, addresses the complex issue of forest governance in the Southern Mediterranean, with particular attention to the role of participation in sustainable development. The chapter analyzes institutional arrangements that govern forest resources and evaluates the degree to which participatory governance models have been incorporated into management practices. Through an examination of policy frameworks and local management experiences, the author highlights both the opportunities and challenges associated with involving rural communities in forest decision-making processes. The chapter ultimately underscores that effective forest governance requires institutional innovation capable of balancing state authority, community participation, and ecological sustainability.

In **Chapter 3**, Mustapha Naggar explores silvopastoral management systems and their potential role in participatory forest management across North Africa. Recognizing that livestock grazing has historically been an integral component of rural livelihoods in Mediterranean landscapes, the chapter examines how integrated forestry-pastoral systems can be designed to support both ecological regeneration and socio-economic resilience. Drawing on regional experiences, the author discusses management approaches that combine forest conservation with controlled grazing practices, community engagement, and local knowledge. The chapter demonstrates that silvopastoral systems, when carefully managed, can serve as a practical pathway toward more inclusive and sustainable forest management strategies.

Section II

Ecosystem Services, Biodiversity and Climate Mitigation

The second section of the book shifts the focus toward the ecological functions of sub-Mediterranean forest ecosystems and their contribution to biodiversity conservation, ecosystem services, and climate change mitigation. The chapters explore a range of topics, from the ecological significance of small and often overlooked habitats to the economic valuation of forest services and the role of forests in carbon sequestration and climate policy.

In **Chapter 4**, Mohammed El Madihi and colleagues examine temporary ponds within southern Mediterranean forest landscapes, ecosystems that, despite their small spatial extent, play a disproportionately important role in maintaining biodiversity and ecological functioning. The chapter provides a detailed overview of the ecological characteristics of these seasonal wetlands, their hydrological dynamics, and the unique plant and animal communities they support. At the same time, the authors draw attention to the growing threats facing these habitats, including land-use change, hydrological alteration, and climate variability. By highlighting their ecological significance, the chapter makes a compelling case for the inclusion of temporary ponds in forest conservation strategies.

Chapter 5, authored by Hamed Daly Hassen and colleagues, addresses the economic valuation of forest ecosystem services, focusing particularly on cultural and environmental benefits in South and East Mediterranean contexts. The chapter reviews methodological approaches used to estimate the economic value of non-market ecosystem services and illustrates how such valuation can inform policy decisions and forest management strategies. Through case studies and comparative analysis, the authors demonstrate how economic assessment tools can help decision-makers better understand the broader societal benefits of forests, strengthening the case for conservation and sustainable management investments.

In **Chapter 6**, Mohamed El Mderssa and collaborators present a comprehensive synthesis of carbon sequestration in Moroccan forest ecosystems, integrating data on both biomass and soil carbon stocks. The chapter highlights the significant role that Mediterranean forest systems can play in climate change mitigation and adaptation strategies. By analyzing carbon storage across different forest types and ecological conditions, the authors provide valuable insights into the potential of Moroccan forests to contribute to national and international climate objectives. The chapter also underscores the importance of improved monitoring and management practices to enhance carbon sequestration capacity in these ecosystems.

Chapter 7, by Saliha Diarra and colleagues, focuses specifically on the carbon sequestration potential of eucalyptus plantations established on restored sub-Mediterranean lands in Morocco. Through an assessment of different eucalyptus clones, the chapter evaluates growth performance, biomass accumulation, and carbon storage capacity in restoration contexts. The findings contribute to ongoing debates about the role of fast-growing species in climate mitigation and landscape restoration strategies. While acknowledging ecological considerations associated with plantation forestry, the chapter demonstrates that carefully selected and managed eucalyptus clones may offer significant opportunities for carbon sequestration and land rehabilitation in degraded environments.

Section III

Monitoring and Innovation in Forest Management

The final section of the book examines the technological and methodological innovations that are transforming forest monitoring, analysis, and governance. As digital tools become increasingly integrated into environmental management, advances in remote sensing, artificial intelligence, and geospatial analysis are opening new possibilities for understanding and managing forest ecosystems at multiple scales.

In **Chapter 8**, Ali Azedou explores the emerging role of remote sensing technologies and artificial intelligence in forest governance in Morocco. The chapter reviews recent advances in satellite observation, machine learning, and geospatial data analysis and discusses how these technologies are contributing to what is increasingly described as "Forestry 4.0 and 5.0." By integrating digital monitoring tools with forest management systems, these innovations offer the potential to significantly enhance the accuracy, efficiency, and responsiveness of forest governance. The chapter highlights practical applications in forest inventory, wildfire monitoring, and ecosystem assessment, illustrating how technological transformation is reshaping the practice of forestry.

In **Chapter 9**, Abderrazak Bannari and his co-authors present an innovative application of drone-based multispectral remote sensing and machine learning techniques to map carob tree dieback associated with damage caused by Barbary macaques. Using UAV-derived data, biomass and chlorophyll indices, and advanced classification algorithms, the study demonstrates how high-resolution geospatial technologies can be employed to detect and monitor forest health disturbances with remarkable precision. The chapter exemplifies the growing role of digital tools in diagnosing ecosystem stress and supporting timely management interventions, highlighting the potential of these technologies to improve forest monitoring and conservation strategies in complex Mediterranean landscapes.

CLOSING REMARKS

The chapters assembled in this volume collectively underscore a central reality: the future of sub-Mediterranean forests will be determined not by a single discipline, policy instrument, or management technique, but by the capacity of the scientific and policy communities to integrate knowledge across ecological, social, economic, and technological domains. These forest systems occupy landscapes where environmental processes and human livelihoods have been intertwined for centuries. As a result, their conservation and sustainable management demand approaches that are at once scientifically rigorous, institutionally adaptable, and socially grounded.

Several themes emerge clearly across the contributions presented in this book. First, the ecological fragility of sub-Mediterranean forests is becoming increasingly evident in the context of accelerating climate change. Rising temperatures, intensifying drought cycles, and changing disturbance regimes, particularly wildfire, are altering ecosystem dynamics in ways that challenge long-standing assumptions about species resilience and forest stability. At the same time, the chapters devoted to biodiversity, ecosystem services, and carbon sequestration remind us that these forests remain among the most valuable ecological assets of the Mediterranean region. Their capacity to regulate water resources, conserve biodiversity, store carbon, and sustain rural economies gives them a central role in both regional sustainability and global climate mitigation efforts.

Second, the book highlights the importance of governance and institutional innovation. Forest management in the Southern Mediterranean cannot be understood solely as a technical exercise; it is equally a question of social organization, public policy, and the equitable distribution of resources. The integration of participatory approaches, the recognition of local knowledge systems, and the development of governance frameworks that reconcile conservation objectives with rural livelihoods are essential conditions for long-term success. Without such institutional alignment, even the most sophisticated ecological or technological solutions are unlikely to achieve lasting impact.

Third, the volume illustrates the growing importance of technological innovation in contemporary forest science and management. Advances in remote sensing, geospatial analysis, and artificial intelligence are rapidly transforming our capacity to monitor forest conditions, detect disturbances, and model ecosystem dynamics. These tools offer unprecedented opportunities for evidence-based management and adaptive governance. Yet their effectiveness will ultimately depend on the extent to which they are integrated into practical management systems and made accessible to practitioners working on the ground.

Taken together, the contributions presented here suggest that the sustainable management of sub-Mediterranean forests requires a shift toward what might be described as a truly integrated forest knowledge system, one that connects ecological research, economic valuation, community participation, and digital monitoring technologies within a coherent framework for decision-making. Such integration will not emerge automatically; it requires sustained investment in research infrastructure, training, institutional capacity, and international collaboration.

As editors, we hope that this volume contributes to that broader effort. Our aim has been to assemble a body of work that not only synthesizes current knowledge but also stimulates further inquiry, dialogue, and innovation. The diversity of disciplines represented in these chapters reflects the diversity of challenges facing sub-Mediterranean forests, but also the richness of the scientific and professional communities committed to their stewardship.

The Southern Mediterranean stands at a critical ecological juncture. Decisions taken in the coming decades, regarding land use, climate adaptation, forest governance, and conservation investment, will shape the trajectory of these ecosystems for generations. If informed by sound science, inclusive governance, and forward-looking management strategies, those decisions can still ensure that sub-Mediterranean forests remain resilient landscapes capable of sustaining both biodiversity and human well-being.

Said Moukrim
Mohammed V University in Rabat, Morocco

Said Lahssini
National School of Forest Engineers, Salé, Morocco

REFERENCES

Agrawal, A., Chhatre, A., & Hardin, R. (2008). Changing governance of the world's forests. *Science, 320*(5882), 1460–1462. DOI: 10.1126/science.1155369 PMID: 18556552

Costanza, R., d'Arge, R., de Groot, R., Farber, S., Grasso, M., Hannon, B., Limburg, K., Naeem, S., O'Neill, R. V., Paruelo, J., Raskin, R. G., Sutton, P., & van den Belt, M. (1997). The value of the world's ecosystem services and natural capital. *Nature, 387*(6630), 253–260. DOI: 10.1038/387253a0

FAO. (2022). FAO Forestry Paper: Vol. 187. *Grazing with trees: A silvopastoral approach to managing and restoring drylands*. Food and Agriculture Organization of the United Nations., DOI: 10.4060/cb9678en

Giorgi, F., & Lionello, P. (2008). Climate change projections for the Mediterranean region. *Global and Planetary Change, 63*(2–3), 90–104. DOI: 10.1016/j.gloplacha.2007.09.005

IPCC. (2021). *Climate Change 2021: The Physical Science Basis*. Cambridge University Press., DOI: 10.1017/9781009157896

Messier, C., Puettmann, K., Chazdon, R., Andersson, K. P., Angers, V. A., Brotons, L., Filotas, E., Tittler, R., Parrott, L., & Levin, S. A. (2015). From management to stewardship: Viewing forests as complex adaptive systems in an uncertain world. *Conservation Letters, 8*(5), 368–377. DOI: 10.1111/conl.12156

Miranda, A., Lara, A., Altamirano, A., Di Bella, C., González, M. E., & Camarero, J. J. (2023). Widespread synchronous decline of Mediterranean-type forest driven by accelerated aridity. *Nature Plants, 9*(11), 1810–1817. DOI: 10.1038/s41477-023-01541-7 PMID: 37845335

Moukrim, S., Lahssini, S., Rhazi, M., Alaoui, H. M., Benabou, A., Wahby, I., El Madihi, M., Arahou, M., & Rhazi, L. (2019). Climate change impacts on potential distribution of multipurpose agro-forestry species: Argania spinosa (L.) Skeels as case study. *Agroforestry Systems, 93*(4), 1209–1219. DOI: 10.1007/s10457-018-0232-8

Moukrim, S., Lahssini, S., Rhazi, M., Menzou, K., El Madihi, M., Rifai, N., Bouziani, Y., Azedou, A., Boukhris, I., & Rhazi, L. (2022). Climate change impact on potential distribution of an endemic species Abies marocana Trabut. *Ekologia (Bratislava), 41*(4), 329–339. DOI: 10.2478/eko-2022-0034

Acknowledgment

The completion of this collective volume stands as a testament to the dedication, expertise, and collaborative spirit of a community of scholars united by a common commitment to advancing knowledge in this field.

The editors wish to express their profound gratitude to all who contributed to this undertaking. Above all, our deepest appreciation is owed to the chapter authors, whose intellectual rigor, scholarly generosity, and sustained engagement made this work possible. Their contributions reflect not only scientific excellence but also a genuine willingness to share knowledge for the benefit of the wider academic community.

The editors also wish to pay tribute to the reviewers, whose exacting and constructive assessments were essential in upholding the standards of quality, coherence, and scientific integrity that a publication of this nature demands. Many of the authors graciously assumed the additional responsibility of peer review, a dual contribution that speaks to their exemplary commitment to the collective endeavor.

We extend our sincere appreciation to colleagues from various academic departments and research institutions for their support in disseminating the call for contributions and for the encouragement they offered throughout the editorial process.

The editors further wish to acknowledge, with particular gratitude, the **CEMED** "*Centre des Experts "**Mastery**" pour les Études et le Développement*", whose institutional support and steadfast accompaniment throughout this project have been invaluable. Their engagement reflects a commendable commitment to the promotion of research and scholarly excellence.

It is our hope that this volume will serve as a meaningful contribution to its field, and a worthy reflection of the collective effort that brought it into being.

Said Moukrim
Mohammed V University in Rabat, Morocco

Said Lahssini
National School of Forest Engineers, Salé, Morocco

Section 1
Sub-Mediterranean Forests: Characteristics and Dynamics

Chapter 1
Sustainable Management and Conservation of Forest Ecosystems:
Evolution of Paradigms and Adaptation Perspectives

Aissa Chkhichekh
http://orcid.org/0009-0003-8951-4744
National Forestry Engineering School at Sale, Morocco

ABSTRACT

The forest ecosystems of Morocco present a great diversity shaped by the interactions between biotic and abiotic factors. Their management has moved from a traditional community system to a productivist model established during the French Protectorate and today ensured by the Moroccan Water and Forests administration. This model has generalized the regular high forest and simple coppice, whose prolonged application has reduced the ecological resilience of stands in the face of growing climatic and biotic disturbances. Moreover, since the 1990s, a paradigm shift has been occurring in Europe, favoring other models such as irregular high forest and continuous cover forestry, which present increased resilience and support superior structural biodiversity. In Morocco, and in the context of climate change, their application requires the integration of three dimensions: control of grazing pressure to secure natural regeneration, human development at the level of forest territories, and adoption of adaptive management with scientific monitoring.

DOI: 10.4018/979-8-3373-6239-7.ch001

INTRODUCTION

The results of the mapping of the National Forest Inventory (NFI) show that the different Moroccan forest formations, including steppes, cover an area of 9,279,100 ha (NFI, 2024). These spaces present a great diversity of natural Mediterranean and Saharan ecosystems (Benabid, 2000) having known two distinct modes of management: i) a community management practiced for centuries, then ii) a modern management entrusted to the State through the Forest Administration, established with the advent of French colonialism during the second decade of the 20th century (Puyo, 2001).

This paradigm shift has not been without consequences. Modern management, although based on rational principles, at least what is declared, has unfortunately led to a dysfunction of natural ecosystems, a reduction of their areas and a loss in biodiversity, under the combined effect of multiple factors: sociological, anthropological, technical, political and climatic. Conversely, traditional community management, although imperfect, disorderly and sometimes brutal, proved more effective in terms of biodiversity conservation. Indeed, the last Atlas lion (*Panthera leo leo*) and the last panther/leopard (*Panthera pardus pardus*), representing the large felines of Morocco, only disappeared from the wild state during the twentieth century (the first decades of the twentieth century for the lion and 1990 for the panther).

Almost all components of modern forestry in Morocco, such as the delimitation of forests, management studies (of forests, rangelands and watersheds) as well as regulations relating to forests, hunting and fishing, were born and developed during the 20th century.

However, the context has profoundly changed with the massive intervention of man, whether forest technicians (engineers and technicians) or neighboring populations. The actions of technicians resulted in management studies favoring production to the detriment of other objectives, and by fixing withdrawals on the basis of observed production, while deliberately neglecting the impact of illegal uses by local residents. This omission was based on the assumption that illegal withdrawals would be settled by the application, by managers, of the legislation in force (Repression). However, in reality, the neighboring population draws the major part of its needs from forest products, whether woody or non-woody.

Moreover, in order to avoid complications, managers continued to apply classical management methods: regular high forests (on stumps for broadleaves and thuya) and simple coppices. Mediterranean forest species, over time, have developed a genetic adaptation and diversity, through sexual reproduction, allowing them to cope with the complex climatic conditions of the region (Ramírez-Valiente et al., 2022; Aurelle et al., 2022; Fady et al., 2003; Cheddadi et al., 2009; Fady et al., 2022; Matesanz & Valladares, 2014). On the other hand, regeneration based on asexual reproduction (stump sprouts), used in the management methods of regular high forest on stump and simple coppice, as well as regular high forest treated with long rotation, allow the persistence of a stable and non-evolutionary genotype, limiting genetic diversity, therefore, the capacity for adaptation in the face of global changes (climate, diseases, disturbances) (Barrett, 2015; Silvertown, 2008; Reusch & Wood, 2007; Honnay & Bossuyt, 2005; Otto & Gerstein, 2006). The maintenance of said treatment modes only perpetuates ancient genetic material, inherited from several decades, even centuries ago, while climatic conditions are evolving towards marked warming, accompanied by drought and increased parasitic attacks.

Environment specialists and more particularly in forestry of the northern Mediterranean, have already begun to review the management and administration methods of forest spaces. Thus new concepts of modern forest management, particularly in Mediterranean context, and facing the challenges of climate change, have emerged and are becoming more and more concrete on the ground.

The officials who have succeeded at the head of the different structures of the Water and Forests Administration, have, at each era, attempted to implement diversified plans and programs aimed at developing the forestry sector. Among the notable achievements, we can cite the National Reforestation Plan (NRP) (1970), developed by an elite of engineers, national and foreign. A valuable document entitled "practical guide for reforestation in North Africa" was published for the occasion. It was used as a reference by foresters and also served as a basic document for the training of foresters, whether engineers or technicians. The NRP was successively replaced by the Reforestation Master Plan (RMP) (1997), then by the National Forestry Plan (NFP) (1998).

Other programs have also been implemented, such as the National Protected Areas Plan (1996), granting a special status to certain zones in order to strengthen the conservation of forest ecosystems and biodiversity and the National Plan to Combat Desertification (NAP-CCD) (2001) which is a strategic framework put in place to respond to the issues of land degradation, desertification and the impacts of climate change, particularly in arid, semi-arid and dry subhumid zones of the country. These efforts were crowned by the development of the national strategy "Morocco Forests 2020-2030", which traces the major orientations for sustainable management and valorization of national forest heritage. However, halfway through its implementation, this strategy already shows its limits and proves incapable of achieving the fixed objectives.

Moroccan forests are increasingly exposed to risks induced by climate change, notably droughts, water erosion, increasingly frequent and severe forest fires, the emergence of new pests that decimate forest biodiversity, and growing desertification. These difficulties are explained in part by the fact that classical forest management and planning, following approaches inspired and imposed by various international conventions and donors, in reality only prolong failures in terms of conservation of our forest ecosystems. Said approaches were born and developed in a sociocultural context incommensurable with that of the southern Mediterranean. The inadequacy of these imported models, combined with the intensification of climatic and anthropogenic pressures, makes a profound renewal of forest governance practices imperative. In this context, a paradigm shift in the governance of forest, pre-forest, pre-steppe and steppe spaces is more imperative than ever, in order to respond to ecological, socio-economic and climatic challenges.

This chapter focuses on the discussion of perspectives linked to sustainable planning of southern Mediterranean forest spaces, integrating both local specificities and imperatives of resilience in the face of climate changes and anthropogenic pressures. The reflection will focus mainly on the dominant species constituting the different Moroccan forest ecosystems, analyzing the planning modes and silvicultural practices that have been applied to them until now, then proposing emerging, adapted and realistic alternatives, likely to better respond to current and future issues of sustainable management.

APPLIED SILVICULTURE

Tree formations in Morocco cover an area of 6,710,400 ha. These expanses are composed of broadleaved species with 60% including notably holm oak, cork oak, zeen oak, argan tree and Saharan acacias, and coniferous species with 24% composed of cedar, thuya, junipers, pines and fir. Artificial plantations cover 9% of this area while matorrals, generally resulting from degradation of forest formations, occupy the rest, i.e. 7%. As for steppe formations consisting of esparto grass areas, wormwood and rosemary, they cover an area of 2,568,700 ha and are generally concentrated in eastern Morocco (NFI, 2024). This diversity of forest ecosystems has historically been subjected to differentiated silvicultural treatment modes, inherited essentially from Mediterranean European forestry practices.

The management of this diversified forest heritage is based on the application of different silvicultural treatment modes that have been progressively introduced and generalized since the colonial era. These management methods, mainly regular high forest and simple coppice, were conceived and developed in the northern Mediterranean then transposed to Moroccan forests according to the production objectives sought and the biological characteristics of species. The examination of these classical practices allows better understanding of their limits in the face of current challenges and identifying necessary improvement pathways.

If forest management develops strategies that orient the structure of the forest towards a target structure, ideal and theoretical, silviculture, for its part, consists of elaborating stand management scenarios, through a sequence of interventions manipulating the structure of stands with the aim of maintaining the desired composition in forest species in cases of mixtures, obtaining woody products of desired dimensions and ensuring the presence of sufficient regeneration at the required time. The different treatment modes were invented and developed in the northern Mediterranean for centuries then introduced in the countries of the southern Mediterranean; first in Algeria then in Morocco and Tunisia (Puyo, 2001).

The development of forestry has primarily emphasized, over the past two centuries, intensive management operations focused on increasing wood production. This "conventional" forest management approach has generally favored the production of monospecific even-aged stands (Bauhus et al., 2009). However, this situation has evolved over the past three decades, particularly in developed countries, due to a combination of new information and concerns regarding the ecological consequences of intensive forestry practices, as well as a willingness on the part of many forest owners and society to adopt a broader set of management objectives (Puettmann et al., 2015).

Regular High Forest Treatment Mode

Characteristics and Objectives

The regular high forest treatment is one of the classical methods of forest management. It consists of managing a stand where all trees belong to the same age class (or to very close age classes), as opposed to irregular high forest where ages and dimensions are mixed. In Morocco, we speak of high forest when we want to obtain old trees capable of giving certain products (timber, cork, fruits...). Whereas by definition, this regime is linked to reproduction by seeds, calling upon, depending on the case, seeds or sprouts and seeds (high forest on stumps). In regular high forest, regeneration cuts practiced on entire plots give rise to a succession of regular stands. Seedlings and the most beautiful trees are favored by improvement cuts. Cedar, maritime pine, Aleppo pine, zeen oak, mountain cork oak are treated in regular high forest, while plains cork oak, mountain holm oak and argan tree are treated in regular high forest on stumps (generally we dispense with specifying "on stumps").

Forest managers of the past have always favored the treatment of regular high forest, compared to other treatment modes, during forest planning for a certain number of species for several reasons. The development of forestry as a scientific and management discipline over the past two centuries has primarily emphasized intensive management operations focused on increasing wood production. This "conventional" forest management approach has generally favored the production of monospecific even-aged stands (Puettmann et al., 2015). First, regular stands allow obtaining straight, homogeneous stems of large diameter (for example for cedar in Morocco, beech in France...), with a planned production cycle (fixed rotation) that allows better control of volumes and wood qualities. Age regularity simplifies interventions (releases, thinning, regeneration cuts), making operations often mechanizable and less costly. Long-term planning constitutes another major advantage of this treatment mode. By fixing a rotation duration (e.g. 72 years for plains cork oak forests, 100 to 120 years for mountain cork oak forests, 100 to 150 years for cedar, 40 to 50 years for pine), it becomes possible to predict economic profitability and forest yields. Clear cuts followed by reforestation and/or controlled regeneration ensure uniform stand renewal, which also allows introducing or selecting species adapted to economic needs or ecological conditions. Also, adaptation to production forestry was a mode often favored in intensive production forests (conifers (cedar for Morocco), quality oaks), allowing maximizing exploitable wood volumes within known timeframes.

Forest Species Treated in Regular High Forest in Morocco

Cedar (Cedrus atlantica)

Ecological Characterization and Distribution

Atlas cedar is the noblest species of the country. It covers approximately 131,470 ha in the Rif, Middle and High Atlas (NFI, 2024). Morocco constitutes the first reserve (75% of its distribution area) worldwide of this endemic species of northwest Africa, classified in 2017 on the IUCN red list as endangered species. Conifer of first magnitude, it can reach 65 m in height (Benabid, 2000). The slender trunk often has a beautiful shape. The bark is cracked. The youthful crown is conical with a very flexible spire, becoming tabular with age. The branches are horizontal. The trees, having a fairly developed but non-tap root system, are subject to windthrow. The persistent needles are grouped in clusters at the end of a short branch. The cover is quite thick. This is a monoecious species, which flowers at the end of summer to give mature cones only in the third winter. It fruits abundantly from 40 years. The resinous seeds do not keep well outside the cone. They germinate very easily, but seedlings only develop under certain shelter conditions and they do not support vital competition. Cedar is a very long-lived species. Some works have confirmed the presence of individuals having ages close to 800 years (Linares et al., 2011), some of which would presently still be alive.

Cedar is a strictly altitude species, of luminous, humid and cold climate (Terrab et al., 2006). It resists temperatures of -25°C as well as the weight of snow (Quézel & Médail, 2003). It is found in the mountains of the Middle Atlas from 1,600 m and reaches its optimum at 1,800 m, and from 1,500 m in the Rif. Benabid (2000), gives the altitudinal distribution of cedar between 1,300 to 2,600 m, being able to reach 2,800 m in the High Atlas. It is indifferent to the chemical character of the soil, but fresh and loose soils are indispensable for its regeneration.

Structure of Stands and Silvicultural Management

Cedar stands can be classified according to the nature of the parent rock:

- The high forest on basalt (example Sehab forest) develops on deep and fresh soils and constitutes a magnificent stand, very slender and presenting good growth varying between 2 and 5 m^3/ha/year (Quézel & Médail, 2003; Sbay & Zine El Abidine, 1994). It occupies only a very limited area where regeneration poses no problem.
- The high forest on limestone (example Itzer forest) forms a stand dominating an understory of holm oak. The trees are less slender with relatively low growth of the order of 1.5 m^3/ha/year (Sbay & Zine El Abidine, 1994) where regeneration is very irregular. The mixed high forest cedar and holm oak (example Ifrane forest) is an old forest of the preceding type that fires have transformed.

Cedar is obligatorily regenerated by seed and, producing timber, it is therefore treated in regular high forest. The development of cedar seedlings, especially on limestone substrate, is difficult. Ground floor herbaceousness of the cut is to be avoided. This species is regenerated by progressive cuts around the age of 90 years on basalt, 120 to 160 years on limestone. The seed cut is relatively light (1/3); seedlings are released and extended by 1 to 3 secondary cuts. A final cut leaves 10 to 15 seed trees per hectare as insurance against fires. Cuts succeed each other approximately every 10 years. Plots in regeneration are prohibited to grazing during this entire period (20 to 30 years). The success of regeneration depends especially on the condition of the soil which must be loose and fresh. The herbaceous layer is sometimes destroyed on patches by burning slash. Young seedlings are possibly released from holm oak sprouts, then young cedars are thinned at the rotation of 10 to 15 years.

Threats, Valorization and Perspectives

Cedar wood is an excellent quality and fragrant timber, although a bit brittle. Depending on the forests of origin, it has very special qualities. The wood of forests on limestone, with very slow growth, is very homogeneous and particularly appreciated by cabinetmakers. Wood with wider rings is used in joinery, framework and crating. When it is too altered by fungi, it is used as filling (M'Hirit, 1999; Quézel & Médail, 2003). Thinning products, after debarking, gave telephone poles and mine timber. Cedar wood is widely used in local constructions: poles, beams, planks. Distilled, it gives tar (Benabid, 2000). It is a mediocre firewood, used mainly for lighting holm oak wood more prized for these types of uses.

The cedar forest, with its holm oak undergrowth, is quite insensitive to fires. Processionary caterpillars that develop in nests, as well as the cedar tortrix, devour the foliage certain years. The great enemies of cedar are very widespread fungi, the "Mjeuj" (*Phellinus chrysoloma*) and the "Saboune" (*Polyporus officinalis*) which attack springwood and often make the wood unusable. However, the most serious threats today come from climate change and anthropogenic pressure. Since the 1980s, severe droughts have influenced the mortality of Atlas cedar in Morocco (Linares et al., 2011), with massive dieback observed after the 1980s, particularly in the xeric region of the High Atlas (Camarero et al., 2021). Future distribution modeling predict a significant reduction of the current potential distribution area of Atlas cedar under future climate change scenarios (Moukrim et al., 2020). Holocene paleo-ecological data reveal that cedar forests have known complex dynamics of migration and contraction over the last millennia, with forest reorganization processes marked by strong spatial and temporal variability across Moroccan mountain massifs (Muller et al., 2022). In the current context of global warming, the transition of cedar forests towards open and stress-resistant assemblages, through materialization processes, raises urgent conservation concerns, mountain refugia possibly representing the last places for this heat-sensitive conifer and vulnerable to prolonged water stress.

Aleppo pine (Pinus halepensis)

Ecological Characterization and Distribution

Aleppo pine covers an area of approximately 89,800 ha as natural stands distributed between the Rif, Middle Atlas and High Atlas (NFI, 2024), to which are added the areas established by plantations. According to Benabid (2000), Quézel & Médail (2003), Barbero et al. (1998), Aleppo pine is a tree of 2nd magnitude, with a trunk of 6 to 8 m. The crown, conical in youth, spreads with age. The wood contains resin ducts. The bark remains membranous for a long time then cracks on old trees. The rooting is spreading or taproot depending on the depth of the soil. Needles of 5 to 10 cm grouped in pairs, are light green and the cover is light.

Monoecious species, Aleppo pine flowers in spring abundantly and almost every year from 20 years. The cones are only mature in the autumn of the 2nd year and open by dryness the year following maturation while remaining on the tree. The samaras are disseminated within a certain radius by the wind. In Morocco, it is considered as a short-lived species: 120 to 130 years (Benabid, 2000; M'Hirit, 1994). Aleppo pine does not tolerate excess moisture and poorly resists the weight of snow. It is found in the cold semi-arid stage in middle mountains. It is a light-demanding species, indifferent to soil, but prefers marly or limestone soils (Benabid, 2000; M'Hirit, 1994; Quézel & Médail, 2003). The species demonstrates a remarkable capacity for adaptation to arid environments and great growth plasticity.

Structure of Stands and Silvicultural Management

Aleppo pine stands often have the appearance of old high forest of selection form or young regular high forests resulting from fires (Barbéro et al., 1992). Mixed with holm oak, it has been reduced by fires to a few scattered groups but tends each time moreover, to regain ground thanks to its colonizing power. Depending on conditions, the growth in circumference is 1 to 2 cm per year (M'Hirit, 1999). This species is treated in regular high forest by regenerating it around 70-80 years by single cut leaving some standards. Thinning will bring the stand to a density of 150 to 200 subjects per hectare at the time of regeneration (M'Hirit, 1999). Aleppo pine is widely used in reforestation by planting seedlings raised in nursery. Trees at the southern limit of their distribution, as in the Moroccan High Atlas, present reduced growth, low relative resilience and more missing rings in response to increasing temperatures and droughts (Vieira et al., 2022).

Threats, Valorization and Perspectives

Aleppo pine gives wood which, in mountains, can be of a certain quality, but mediocre elsewhere. Trees of large dimensions are rare in Morocco. Thinning products give mine timber and poles. It is not appreciated for local constructions because it is attacked, being attacked by insects (M'Hirit, 1999). This species is particularly sensitive to fires. Wood impregnated with resin, empty cones remaining on the tree and the abundance of live cover, are the factors of this sensitivity (Rigolot, 2004). It is attacked by the fungus Mjeuj (*Phellinus chrysoloma*) and processionary caterpillars (*Thaumetopoea pityocampa*) but without severity. But, weakened trees are easy prey to xylophagous insects such as bark beetles (Lieutier et al., 2004). Despite its vulnerability to fires, Aleppo pine is considered as a “fire-embracer” species that bases its strategy on post-fire regeneration by seedling production (Romero et al., 2023). However, with the observed and predicted increase in the frequency, duration and intensity of droughts, as climate models predict, the recovery capacity of these trees will be compromised, leading to dieback and potential contraction of the southern range of the species (Campelo et al., 2022).

Maritime pine (Pinus pinaster)

Ecological Characterization and Distribution

Maritime pine covers an area of approximately 20,900 ha distributed over the Rif, Middle Atlas and High Atlas. It is represented by two varieties: plains maritime pine (*Pinus pinaster var. iberica*) and mountain maritime pine (*Pinus pinaster var. maghrebiana*) (Benabid, 2000). It is a species of first magnitude (20 to 40 m) having a very straight trunk. The wood contains resin ducts. The bark is very scaly. The crown spreads with age. The rooting is spreading or taproot depending on the depth of the soil. Needles grouped in pairs are long and dark green. They provide quite thick cover. Monoecious species, maritime pine flowers in spring to give large mature cones in the autumn of the 2nd year. Fructification, from 20 years, is frequent and abundant. The cones open by dryness and release large winged seeds and remain adherent to the branches. Maritime pine can live up to 120 – 130 years (Benabid, 2000; M'Hirit, 1999). Light-demanding species, maritime pine requires a certain humidity and resists snow and cold. It is found in mountains in the subhumid and humid stages on limestone and dolomitic limestone substrates (Benabid, 2000).

Structure of Stands and Silvicultural Management

Maritime pine in Morocco forms regular high forests resulting from post-fire regeneration, and older stands can give the appearance of old high forests with selection character (M'Hirit, 1999). Its growth is more active compared to Aleppo pine (2 cm/year in circumference). Management in regular high forest with short rotation (70–80 years) is the classical method in Mediterranean forests for maritime pine, with some standards left for seed production and ensure seeding (Quézel & Médail, 2003).

Threats, Valorization and Perspectives

Maritime pine gives coarse-grained but light timber that can receive many uses, particularly in crating. Young woods give good poles as well as mine timber. Like Aleppo pine, maritime pine is sensitive to fires because of its abundant resin, persistence of serotinous cones and thick vegetation cover that favors fire propagation (Rigolot, 2004; Barbero et al., 1998). However, like Aleppo pine, maritime pine has developed adaptations to fire, allowing effective post-fire regeneration, although this capacity may be compromised under the effect of recurrent droughts and fire regimes altered by climate change.

Cork oak (Quercus suber)

Ecological Characterization and Distribution

It is the species that has the most importance from an economic point of view. It covers approximately 280,000 ha distributed over the Rif, the northwestern coastal zone (Atlantic cork oak forest), the Central Plateau (Oulmes, Sidi Hassine regions) and Bab Azhar (Taza region). The Mamora forest constitutes the largest cork oak forest in one piece in the world (Benabou et al, 2022) and presents the greatest genetic diversity compared to other similar forests of Morocco. Tree of 2nd magnitude (can reach 25 m) with short trunk, cork oak has a globular crown when isolated. It has a developed and taproot system. Growth in circumference is 1.5 to 2 cm per year. The bark is constituted solely by cork. Leaves are simple and persistent, however the tree can be entirely defoliated (drought or attack by *Limantria dispar*). The cover is light (Benabid, 2000). Monoecious species and fruits more or less regularly up to an age of 95 to 100 years. Acorns are ripe in autumn. At the beginning, the seedling mainly develops its root system. It sprouts vigorously from stumps up to an advanced age: 80 years in plains. Sprouts have very rapid elongation during the first years. It is quite long-lived: 150 years on average (M'Hirit, 1999). Cork oak requires heat, humidity and light (Boudy, 1952). Temperatures of -10°C are fatal to it (Gil & Varela, 2008). The species is present in semi-arid (Mamora), subhumid (Harcha), humid (Bab Azhar, Chefchaouen) and even per-humid (Bouhachem) climates (Benabid, 2000). It does not tolerate limestone and clay. Sand, sandstone and schist most often form the subsoil of cork oak forests (Sauvage, 1961).

Structure of Stands and Silvicultural Management

Cork oak stands are different depending on existence conditions and can be divided into two types:

- The plains type (example Mamora) forms pure open stands with diversified and abundant undergrowth, presenting difficulties of regeneration by seed (Lepoutre, 1965). Regeneration by seed being rare, it is mainly ensured by stump sprouts and carried out by single clear cut. The exploitation age is fixed at 72 years. 6 years of grazing prohibition are sufficient for sprouts to become self-defensible, since goats are not authorized there. Around 6 years, cleaning leaves 2 to 4 pruned shoots on each stump; then 3 thinning gradually bring the stand to the density of 200 trees per hectare. Eliminated trees will be first subjects injured at harvest or non-producers of cork. Twelve years before the clear regeneration cut, the plot will be closed again to grazing, in order to allow the development of some seedlings in this stand which is sufficiently open. Valorization is done when the tree reaches 7 dm in girth (26 years), and cork is harvested 5 times at the rotation of 9 years.

- The mountain type constitutes a more compact stand, sometimes mixed with zeen oak or holm oak (Chkhichekh, 1993; Chkhichekh et al., 2015). Cork production requires conserving trees up to an advanced age; also the cork oak forest is treated in regular high forest generally on stump. Regeneration can count on seedlings, without neglecting, however, sprouts. The coming and development of seedlings are ensured by progressive cuts. Around 100 to 120 years, the old high forest is opened by a light seed cut, followed by a final with sometimes a secondary between the two. On average the interval between two cuts is 6 to 10 years. Obviously, these plots in regeneration are prohibited to grazing. Seedlings are released. Thinning have the same characteristics as in plains stands and in addition, they ensure suitable mixing of species. Valorization is done on trees of 7 dm in circumference at 1.30 m in height (at an age of approximately 40 years) and up to 6 harvests are done every 12 years. Recent studies have highlighted the loss of natural regeneration in cork oak stands and climate change could further aggravate the sustainability of this ecosystem (Benabou et al., 2022).

Threats, Valorization and Perspectives

Fire remains the greatest enemy of cork oak. The abundance of undergrowth makes these stands very sensitive. The caterpillars of a butterfly (*Lymantria dispar*) devour the foliage and thus cause losses in wood and cork growth. They can even cause the death of certain trees (Boudy, 1950; Ben Jamâa & Mnara, 2000). A fungus (*Hypoxylon mediterraneum*) destroys the phloem and kills affected subjects. It manifests by black patches in cork cracks (Costa & al., 2020).

Cork is the main product, due to its great value. The wood is heavy and has a tendency to split. It gives good firewood: heating and carbonization. The harvest of edible acorns is tolerated in plots that are not in regeneration. However, progressive degradation results from a combination of climatic and anthropogenic factors that continue to intensify; the lack of natural regeneration being one of the most worrying problems requiring intensification of regeneration programs by involving more and more local populations (Laaribya, 2023).

Argan tree (Argania spinosa)

Ecological Characterization and Distribution

The argan tree is the most original and emblematic species. Due to its unique ecological, economic and social importance, the Argan forest was recognized as a

Biosphere Reserve by UNESCO in 1998. It covers approximately 1,000,000 ha in southwestern Morocco, Souss and Essaouira. Tree of third magnitude being able to reach 10 to 12 m, with short and twisted trunk. It has a globular and dense crown. The light gray bark is cracked in snake skin. The branches are thorny. The rooting is very developed and spreading (Benabid, 2000). Leaves are persistent and fasciculated, small, grouped in clusters. In drought years, certain districts can defoliate during summer (Benabid, 2000). Complete flowers appear in spring. Argan nuts, in drupe form (Achour, 2014) of very variable shapes are only ripe the following year: June to August. Fructification, beginning on 5-year sprouts, is abundant and frequent in argan forests. This species sprouts powerfully from stump up to 200 years, with very rapid height growth in the first years (3 m in 5 years). Its longevity would be more than 250 years. The argan tree develops a deep and spreading root system, often described as spreading, which allows it to anchor solidly in very varied soils, including compact or stony ones. This dense root network contributes strongly to soil stabilization and fight against erosion (Quézel & Médail, 2003; Bani-Aameur & Ferradous, 2001).

It is a species loving heat and fearing frost. The argan tree can withstand high temperatures exceeding 50°C and those below 0°C for short periods. It is content with low rainfall, between 160 and 400 mm (Benabid, 2000). Reputed as thermophilic and xerophilic, it is found in arid and semi-arid bioclimates and little in the Saharan. From an altitudinal point of view, it exists from the sea shore up to 1500 m in the Anti-Atlas (Benabid, 2000).

Structure of Stands and Silvicultural Management

Human action has so marked its stands which are characterized by two main types. The argan forest, a high forest with relatively dense selection appearance with continuous thorny live cover (spurges, thorny broom, etc.). The cultivated argan forest, an incomplete regular high forest with clearings (80 to 100 trees/ha) where argan trees are of large circumferences, with developed and rounded crown, the soil being cleared of live cover and it is cultivated certain years.

Due to the special legal status, these stands are not yet subjected to a well-defined treatment. The first cuts practiced during the war of 1939-1945 allowed determining the following bases: i) Argan nuts constitute the main product that the treatment must favor. This production is maximum around 60 years (Orwa et al., 2009): hence the treatment in regular high forest; ii) Regeneration by seedling being exceptional due to the lack of favorable conditions for seedling development, vegetative multiplication is provoked by a single cut (clear cut) carried out around 140 years; iii) A cleaning removes numerous shoots in the thorny and bushy stump, leaving 3 to 4 on average. This intervention is done at a variable age and which still needs to be specified. Subsequently, 1 or 2 thinning bring the stand to the density of 150 trees/ha (best conditions) around 60 years, time of full fruit production.

Threats, Valorization and Perspectives

This vigorous and robust species in its environment has practically no enemy except man and his herd. Without this providence-tree, a large part of these regions would be uninhabitable. It provides populations with very heavy and homogeneous wood that serves to make everything that is necessary for users (framework, roofing, plow, household objects, etc.), also constituting an excellent firewood and carbonization wood. The almond of argan nuts gives oil of very good quality having several uses (food, cosmetology, pharmaceutical products, etc.), with relatively low argan oil yields, of the order of 3 liters per 100kg of dry fruits (El Yousfi, 1988). The dried pulp and the residue from oil manufacturing are concentrated foods for domestic animals. The argan forest-forest is subjected to grazing which is often excessive and which sometimes threatens its existence. Animals consume nuts, young branches and foliage (Aafi et al., 2007; Bani-Aameur & Ferradous, 2001; M'Hirit et al., 1998). It is the argan tree that allows these numerous herds of goats to get through certain bad years. However, the argan forest faces today growing threats including recurrent drought, overgrazing, demographic pressure and agricultural extension, requiring urgent measures of conservation and sustainable management to preserve this unique ecosystem recognized worldwide for its ecosystem services and its exceptional biodiversity. Also, future distribution modeling predict a significant reduction of the potential distribution area of this species (Moukrim et al., 2019).

Zeen oak (Quercus faginea, Quercus canariensis)

Ecological Characterization and Distribution

It covers approximately 15,900 ha in the Rif and the Middle Atlas, including 7,400 ha of pure stands and 8,500 ha of mixed stands (IFN, 2024). It is a beautiful tree of 1st magnitude (20 to 30 m high), with a slender trunk and finely fissured bark (Benabid, 2000). The rooting is deep (https://mediterraneanforest, 2025). The simple, cut, deciduous and marcescent leaves are wide and constitute a thick canopy (Benabid, 2000). Monoecious tree, fruiting abundantly from 30 years. It resprouts and suckers until 150 years. The height growth of seedlings and especially of sprouts is very rapid. Its longevity exceeds 300 years (https://agris.fao.org, 2025). It requires humidity as well as fresh and deep soil from sea level up to 1,800 m altitude (IFN, 2024) (basalt in the Middle Atlas, Numidian sandstone at Bouhachem and Jbel Allam in the Rif). It has no plasticity.

Structure of Stands and Silvicultural Management

Zeen oak constitutes magnificent dense pure stands or mixed especially with cork oak in the Rif (case of Bouhachem) (Chkhichekh, 1993) and in the surroundings of Ifrane. Elsewhere, it is mixed in patches with cork oak (Chkhichekh, 2015), cedar and holm oak.

Zeen oak stands are treated in regular high forest (generally coppice with standards) by single cut (clear cut) at the age of about a hundred years. Seedlings, sprouts and suckers contribute largely to ensuring its regeneration. Thinning then progressively reduce the density.

Threats, Valorization and Perspectives

Zeen oak is not very sensitive to fires, because its thick canopy drives out the undergrowth. The heavy wood has a beautiful appearance, unfortunately, it warps, splits and dries with difficulty (Mechergui, 2019). Also was it used only for the manufacture of railway sleepers. Thinning products can give mine timber.

Treatment method in Simple *Coppice*

Characteristics and Objectives

The treatment in simple coppice constitutes one of the traditional methods of forest management based on the exploitation of the vegetative regeneration capacity of species. It consists of periodically cutting the trees at ground level, causing the emission of coppice shoots which form a new stand. Unlike regular high forest which relies on regeneration by seeds, simple coppice exclusively exploits vegetative reproduction, allowing shorter rotations and continuous production of service wood, firewood and poles.

In Morocco, this treatment method has been widely adopted for its simplicity of management and its adaptation to species with high resprouting capacity. Forest managers of yesteryear favored simple coppice for several reasons: ease of exploitation with short rotations (45 to 60 years), regular production of small and medium-sized wood meeting local needs, low investment cost since regeneration is naturally ensured by sprouts, and resilience to disturbances (fires, overgrazing) thanks to the vegetative regeneration capacity. This management method also made it possible to maintain forest cover on difficult terrains and to provide fuel to rural populations.

Forest species treated in Simple Coppice in Morocco

Thuya (Tetraclinis articulata)

Ecological Characterization and Distribution

Thuya occupies the fourth place from the point of view of area after holm oak, argan tree and acacias with 694,380 ha. It presents three main blocks: the northern and eastern block; the block of the Central Plateau and the western Meseta, and the block of the western Middle Atlas and the High Atlas (IFN, 2024). Thuya is a conifer of third magnitude (15 to 25 m), with a pyramidal habit in youth and spreads with age. The bark is finely furrowed lengthwise and has resin canals. The rooting is spreading and ensures a solid fixation even on very compact soils. The scale-like leaves are placed on the articulated branches and the canopy is light. Thuya is a monoecious species, fruiting from 15 years in a fairly abundant manner. The small bluish cones with 4 scales open with heat and release a few winged seeds. The harvest and conservation of thuya seed are delicate. It resprouts abundantly and for a long time from stumps. The height growth is slow, as well as that in circumference (M'Hirit, 1999; Benabid, 2000; Quézel & Médail, 2003; Hadjadj & Letreuch Belarouci, 2017).

This tree requires a hot and dry climate; it is a light-demanding species. It is characteristic of the semi-arid mild stage, but it can be found at the limit of the arid (in the south of the country) and the subhumid (Middle Atlas) (Benabid, 2000). Indifferent to soil, but it is often found on more or less warm calcareous soils. Heavy and humid soils do not suit it (clayey bottomlands). Thuya is long-lived: about 400 years (Quézel & Médail, 2003).

Structure of Stands and Silvicultural Management

It forms stands of varied types: i) old coppice with standards, in the state of islets in the coppice (aspect currently increasingly rare); and ii) coppice which is the most frequent, coming from fires or coppicing whose undergrowth varies inversely with density. This coppice can share the ground, depending on exposure, with cork oak (El-Harcha plateau – Oulmès) (Chkhichekh, 2015) or with argan tree (at Tamanar). Thuya gives poles which must be at least 60 cm in circumference (circumference at 1.30 m). For this, it is treated in regular simple coppice at the exploitation age of 50 – 60 years. In order to maintain the soil covered, we seek to maintain a dense stand until 25 to 30 years where a thinning leaves 2 poles per coppice. The height growth being low and these stands being grazed by goats, the protection can last 10 years. To obtain timber (planks) or cabinet-making wood for crafts, 30 to 40 very vigorous reserves are sometimes spared per hectare.

Threats, Valorization and Perspectives

The old thuyas give a beautiful red timber with fine grain. It is however very brittle. Thuya burl is constituted by the collar of old thuyas mutilated by fire and by repeated cuts. It is highly sought after by artisans (Essaouira). Poles are used as service wood, posts and mine timber. Old thuyas are appreciated as firewood. Its charcoal is mediocre. The distillation of shavings from the collar gives tar. Sandarac gum is extracted by superficial wounds of the bark.

This species is the most sensitive to fires, especially if the undergrowth and the herbaceous layer are developed. Also, repeated treatment in simple coppice for several decades has led to a genetic impoverishment of thuya stands, regeneration being almost exclusively ensured by vegetative reproduction. This situation reduces the species' capacity for adaptation to increasing climatic stresses and diseases, requiring reflection on the integration of regeneration phases by seedlings to maintain genetic diversity and long-term resilience of these ecosystems.

Holm oak (Quercus rotundifolia)

Ecological Characterization and Distribution

Holm oak, with 1,339,200 ha, is the most represented species in Morocco by its great capacity to adapt to various ecological conditions (IFN, 2024). Tree of 2nd magnitude, it exceeds 20 m high in humid high forests and can reach 25 m according to Benabid (2000). The trunk with fissured bark. The growth is relatively slow. The rooting is tap with powerful lateral roots. The young seedling extends its roots during the first years (Quézel & Médail, 2003). The leaves are simple, persistent, shiny green above, gray-green below, rigid, more or less large depending on the stations. The canopy is thick. Holm oak is a monoecious species and fruits quite often from 15 years. It resprouts from the stump and suckers abundantly until an advanced age: 150 – 200 years. Its growth is quite slow. It is long-lived and can reach, even exceed, 300 years in good conditions (FAO, 2012). It adapts to all climate, soil and human action conditions. It is robust and plastic. It resists temperatures of -20°C. It tolerates canopy cover and then forms the undergrowth dominated by other species. It is found between 500 and 2,700 m altitude; from the semi-arid to per-humid bioclimatic stage (Benabid, 2000).

Structure of Stands and Silvicultural Management

A species so widely distributed gives stands with different aspects, which can be arranged in 3 types of regimes: i) Dry high forest (at Debdou): open stands, isolated trees with short trunks and globular crowns mixed with Aleppo pine. Some of these stands are pure in the state of coppice with standards from coppice having undergone

Thinning and Thinning; ii) Humid high forest (Azrou, Aïn Leuh): trees in massif with slender trunk and candle-like crown, mixed with cedar and zeen oak; and iii) Coppice at the semi-arid stage, more or less dense and vigorous according to climate and soil conditions. Some of these stands develop at the level of the subhumid stage in the surroundings of Azrou and Khénifra.

From a biogeographic point of view, holm oak forests are organized into four types (IFN, 2024): the holm oak forest of the Rif and the plateaus of the Oriental; the holm oak forest of the Middle Atlas and the Central Plateau; the holm oak forest of the Central and Oriental High Atlas; and the holm oak forest of the Western High Atlas and the Anti Atlas.

The most widespread treatment is regular simple coppice coppiced from 45 to 60 years. Thinning is postponed until 20 years in order to obtain merchantable wood. A thinning is sometimes considered in stands maintained beyond 45 years. The massifs of the humid stage are treated in regular high forest by single cut (clear cut) carried out around 120 years. Thinning also occurs at 20 years and a thinning 20 years later which brings the stand to the suitable density. Subsequently, cuts will remove only defective trees. Currently laudable decisions have been made to abolish clear cuts by substituting them with "progressive cuts" to bare stump.

Threats, Valorization and Perspectives

Holm oak gives excellent firewood, very heavy and very dense[1]. It is used as mine timber, service wood and firewood, and it gives excellent charcoal (Benabid, 2000). The butt logs of trees from humid high forest have long been used as railway sleepers. In periods of drought or snow, the branch and top cutting of holm oak trees serve as emergency feed for herds (pollard cuts). Formerly, it was widely used for the extraction of tannin (bast). Today, this practice has been fortunately abandoned due to the availability of cheap alternative products.

Apart from *Catocalas* (defoliating caterpillars) (*Catocala nymphagoga*), holm oak has few enemies. It is not very sensitive to fires and is always maintained by sprouts and suckers. However, the anthropozoogenic factor remains the main actor in the degradation of holm oak forest ecosystems in Morocco.

NECESSITY OF PARADIGM SHIFT IN MOROCCAN FOREST MANAGEMENT

Limits of Conventional Approaches

Conventional approaches to forest management, whether regular high forest or simple coppice, today show their limits in the face of current ecological and climatic challenges. Numerous studies have reported increased multifunctionality and financial profits through the transition from even-aged management to uneven-aged management (Diaci et al., 2011). In the Mediterranean region, particularly vulnerable to climate change, ecosystem services provided by forests should decrease on average by the end of the century, with increased dieback and mortality phenomena in many forest species, lack of management and aging of stands constituting additional vulnerability factors (European Commission, 2023). The prolonged application of simple coppice presents specific limits: exclusively vegetative reproduction prevents genetic renewal of stands, reducing their adaptive capacity; repeated cuts lead to progressive aging of stumps and a decrease in the vitality of sprouts; the uniform structure limits forest biodiversity; in the long term, intensive exploitation can lead to soil impoverishment. These findings underline the need to adapt traditional silvicultural practices. Alternative approaches, such as irregular high forest and mixed silviculture with continuous cover, are gradually emerging as solutions more adapted to the contemporary Mediterranean context (Puettmann et al., 2015).

Current Situation in Morocco

Almost all managed forests in Morocco are currently treated in regular high forest (Figure 1-a) or in simple coppice (Figure 1-b). The rare forests where irregular or selection-type stands (Figure 1-c) can be observed are not yet subject to formalized management. These formations are located mainly in the north of the country, particularly in the Rif, where the land situation remains partly unresolved. It should also be emphasized that one of the particularities of this region resides in a relatively less important pastoral pressure compared to other regions of the country. This situation offers favorable conditions for the installation and continuous development of natural seedlings, within certain restricted spaces still little or no frequented by herds.

Figure 1. Illustrations for examples of silvicultural treatment of Moroccan forest stands. a) Atlas Cedar in Regular High Forest (Taffert forest, 2025); b) Holm oak in simple coppice (in the Middle Atlas, 2018); and c) Zeen oak in irregular high forest with selection tendency (Bouhachem forest, 2016) (@ Chkhichekh A.)

Although the current governance of forest resources is, to a large extent, behind the problems of dysfunction, dieback and dedensification that the different forest and steppe ecosystems of Morocco are experiencing (Figure 2), the silvicultural techniques and management methods applied have considerably impacted their resilience.

Figure 2. Global warming, with increased drought and soil erosion, has gradually made the habitat unsuitable for some species. a) Old regular high forest of Atlas Cedar declining, evolving on a completely eroded soil in the Oriental Middle Atlas; b) Decline of cork oak at the edge of its ecological range in an area of contact with thuya (SIBE El Harcha, Central Plateau, Morocco. @Chkhichekh A., 2025)

In strictly protected Mediterranean forests, unmanaged areas were found to be about 1°C cooler than productive forests, even after accounting for topographic and stand attributes, highlighting the specific emergent properties that develop when stands evolve toward more natural stages (Solano et al., 2025).

Impacts of Conventional Methods on Resilience

The productivist conception of forest management, inherited from the West and perpetuated by our managers and planners, should be substituted by a much more open conception, leaving an important place to the other functions that these ecosystems ensure: functions of maintaining biodiversity, regulation of biogeochemical cycles, hydrological and climatic regimes, carbon storage and sequestration, without however omitting the social functions which are taking on growing importance. Silvicultural treatments based on fixed and more or less long rotations have hindered the natural processes of selection and genetic adaptation of stands. Sexual reproduction generates genetic variability, essential for evolutionary adaptation to environmental changes, diseases and disturbances. It also allows long-distance dispersal and colonization of new habitats (Barrett, 2015), especially since forests managed according to uneven-aged approaches present increased resilience to climatic and biotic disturbances, with superior adaptive capacity compared to homogeneous even-aged stands (Messier et al., 2019). Reduced biodiversity due to stand uniformity also constitutes a notable disadvantage of conventional treatments.

Toward a Pluralistic and Adaptive Forest Management Paradigm

Treatments in regular high forest and simple coppice are the fruit of the evolution, for several centuries in Europe, of forestry techniques of rationality reducing a plant grouping dominated by trees to a product of uniform and industrial type. Production cycles that are too long have only recently revealed an awareness of the limits of the system (AFI, 2009). Over the past decades, several major developments have profoundly modified the perception of the forest, its functions and its management methods. Various meteorological catastrophes (storms, snowfalls, episodes of freezing rain) and biotic (attacks by insect pests or pathogenic fungi), have prompted reflections aimed at strengthening the stability and resilience of forest ecosystems. Furthermore, the need to maintain biodiversity, the evolution of knowledge in ecology and climate change direct silvicultural practices toward management methods closer to nature. Finally, climate change requires the adoption of management practices favoring rapid and sustainable adaptation of forest species (AFI, 2009).

As a result, new concepts arising from a renewed management paradigm have emerged. This is part of the vision developed by Paul Feyerabend in Against Method (1975), which radically questions the existence of a universal and normative scientific method. In this perspective, science appears as a complex human activity, influenced by a multitude of factors, rather than as a strictly rational enterprise guided by a single and universal method. In forest management, the multifunctionality of ecosystems requires full inscription in this pluralistic and adaptive vision. Forests, as complex adaptive systems, require management approaches that recognize their dynamic nature, their heterogeneity and their capacity for self-organization, rather than simplified interventions based on deterministic models (Messier et al., 2013).

On the basis of these observations, the Association *'Futaie Irrégulière'* (AFI) is created at the initiative of experts and forest owners in 1991, in order to study and develop the treatment of irregular high forest in forest stands. This reflection group has given itself the mission of studying and disseminating forest management methods that would ensure income while agreeing with the natural functioning of the forest (AFI, 2009).

The Management Method of Irregular High Forest

Irregular high forest is a forest structure resulting mainly from regeneration by seeds, characterized by the coexistence of all age classes, without obeying a strictly defined mathematical model. However, a structure with selection tendency, in which the number of individuals decreases with increasing age (or diameter), appears more relevant to translate a balanced dynamic of stands and ensure their long-term stability as well as their ecological sustainability. This treatment method presents numerous ecological and economic advantages: increased resilience to climatic and biotic disturbances, continuity of forest cover limiting erosion and favoring water regulation, superior structural and functional biodiversity, as well as regular and diversified forest income. Said treatment method would lead, in the medium and long term, to the progressive installation of a continuous forest cover, favoring the redynamization of the ecosystem. This evolution would allow the reinstallation of plant species contributing to the reconstitution of vertical stratification, which would in turn favor the development and maintenance of animal biodiversity, both microscopic and macroscopic.

It is appropriate however to recognize that the implementation of irregular high forest, in the national sociocultural context, would pose notable difficulties, contrary to what is observed in several countries of the north of the Mediterranean. The extensive livestock system widely practiced in Morocco, combined with the use rights recognized to populations bordering forests, constitutes a major constraint. Indeed, irregular high forest, based on the simultaneous presence of all age and height classes within the forest, requires rigorous protection of natural regeneration. However, the success of such an approach presupposes a profound transformation of the lifestyles of rural populations, through the implementation of public policies capable of reducing their vital dependence on the forest. This dependence is expressed both by the pastoral use of forest ranges and by illicit cuts for commercial purposes, practiced to satisfy essential economic needs.

Approaches and Determinants of Successful Implementation of Irregular High Forest

The implementation of irregular high forest in the national forestry context cannot be considered without in-depth reflection integrating both ecological, silvicultural and socio-economic dimensions. Several preconditions appear essential to increase the chances of success of this treatment method. First, securing natural regeneration constitutes a fundamental lever. It necessarily passes through control of pastoral pressure, either by temporary or permanent protection devices, or by strict regulation of pastoral load adapted to the production capacity of stands. Next, the success of irregular high forest presupposes adherence of neighboring populations, which cannot be obtained without a tangible improvement of their living conditions. Reducing direct dependence on forest resources requires the establishment of integrated rural development policies, favoring alternative sources of income (agroforestry, intensified livestock farming, ecotourism, valorization of non-timber forest products).

Moreover, the adoption of this management method requires a paradigm shift in forest management, moving away from rigid normative models in favor of adaptive management, based on continuous observation of natural dynamics of stands and on progressive experimentation at the local scale. Finally, irregular high forest must be considered as a long-term process, requiring rigorous scientific monitoring, continuous training of forest managers and the development of decision support tools adapted to local ecological and social realities. From a technical point of view, the introduction of irregular high forest could be considered as a priority in the most vulnerable forests, where the survival duration of trees does not allow waiting for prolonged delays before intervening. However, given use rights and usual pastoral practices, this approach should be based on small-scale interventions, in the form of closed, discontinuous and spatially dispersed surfaces, representing globally 30 to 40% of the area of said forests. The rest of the area would remain open to pastoral grazing, thus allowing a progressive compromise between conservation, production and local use objectives. This approach would make it possible to partially achieve the objectives of irregular high forest, while awaiting the progressive emergence of social acceptability. This could be built gradually, in parallel with the implementation of an integrated rural development policy, likely to sustainably reduce the dependence of local populations on forest resources.

Assisted Migration

Climate change rapidly modifies the ecological conditions to which forest species are adapted, particularly through increasing temperatures, intensification of drought and increased frequency of extreme events (IPCC, 2023). In many cases, the natural migration capacity of species is no longer sufficient to follow the speed of climate change, particularly for species with slow dispersal or for those whose habitats are fragmented. In the case of Morocco, it would be almost unimaginable to count on the natural migration of species mainly because of the uncontrollable movement of livestock herds (Figure 2-b). In this context, assisted migration (or directed migration) is considered as a proactive adaptation strategy. It consists of intentionally transferring species, provenances or populations toward geographically or ecologically more favorable areas to future climatic conditions. In forestry, this approach aims to maintain productivity, resilience and ecological functions of forest ecosystems. Short-distance assisted migration experiments have been conducted successfully in several Mediterranean regions, particularly through the transfer of provenances adapted to drier conditions toward sites anticipating future aridification (Aitken & Whitlock, 2013).

However, assisted migration raises important ecological and ethical issues. It involves risks linked to the introduction of species or genotypes likely to become invasive, to disturb local ecological balances or to reduce the genetic diversity of native populations (McLachlan et al., 2007; Hoegh-Guldberg et al., 2008; Ricciardi & Simberloff, 2009; Pedlar et al., 2012). It also questions the traditional concepts of naturalness and conservation. Therefore, assisted migration can only be considered within the framework of adaptive management, based on solid scientific knowledge (ecophysiology, genetics, climate modeling), small-scale experiments, and long-term monitoring. It must be considered as a complementary tool, and not as a universal solution, in the same way as species diversification, management close to nature and conservation of forest genetic resources. Also, to avoid risks linked to the introduction of new species, these should be characterized by ecological and geographical proximity.

CONCLUSION

The forest, pre-forest and steppe ecosystems of Morocco are characterized by great diversity, resulting from complex interactions between biotic and abiotic factors. These ecosystems today face unprecedented threats, with massive dieback of certain forest species and alarming projections of contraction of distribution areas for all main species by 2100.

Historical analysis reveals that management methods inherited from the north of the Mediterranean, regular high forest and simple coppice, have been generalized according to a productivist paradigm. However, the prolonged application of these treatments has considerably reduced the ecological resilience of stands by hindering natural processes of genetic adaptation. Whereas forests managed according to uneven-aged approaches and the adoption of mixed silviculture with continuous cover present significantly increased resilience to climatic and biotic disturbances.

The transposition of these approaches to the Moroccan context requires the integration of three dimensions: control of pastoral pressure, improvement of living conditions of local populations through rural development policies, and adoption of adaptive management based on progressive experimentation. Also, the gap between the speed of climate change and the natural migration capacity of trees also imposes proactive strategies such as assisted migration, provided there is rigorous scientific supervision.

REFERENCES

Aafi, A., Cuzin, F., & Tahiri, A. (2007). La faune sauvage et l'arganeraie: interactions et pressions. In: Actes du Premier Congrès International de l'Arganier. Essaouira.

Achour, A. (2014). *Contribution à l'étude de la phénologie de l'arganier et de la biodiversité au sein d'une parcelle clôturée. Evaluation d'un essai de régénération artificielle*. Thèse de Doctorat National. PP. 204. Université Ibn Zohr; Faculté des Sciences. Agadir.

AFI (Association Futaie Irrégulière) (2009). *Les traitements des Futaies Irrégulières. Valoriser les fonctions multiples de la forêt*. Convention France Bois Forêt/AFI 2009.

Aitken, S. N., & Whitlock, M. C. (2013). Assisted gene flow to facilitate local adaptation to climate change. *Annual Review of Ecology, Evolution, and Systematics, 44*(1), 367–388. DOI: 10.1146/annurev-ecolsys-110512-135747

Aurelle, D., Thomas, S., Albert, C., Bally, M., Bondeau, A., Boudouresque, C.-F., Cahill, A. E., Carlotti, F., Chenuil, A., Cramer, W., Davi, H., De Jode, A., Ereskovsky, A., Farnet, A.-M., Fernandez, C., Gauquelin, T., Mirleau, P., Monnet, A.-C., Prévosto, B., & Fady, B. (2022). Biodiversity, climate change, and adaptation in the Mediterranean. *Ecosphere, 13*(4), e3915. DOI: 10.1002/ecs2.3915

Bani-Aameur, F., & Ferradous, A. (2001). Fruit and seed variability in Moroccan argan tree (Argania spinosa). *Forest Genetics, 8*(1), 39–45.

Barbéro, M., Loisel, R., & Quézel, P. (1992). Biogeography, ecology and history of Mediterranean Pinus halepensis and Pinus brutia. *Plant Ecology, 4*(1), 95–115.

Barbero, M., Loisel, R., Quézel, P., & Richardson, D. M. (1998). Pinewoods of the Mediterranean Basin. In *Ecology and Biogeography of Pinus*. Cambridge University Press.

Barrett, S. C. H. (2015). Clonality, sexual reproduction and evolution in plants. *Molecular Ecology, 24*(13), 2777–2796.

Bauhus, J., Puettmann, K., & Messier, C. (2009). Silviculture for old-growth attributes. *Forest Ecology and Management, 258*(4), 525–537. DOI: 10.1016/j.foreco.2009.01.053

Ben Jamâa, M. L. & Mnara, S. (2000) — *Effect of the defoliations of Lymantria dispar on the cork growth.* Article, disponible via IOBC-WPRS.

Benabid, A. (2000). Flore et écosystèmes du Maroc; *Evaluation et préservation de la biodiversité*. Edition Ibis Press, 2000. ISBN: 2-910728-13-7

Benabou, A., Moukrim, S., Lahssini, S., El Aboudi, A., Menzou, K., Elmalki, M., & Rhazi, L. (2022). Impact of climate change on potential distribution of *Quercus suber* in the conditions of North Africa. *Biosystems Diversity, 30*(3), 289–294. DOI: 10.15421/012231

Boudy, P. (1950). *Économie forestière nord-africaine. Tome II: Monographie et traitement des essences forestières.* Larose, Paris.

Boudy, P. (1952). *Guide du forestier en Afrique du Nord.* La Maison Rustique.

Camarero, J. J., Sánchez-Salguero, R., Sangüesa-Barreda, G., Lechuga, V., Viñegla, B., Seco, J. I., Taïqui, L., Carreira, J. A., & Linares, J. C. (2021). Drought, axe and goats. More variable and synchronized growth forecasts worsening dieback in Moroccan Atlas cedar forests. *The Science of the Total Environment, 765*, 142752. DOI: 10.1016/j.scitotenv.2020.142752 PMID: 33082041

Cheddadi, R., Fady, B., François, L., Hajar, L., Suc, J.-P., Huang, K., Demarteau, M., Vendramin, G. G., & Ortu, E. (2009). Putative glacial refugia of Cedrus atlantica deduced from Quaternary pollen records and modern genetic diversity. *Journal of Biogeography, 36*(7), 1361–1371. DOI: 10.1111/j.1365-2699.2008.02063.x

Chkhichekh, A. (1993). Cartographie de la forêt de Bouhachem dans un objectif d'aménagement. Mémoire de 3ème cycle; ENFI. Salé. Maroc.

Chkhichekh, A. (2015). *Cartographie et étude de la diversité des groupements végétaux du Plateau Central marocain: cas du Site d'Intérêt Biologique et Ecologique (SIBE) d'El Harcha, Maroc.* Thèse de Doctorat National. PP. 103. Université Mohammed V; Faculté des Sciences. Rabat. Maroc.

Chkhichekh, A., El Aboudi, A., Aafi, A., Wahid, N., & Benabid, A. (2015, December). A contribution to the knowledge of the Moroccan forest ecosystems: Association of Quercetum rotundifolio-suberis in the Central Plateau of Morocco. *Plant Sociology, 52*(2), 57–68.

Costa, D., Tavares, R. M., Baptista, P., & Lino-Neto, T. (2020). Cork Oak Endophytic Fungi as Potential Biocontrol Agents against Biscogniauxia mediterranea and Diplodia corticola. [— revue] [Open Access, PMC]. *Journal of Fungi (Basel, Switzerland), 6*(4), 287. DOI: 10.3390/jof6040287 PMID: 33202643

Diaci, J., Kerr, G., & O'hara, K. (2011). Twenty-first century forestry: Integrating ecologically based, uneven-aged silviculture with increased demands on forests. *Forestry, 84*(5), 463–465. DOI: 10.1093/forestry/cpr053

El Yousfi, S. M. (1988). *La dégradation forestière dans le Sud marocain. Exemple de l'arganeraie d'Admine (Sous) entre 1969 et 1986*. Mémoire de 3ème cycle [P; Rabat.]. *IAV Hassan, II*, 117.

Fady, B., Esposito, E., Abulaila, K., Aleksic, J. M., Alia, R., Alizoti, P., Apostol, E.-N., Aravanopoulos, P., Ballian, D., Bou Dagher Kharrat, M., ... Vendramin, G. G. (2022). Forest genetics research in the Mediterranean Basin: Bibliometric analysis, knowledge gaps, and perspectives. Current Forestry Reports, 8, 277–298. DOI: , B., Lefèvre, F., Reynaud, M., Bou Dagher-Kharrat, M., Vendramin, G. G. & Bariteau, M. (2003). Gene flow among different taxonomic units: evidence from nuclear and cytoplasmic markers in Cedrus plantation forests. Theoretical and Applied Genetics, 107, 1132–1138.DOI: 10.1007/s40725-022-00169-Fady

FAO (2012). *Les chênaies méditerranéennes: écologie, dynamique et gestion.*

Feyerabend, P. (1975). *Against method: Outline of an anarchistic theory of knowledge*. Verso.

Gil, L., & Varela, M. C. (2008). *EUFORGEN Technical Guidelines for genetic conservation and use for cork oak (Quercus suber)*. Bioversity International.

Hadjadj, K., & Letreuch Belarouci, A. (2017). *Synthèse bibliographique sur le thuya de berbérie (Tetraclinis articulata* (Vahl) Mast.). *Geo-Eco-Trop*, *41*(1), 13–27.

Hoegh-Guldberg, O., Hughes, L., McIntyre, S., Lindenmayer, D. B., Parmesan, C., Possingham, H. P., & Thomas, C. D. (2008). Assisted colonization and rapid climate change. *Science*, *321*(5887), 345–346. DOI: 10.1126/science.1157897 PMID: 18635780

Honnay, O., & Bossuyt, B. (2005). Prolonged clonal growth: Escape route or route to extinction? *Oikos*, *108*(2), 427–432. DOI: 10.1111/j.0030-1299.2005.13569.x

IPCC (Intergovernmental Panel on Climate Change). (2023). AR6 Synthesis Report. *Climatic Change*, •••, 2023.

Laaribya, S. (2023). A review analysis of the degradation of cork oak forests in North Atlantic, Morocco. *Journal of Biometry Studies*, *3*(2), 54–59. DOI: 10.61326/jofbs.v3i2.05

Lepoutre, B. (1965). Régénération artificielle du chêne-liège et équilibre climacique de la subéraie en forêt de la Maâmora. *Annales de la Recherche Forestière du Maroc*, *9*, 1–188.

Lieutier, F., Day, K. R., Battisti, A., Grégoire, J. C., & Evans, H. F. (2004). *Bark and wood boring insects in living trees in Europe, a synthesis*. Springer. DOI: 10.1007/978-1-4020-2241-8

Linares, J. C., Taïqui, L., & Camarero, J. J. (2011). Increasing drought sensitivity and decline of Atlas cedar (Cedrus atlantica) in the Moroccan Middle Atlas forests. *Forests*, *2*(3), 777–796. DOI: 10.3390/f2030777

M'Hirit, O. (1994). *Le pin d'Alep (Pinus halepensis Mill.): Ecologie et sylviculture en zones arides et semi-arides*. FAO.

M'Hirit, O. (1999). *La forêt Marocaine: Enjeux et perspective d'une gestion durable*. Ministère Chargé des Eaux et Forêts.

M'Hirit, O., Benzyane, M., Benchekroun, F., El Yousfi, S. M., & Bendaanoun, M. (1998). *L'arganier, une espèce fruitière-forestière à usages multiples*. Éditions Mardaga.

Matesanz, S., & Valladares, F. (2014). El papel de la plasticidad fenotípica en la respuesta de la vegetación mediterránea al cambio global. In Herrero, A., & Zavala, M. A. (Eds.), *Impactos, Vulnerabilidades y Adaptación de los Bosques y la Biodiversidad de España frente al cambio climático (chapitre, pages N/D)*. MAGRAMA.

McLachlan, J. S., Hellmann, J. J., & Schwartz, M. W. (2007). A framework for debate of assisted migration in an era of climate change. *Conservation Biology*, *21*(2), 297–302. DOI: 10.1111/j.1523-1739.2007.00676.x PMID: 17391179

Mechergui, T. (2019). Influence of mulching and tree shelters on 4-year survival and growth of zeen oak (Quercus canariensis) seedlings. Journal of Forestry Research.

Messier, C., Bauhus, J., Doyon, F., Maure, F., Sousa-Silva, R., Nolet, P., Mina, M., Aquilué, N., Fortin, M.-J., & Puettmann, K. (2019). The functional complex network approach to foster forest resilience to global changes. *Forest Ecosystems*, *6*(1), 1–16. DOI: 10.1186/s40663-019-0166-2

Messier, C., Puettmann, K. J., & Coates, K. D. (Eds.). (2013). *Managing forests as complex adaptive systems: building resilience to the challenge of global change*. Routledge. DOI: 10.4324/9780203122808

Moukrim, S., Lahssini, S., Rhazi, M., Alaoui, H. M., Benabou, A., Wahby, I., El Madihi, M., Arahou, M., & Rhazi, L. (2019). Climate change impacts on potential distribution of multipurpose agro-forestry species: Argania spinosa (L.) Skeels as case study. *Agroforestry Systems*, *93*(4), 1209–1219. DOI: 10.1007/s10457-018-0232-8

Moukrim, S., Lahssini, S., Rifai, N., Menzou, K., Mharzi-Alaoui, H., Labbaci, A., Rhazi, M., Wahby, I. W., El Madihi, M., & Rhazi, L. (2020). Modélisation de la distribution potentielle de *Cedrus atlantica* Manetti au Maroc et impacts du changement climatique. *Bois et Forêts des Tropiques, 344*, 3–16. DOI: 10.19182/bft2020.344.a31888

Muller, S. D., Daoud-Bouattour, A., Fauquette, S., Bottollier-Curtet, M., Rifai, N., Robles, M., Saber, E.-R., El Madihi, M., Moukrim, S., & Rhazi, L. (2022). Holocene history of peatland communities of central Rif (Northern Morocco). *Geobios, 70*, 35–53. DOI: 10.1016/j.geobios.2021.12.001

Orwa, C., Mutua, A., Kindt, R., Jamnadass, R., & Simons, A. (2009) — Agroforestry Database — *surtout sur les rendements, la longévité, la capacité de rejets.*

Otto, S. P., & Gerstein, A. C. (2006). Why have sex? The population genetics of sex and recombination. *Biological Reviews of the Cambridge Philosophical Society, 81*(4), 647–679. PMID: 16856849

Pedlar, J. H., McKenney, D. W., Beaulieu, J., Colombo, S. J., McLachlan, J. S., & O'Neill, G. A. (2012). Placing forestry in the assisted migration debate. *Bioscience, 62*(9), 835–842. DOI: 10.1525/bio.2012.62.9.10

Puettmann, K. J., Wilson, S. M., Baker, S. C., Donoso, P. J., Drössler, L., Amente, G., Harvey, B. D., Knoke, T., Lu, Y., Nocentini, S., Putz, F. E., Yoshida, T., & Bauhus, J. (2015). Silvicultural alternatives to conventional even-aged forest management-what limits global adoption? *Forest Ecosystems, 2*(1), 8. DOI: 10.1186/s40663-015-0031-x

Quézel, P., & Médail, F. (2003). *Écologie et biogéographie des forêts du bassin méditerranéen*. Elsevier.

Ramírez-Valiente, J. A., Santos-del-Blanco, L., Alía, R., Robledo-Arnuncio, J. J., & Climent, J. (2022). Adaptation of Mediterranean forest species to climate: Lessons from common garden experiments. *Journal of Ecology, 110*(5), 1022–1042. DOI: 10.1111/1365-2745.13730

Reusch, T. B. H., & Wood, T. E. (2007). *Molecular ecology of global change. Molecular Ecology, 16*(19), 3973–3992. DOI: 10.1111/j.1365-294X.2007.03454.x PMID: 17894755

Ricciardi, A., & Simberloff, D. (2009). Assisted colonization is not a viable conservation strategy. *Trends in Ecology & Evolution, 24*(5), 248–253. DOI: 10.1016/j.tree.2008.12.006 PMID: 19324453

Rigolot, E. (2004). Fire, fuel management and landscape dynamics in Mediterranean forests. *Annals of Forest Science, 61*(6), 599–610.

Romero, B., Scotti, I., Fady, B., & Ganteaume, A. (2023). Fire frequency, as well as stress response and developmental gene control serotiny level variation in a widespread pioneer Mediterranean conifer, *Pinus halepensis. Ecology and Evolution, 13*(3), e9919. DOI: 10.1002/ece3.9919 PMID: 36960240

Sauvage, Ch. 1961: *Recherches géobotaniques sur les subéraies marocaines*. Trav. Inst. Sci. Cherif. Serv. Bot., 21: 462p.

Sbay, H., & Zine El Abidine, A. (1994). *Croissance et production du cèdre de l'Atlas au Maroc*. Revue Annales de la Recherche Forestière au Maroc.

Silvertown, J. (2008). The evolutionary maintenance of sexual reproduction: Evidence from the ecological distribution of asexual reproduction in clonal plants. *International Journal of Plant Sciences, 169*(1), 157–168. DOI: 10.1086/523357

Solano, F., Mansi, C., Baliva, M., Canestrelli, D., Chiarucci, A., Manicone, R., Maugeri, M., Modica, G., Palli, J., Quilghini, G., & Piovesan, G. (2025). Mediterranean strictly protected forests are cooler. *Agricultural and Forest Meteorology, 375*, 110858. DOI: 10.1016/j.agrformet.2025.110858

Terrab, A., Talavera, S., Arista, M., Paun, O., Stuessy, T. F., & Tremetsberger, K. (2006). *Cedrus atlantica* (Endl.) Manetti ex Carrière: Biogeography and genetic differentiation across the Moroccan Atlas Mountains. *Plant Ecology, 187*, 133–147.

Vieira, J., Nabais, C., & Campelo, F. (2022). Dry and hot years drive growth decline of *Pinus halepensis* at its southern range limit in the Moroccan High Atlas Mountains. *Trees (Berlin), 36*(5), 1585–1595. DOI: 10.1007/s00468-022-02314-z

ENDNOTES

1 https://www.wood-database.com/holm-oak/, 2025

Chapter 2

Forest Governance in the Southern Mediterranean:

Sustainable Development Challenged by Participation

Kamal Menzou
http://orcid.org/0009-0005-1125-1854
Ecole Nationale Forestière des Ingénieurs, Salé, Morocco & Institut National d'Aménagement et d'Urbanisme, Rabat, Morocco

ABSTRACT

Mediterranean forest issues are increasingly framed through the lens of governance. International reports often portray forest degradation as a structural feature of so-called developing regions, particularly in Africa and South America, promoting participatory governance as a normative response to environmental crises. This paper critically examines forest governance in Mediterranean contexts by questioning the modern representation of nature underlying dominant policy frameworks. Governance principles- participation, transparency, sustainability, equity, and accountability- are analyzed through the notion of a loss of the object, understood as the shift from an appropriated to an administered nature. This procedural normativity reveals conceptual and political limits that weaken the analytical relevance of governance models, especially in the southern Mediterranean. Participation is further examined as a dispositif marked by hybridized social solidarities and expanding power relations that obscure the socio-political nature of forest conflicts.

DOI: 10.4018/979-8-3373-6239-7.ch002

INTRODUCTION

The issues of forest conservation and exploitation and its spaces have clearly structured global action from 1970 to the present day. They are already at the center of States' public policies and constitute the subject of world summits whose resolutions aim to promote governance frameworks for sustainable forest management (UN, 2017). Reference to forest governance and natural resources has become unavoidable for decades when it comes particularly to presenting political responses to environmental issues and sustainable development (Goxe, 2010). This reference is well-founded particularly in contexts where the crisis of classical management models is evident and where stakeholder participation shows limits (Defarges, 2022), but also when the regulation of illegal timber trade requires a public policy change through recentralization toward the State (Montouroy, 2014).

In this perspective, governance in general and forest governance in particular were formulated imperatively to participate in the reconfiguration of international environmental cooperation, a generic category of the new international regime that took shape from the end of the 20th century (Hufty, 2005) based on the dissociation of figures of power and authority. Thus we witness the shift from government to governance to denote the crisis of representative government, and the substitution of subsidiarity for sovereignty, efficiency for justice, regulation for reglementation, norm for rule. (Supiot, 2015).

Calls for forest governance also take on regional dimensions, thus testifying to the normative and non-descriptive aspect of this concept and explaining its equivocal nature. At the Mediterranean level, the promotion of forest governance and land reforms at the territorial scale constitute an important axis in the strategic framework on Mediterranean forests. This is an operational translation of the leaders' will to respond to a situation characterized by the depletion of forest ecosystems caused essentially by anthropogenic activities determining the ecology of said forests (Arano et al., 2016). This rarefaction is all the more concerning as climate change today constitutes a major threat to the spatial distribution of Mediterranean forest ecosystems and their capacity to provide ecosystem services, particularly in the southern Mediterranean where ecological tolerance margins are already narrow (Benabou et al., 2022a; Moukrim et al., 2022).

The promotion of population participation in the forestry process constitutes the leitmotiv of the rhetoric of good forest governance, particularly in the southern Mediterranean where the underlying questions that forest-related concerns cover often seem more assigned to cultural traits. Thus, governance absorbed and incorporated state ownership and was able to assert itself as a legal-technical principle dissociated from its modern origins and capable of serving forest policies, particularly in the southern Mediterranean.

This chapter thus aims to examine population participation through the prism of administered forest rather than the forest as such. This analytical perspective will allow us to understand, firstly, the limits of public action and to apprehend non-participation through the prism of the legitimacy of powers, and secondly, to examine the constructivist approach of conditional participation[1] in relation to the loss of the social subject whose diversity is the essential substance and power sharing (Borrini & Jaireth, 2007) is the essence and the limits not only of the representative system but also of democratic governance (Hermet, 2004). At this stage, the problem of representations and perceptual schemes presents itself as an unavoidable entry point for apprehending the controversy over forest governance and particularly with regard to participation. This apprehension will allow us to rethink the typology of conflicts around forests, their spaces and their ecosystem services and establish correlations not only to interests but also to the legitimacies of governance structures and decision-making.

In the Southern Mediterranean, state ownership, presented by public authorities as a guarantee of forest conservation and an essential element of the spatialization of forest territories and spaces and of forest governance, is conversely perceived and experienced by rural populations as spoliation (Simenel et al., 2016) thus contributing to imbalances in centuries-old land ownership regimes (Arano et al., 2016). Calls for land policy approaches from a heritage perspective (Barrière, 2011; Barrière & Barrière, 1997; Herzenni, 2008; Etienne, 1997) denote not only the limits of forest state ownership to respond to ecological and economic balances but also socio-political balances (Fay, 1986; Nadir, 2008).

Speaking of participation within the framework of forest governance occupies an important place in political and scientific discursive production in Morocco and in sectoral strategies. Strategies and action plans in the southern Mediterranean aim to promote the articulation of user and citizen participation in forestry projects. This articulation translates the will to strengthen democratic capital in order to build a new forest governance model capable of achieving sustainable development. Analysis of the process of constructing participation structures could inform us about the hybridization of local solidarities, part of which, until the early 2000s, conserved forest spaces in accordance with identity considerations involving incommensurable predispositions (Menzou, 2024), and could also examine the nature of conflicts around forestry projects that translate the will to implement participation through the prism of 'acting with' where participants are considered as objects as soon as they are redefined as a legal category (cooperative or association). From then on, conditional participation not only embodies the conversion of subject into reactive object, but it is operated with a view to weakening and dismantling traditional structures.

CONTROVERSIAL TRAJECTORIES OF FOREST DEVELOPMENT

Mediterranean forests present two contrasting trajectories in terms of development (Figure 1). Between the North and South of the Mediterranean, forest area takes on trends calling on States to commit further to stopping forest degradation and supporting the restoration and management of forest ecosystems (FAO, 2022).

Figure 1. Annual net change in forest area, by decade and world region, 1990-2020

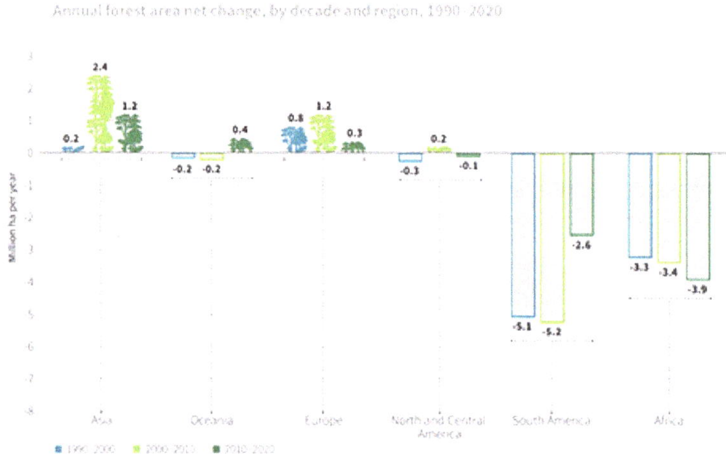

Annual changes in forest areas, including biodiversity and the functions of its species, certainly translate the effect of biotic factors, particularly climate changes over the last century, but also and primarily anthropogenic actions whose representations are to be understood beyond natural universalism (Descola, 2015). Moreover, the demographic balance could easily be correlated with the process of forest degradation (FAO, 2013), however increasing exploitation, threats to biodiversity and pressure on natural resources and their impacts on the environment question the indirect causes of unsustainable trajectories of development models and not only traditional uses of wooded and pastoral spaces.

It is about rethinking the fundamentals of the Human/nature relationship and forest management in modern times. A rereading through the prism of methodological pluralism regarding diversity and the **incommensurability** of social grammars and forms of life to avoid the pitfalls of **determinism**.

The relevance of ecological indicators that testify to i) considerable degradation of the Mediterranean forest, particularly in the South, and functional imbalances of its components, and ii) sociopolitical transformations that denote a remarkable hybridization of local solidarities, has strongly contributed to the choice of the Mediterranean as the subject of this study.

Trends in forest policies testify to a growing awareness of ecological challenges and a will expressed and acted upon during world summits related to forest issues. Bilateral and multilateral international cooperation attempts to establish the foundations of global and regional forest governance.

The Mediterranean Forest: A Heritage in Two Times

The Mediterranean space constitutes one of the world's main centers of forest biodiversity, while representing one of the oldest territories of continuous interactions between human societies and natural environments. Mediterranean forests are the result of a multi-millennial coevolution between ecological dynamics and anthropogenic uses, marked by successive cycles of clearing, overexploitation, reconstitution and, more recently, protection and sustainable management. This long history has shaped complex forest landscapes, profoundly anthropized, but ecologically rich and strategic. However, the Mediterranean forest situation is characterized today by strong disparities between the north and south of the basin, both in terms of area and forest structures and health status. These contrasts cannot be explained solely by climatic factors; they are part of differentiated historical, economic, social and institutional trajectories, linked to state-building, land and forest policies, as well as natural resource management modes.

Forest Area Dynamics: Between Reconquest and Regression

At the scale of the Mediterranean basin, forests and other wooded lands cover approximately 28% of the total land surface, which represents a considerable forest heritage. In absolute value, the Mediterranean forest area is estimated at approximately 99.2 million hectares in 2020 (FAO, 2025), constituting an essential part of regional natural resources and a major lever for biodiversity, climate regulation and soil protection. Nevertheless, this forest cover is unequally distributed between the two shores.

Comparative analysis of forest area, main stand types and trends in health and development highlights a structural opposition between a relatively stabilized, even expanding northern Mediterranean forest, and a more fragmented, vulnerable southern Mediterranean forest under strong anthropogenic pressures (Table 1).

Table 1. Distribution of forest area in the Mediterranean (North / South)

Indicators	**Northern Mediterranean**	**Southern Mediterranean**
Share of Mediterranean forest cover	≈ 49%	≈ 10%
Forest area	High	Low to medium
Share of forest in national territory	Medium to high (often > 30%)	Low (often < 10%)
Recent evolution (1990-2020)	+12% (+14.6 millions ha)	Stable to locally declining
Spatial continuity	Relatively continuous forests	Fragmented forests

source: FAO. Status of Mediterranean Forests (2025)

The concentration of forest area in the north of the basin is explained by a historical dynamic of reforestation and stabilization initiated at the end of the 19th century and reinforced throughout the 20th century. In countries such as France, Spain, Italy or Greece, rural exodus, agricultural abandonment, industrialization and the implementation of proactive forest policies have favored forest reconquest. According to the FAO, Southern Europe recorded a net increase in its forest cover of approximately 12% between 1990 and 2020, or more than 14.6 million additional hectares, despite growing pressures related to urbanization, infrastructure and fires (FAO, 2025).

Conversely, south of the Mediterranean – particularly in the Maghreb countries and in certain regions of the Middle East – forest area remains structurally low in proportion to national territory. Forests there are often discontinuous, embedded in agro-sylvo-pastoral systems and subject to strong ecological constraints, such as aridity, climate variability and low soil fertility (FAO, 2025). While some countries show overall stability in their forest area, this mask local dynamics of degradation, regression or transformation of forests into pre-forest formations (matorrals, garrigues, wooded steppes), translating a progressive weakening of forest ecosystems.

Thus, the asymmetry of area and forest structure between the north and south of the Mediterranean reflects a combination of differentiated ecological constraints and unequally regulated anthropogenic pressures, but also contrasting institutional capacities in terms of forest governance. Where northern forests benefit from structured management policies and significant technical means, those of the south remain more exposed to subsistence uses and the cumulative effects of climate change. This structural divergence constitutes one of the major challenges for the future of Mediterranean forests and calls for differentiated but coordinated policies at the basin scale.

Ecosystem Productivity of Southern Mediterranean Forests: From the Timber Paradigm to Multifunctionality in the Context of Global Change

Beyond forest area, ecosystem productivity today constitutes a central indicator for assessing the sustainability of Mediterranean forests. This productivity is not limited to wood production: it refers to all ecosystem services provided by the forest, including carbon sequestration, water regulation, soil fertility, biodiversity and socio-economic services. This integrative framework allows us to understand that the productivity of a forest ecosystem is not limited to a strictly economic measure, but reflects a complex interaction between ecological processes and human needs, particularly salient in contexts of climate and social vulnerability in the southern Mediterranean.

In the countries of the northern Mediterranean, particularly in Southern Europe (Spain, France, Italy, Greece), forest productivity has historically been dominated by high woody biomass production supported by formalized silvicultural management. The net growth in forest area (+ 12% between 1990 and 2020) reflects proactive forest policies, agricultural abandonment in certain areas and well-managed natural regeneration (FAO, 2025).

Northern Mediterranean stands often present relatively high biomass compared to their southern counterparts, due to relatively more favorable climatic conditions and active silvicultural management practices. This biomass translates into a significant carbon sequestration capacity, thus contributing to regional and global climate regulation. Northern Mediterranean forests are among the main carbon sinks in the basin, although this capacity is now compromised by the increasing frequency and severity of droughts and fires.

In addition to carbon storage, these forests provide essential water regulation and soil protection services, reducing erosion in often mountainous or sloped territories.

They also provide non-timber forest products integrated into local value chains (cork, honey), and support recreational and tourism activities generating income, particularly in peri-urban and rural areas. Despite these assets, northern Mediterranean forests are not immune to ecological tensions. In recent times, we observe an intensification of forest fires, with an average of 1,600 fires per year burning nearly 400,000 hectares, a phenomenon that weakens the expected ecosystem services

In the southern Mediterranean, particularly in the Maghreb, Middle East and North Africa, forests present significantly different spatial and functional configurations. They are characterized by a smaller area, limited timber productivity and a high dependence on not only ecological but also socio-economic services for local communities. Low woody biomass production in southern forests is largely dictated by conditions of climatic aridity, significant water variability and often poor soils. Stands there are generally more open and less dense, which reduces wood production capacity and carbon sequestration per unit area. However, this observation should not be interpreted as an absence of productivity: it rather reflects a functional adaptation of ecosystems to environmental constraints.

In these regions, non-timber ecosystem services play a crucial role for rural populations:

- Aromatic and medicinal plants, melliferous and pastoral species contribute directly to income and food security of local communities.
- Protection against desertification: forest formations, even fragmented, actively participate in preventing wind and water erosion, thus limiting the advance of desert areas
- Local construction materials, besides fuelwood, remain vital resources for households.
- Ecosystem productivity in the southern Mediterranean is therefore strongly linked to the provision of these services, which are often underestimated in approaches strictly focused on wood production (Benabou et al., 2022b).

New services are appearing in the Mediterranean forest service portfolio. Indeed, Mediterranean forests of Europe (Spain, France, Italy, etc.) represent a significant part of spaces accessible for recreation: 78% of forests in Southwest Europe are available for recreational activities, with visit rates that can reach 28.7 visits per hectare per year in some countries (FAO, 2013).

Moreover, a clear differentiation can be observed between practices and policies for managing urban forests in the Northern and Southern Mediterranean. In the north, urban forests benefit from well-established regulatory frameworks, planning integrated into urban policies and sustained investments. Recreational and ecological valorization is considerably consistent with territorial planning, with infrastructure adapted to tourist and local attendance. However, it should be noted that there are few operational tools to systematically and practically evaluate the effectiveness of sustainable management of said spaces (Komarolya & Maleknia, 2025).

However, south of the Mediterranean, despite significant ecological potential, management of urban forests often remains limited by institutional, financial and technical constraints. Programs are often ad hoc, with growing efforts to integrate these green spaces into urban strategies, particularly in Morocco and Tunisia. We thus witness conventional arrangements between deconcentrated services and decentralized authorities with some public-private partnership initiatives in Morocco such as the Bouskoura forest in Casablanca and the Ifrane park.

Opting for forest governance in urban territories calls for a number of indicators and criteria to be considered, particularly citizen participation which is neither secondary nor marginal but a lever for long-term sustainability. Moreover, recent studies on forest governance in urban territories have defined about fifteen priority indicators combining ecology with social and technical aspects (Table 2). This is a matrix that concretely highlights the problem of the technical to the political and puts ecology to the test at the crossroads of the technical and social (Aubert & Sabir, 2014).

Table 2. Urban Forest Governance Indicators (UFGI) (Komarolya & Maleknia, 2025)

Category	Indicator	Definition	Relative Importance
Ecological	Plant species diversity	Richness and variety of tree species promoting ecological resilience	Very high
	Presence of native/ endemic species	Use of species adapted to the local biogeographic context	Very high
	Soil conservation	Protection of structure, fertility and stability of urban forest soils	High
	Tree health and vitality	Physiological and health status of urban stands	Medium
	Ecological management of stands	Silvicultural practices respectful of urban biodiversity	Medium
Social and participatory	Citizen participation	Involvement of residents in planning and management	Very high
	Safety and security	Users' sense of security	High
	Accessibility	Ease of physical and spatial access to urban forests	High
	Environmental awareness	Level of ecological awareness and environmental education	High
	Equity of use	Fair distribution of benefits and uses	Medium

continued on following page

Table 2. Continued

Category	Indicator	Definition	Relative Importance
Planning	Spatial integration	Connection with other urban green spaces	High
	Green paths and trails	Quality of pedestrian and recreational paths	High
	Nature-oriented design	Integration of ecological design principles	Medium
	Multifunctionality	Capacity to fulfill multiple urban functions	Medium
Technical management	Management of damaged trees	Pruning and replacement of at-risk trees	Low

If ecology and biodiversity take on intrinsic value in modern representation, relational value leads to reconsidering the technical approach (silviculture, forest management planning) through the prism of the social. Technology can no longer be purely biocentric (focused solely on the ecosystem) or economist; it must integrate human, cultural and societal dimensions, particularly those relating to i) public health where forest cover is linked to disease reduction, ii) social cohesion where the forest presents itself as a place of meetings and community activities, iii) education and awareness where the forest is presented as an educational support and finally iv) population identity, in other words intangible capital where the emotional and identity value of forests is part of topophilia perspectives and incommensurable forms of life (Altman & Low, 1992; Tuan, 2011).

Forest Policy in the Southern Mediterranean

The history of forestry in African countries offers a striking echo to the diagnosis made for countries in the north of the Mediterranean. It reveals the same modernizing will to impose, via the State, a universalist model of land as an appropriable and administered 'good' (Charles & Kalaora, 2001) according to the modern Anthropocene representation, to the detriment of traditional heritage relations.

Genealogy of Modern Forest Policy: Loss of the Object

In the north of the Mediterranean, forest policies aimed not only to restrict the usage rights of rural populations, as reflected in the experience of the northern Mediterranean (Beau & Larrère, 2018; Boutefeu, 2005), but to fight against all forms of relationships with nature that are incommensurable with the modern paradigm, particularly in the South of the Mediterranean where the first features of forest policy began during the colonization period. This is a political translation of the modern representation of nature whose objective is to construct the forest not only as an object but essentially as a spatialized object.

Faced with ecological and social crises, this model reforms and even hybridizes even at the European and American continent level, progressively integrating sometimes a more heritage and systemic conception of the forest (Etienne, 1997), conceived as a living space at the crossroads of ecological, economic and social issues, sometimes a non-anthropocentric subjectivation (Stone, 1974), of nature breaking from any objectification perspective (Ost, 2003). Contemporary forest policy thus seeks, not without difficulties, to go beyond the modern great divide between man and nature to invent a new, more complex and integrated relationship in a perspective of a new natural contract (Serres, 1990). This is a transposition of 'pure nature' (Schmutz, 2001) non-relational which was sketched in terms of univocally constructive processes according to a Mathesis of rational human artifice and mimicry of the representation of a supposed nature and society reduced to measurable quantities (Supiot, 2015).

Indeed, this project has been doubly nourished by a reductive mimicry:

- Mimicry of the representation of a supposed nature: It often involves recreating an 'ideal,' climax, closed and wood-producing forest, modeled on regular high forests of temperate Europe (such as beech-oak forests). This vision ignores and combats the specific ecological dynamics and mosaics of the Mediterranean basin (matorrals, garrigues, open pine or holm oak forests), perceived as degraded (Boutefeu, 2005, 2009) .
- Mimicry of a society reduced to measurable quantities: Local societies, with their complex practices and vernacular knowledge, are reduced to quantifiable variables (number of livestock heads, fuelwood needs) and often perceived as disturbance factors, 'pests' to be excluded (Marghadi, 2009).

This reading grid allows deciphering centuries of forest and environmental policies. Mountain reforestation projects in 19th century France (1860 law), particularly with RTM (Mountain Terrain Restoration) works, are the archetype: it involved 'repairing' a nature supposedly degraded by peasants and pastoralists, by massively imposing Austrian black pines and Atlas cedar according to geometric layouts, in disregard of ecosystems and local socio-economic balances (Kalaora, 1993). The creation of a strong state Forestry Corps, endowed with police power, embodies this will to subject both nature and societies to a rational and productivism order (Kalaora et al., 2023), denying the historical multifunctionality of Mediterranean forests (place of grazing, gathering, terrace cultivation, hunting, sociability).

Forest policy, like any modern public policy, then becomes an instrument of social discipline in the sense of Michel Foucault (Foucault, 2004), aiming to substitute governance for government and stakeholders for the user-citizen (Supiot, 2015) obeying incommensurable registers of subjectivation. In this regard, the current state of forests in the southern Mediterranean is by no means an ad hoc and cyclical posture, but fundamentally translates that they are the product of a long history (Mhirit & Benchekroun, 2006). As conceived space (Lefebvre, 2000), they are the product of politics at the global level, we speak of 'political forest' in the sense that colonial and postcolonial States have progressively redefined forest spaces, not only as ecological entities, but especially as objects of government, subject to specific legal regimes and control devices (Peluso & Vandergeest, 2001). It is a territory where dwelling and appropriation are defined as resistance (Vidalou, 2017). Also, for African States and the southern Mediterranean in particular, the forest was a political category produced by the State, and not a simple socio-ecological reality. The 'political forest' is redefined as a space where the State claims exclusive authority over access, use and management. 'The African States that became independent all believed in the universality of the civil law regime of property rights and in the panacea of public state ownership. The power of individual exclusivity that transforms land or property into goods, direct management by the State of rural space constitutes, it was thought, an inevitable reference framework for a modern society' (Barrière & Barrière, 1997).

It is in this contingent situation that the premises of modern forest management planning take root in the countries of the southern Mediterranean where the essence of French forestry thinking was deployed from 1913 until independence (Aubert, 2014). Instead of etymologically seeking the meaning to attribute to forest management planning, it is more appropriate to seek the anthropological effect of this new refoundation of modern forest management that François Ost qualifies as the invention of nature (Ost, 1995).

Forest Management Planning: Illusion of Technical Neutrality

From the Ancien Régime, forest management planning imposed itself in France as the normative tool par excellence, establishing a framework of rights and prohibitions and organizing timber exploitation. The Revolution of 1789 did not break with this logic; it radicalized its application. Instrumentalized by a Water and Forests administration seeking authority, management planning became the lever of a policy aimed at abolishing customary usage rights (Larrère, 2002). This confrontation, which opposes the scientific and legal knowledge of foresters to the incommensurable practices of peasants, is at the origin of numerous rural revolts (Boutefeu, 2009).

Management planning and forest sciences are not grasped, as such, as a type of knowledge, or the results are vectors towards a new way of life, they are apprehended as 'signified' that is to say reality in itself. Underlying this apprehension is a certain logic of cumulative knowledge that Lionel Charles and Bernard Kalaora qualify as determinant whose temporality is linear and homogeneous canceling all other experiences with nature (Charles & Kalaora, 2003).

What interests us at this stage is the observation around the nature of forest sciences, which parallel to their claim to universality, claim their legitimacy through the performativity of their procedures. Thus we can underline legitimation by procedure as a characteristic of modern societies where we clearly note the replacement of the normativity of laws by the performativity of procedures is the rule (Luhmann, 2001). This remark leads us certainly to question the relationship between science and politics, but at the same time it will also allow us to choose the appropriate method for our hypothesis on the conflict of legitimacy between scientific knowledge and knowledge (Menzou, 2024).

If Luhmann's work demonstrates, among other things, the utility of procedures for Politics in modern societies, it also reveals the normative aspect behind it. The form of interaction specific to the procedure therefore does not have the sole function of filtering usable decision-making perspectives; it also serves, directly, to absorb conflicts, to weaken and exhaust participants, to transform and neutralize their motivations during a history in which presentations and commitments transform into presentations serving the elimination of alternatives (Luhmann, 2001).

If we thus attack forest management planning, it is that we see in it the third pillar of the will to produce conceived space where the relationship between space and politics emerges. This leads us to shed light on the political instrumentalization of space through forest management planning which is defined through the prism of economics. This abstraction in the definition of management planning, like other concepts, hides behind it the modern normative aspect, and eliminates not only other alternatives (Luhmann, 2001), but also other experiences that do not refer to the same cognitive reference framework. It is in this sense that in North America traditional knowledge was long excluded not only for its epistemological incompatibility but also for its spiritual foundations (Chiasson et al., 2005).

In this perspective, how can we speak of evolution or change in forest policies in the countries of the southern Mediterranean, in a context where we keep the same foundations of the policy initially designed before independence, despite the introduction of new intervention logics and approaches that claim to be inclusive (Aubert, 2014).

The inclusiveness expected by forest policies in the southern Mediterranean translates the limits of the forest management model in terms of sustainability and participation. The deforestation syndrome should by no means be reduced to a population problem and usage rights recognized for neighboring tribes. It is more judicious to dwell on the causes of exclusion, but also on the conflict of territorial legitimacy as well as the sensitivities and attitudes generated by modern concepts in an incommensurable context.

In the southern Mediterranean, where state ownership was the solution invented to conserve forests and allow forest engineers to carry out development projects in a linear sustainability, the syndrome of forest ecosystem degradation. The deforestation syndrome should not be reduced to a population problem and usage rights recognized for neighboring populations. It is more judicious to dwell on forms of exclusion from local development projects, but also and mainly on the conflict of territorial legitimacy as well as sensitivities and perceptions generated by modern concepts in an incommensurable context where local representations are not recognized in projects.

Thus correlation could be established between sociopolitical imbalance and degradation of nature and its forest spaces (Aspe & Genin, 2014; Herzenni, 2008; Romagny et al., 2008, 2018). Analysis of perceptual schemes helps us not only to establish their links with appreciation and judgment schemes and those of action in favor of or against forest policies, but also to question the paradoxes of modern representation of forests as a political object, particularly its inalienability.

Such a reconfiguration exposes the incommensurable political nature of the technical and challenges any politically neutral claim of forest technique (Inventory, Silviculture, Forest Management Planning, Dendrometry...) (Marghadi, 2009). Wanting to do modern forestry in this type of incommensurable territories therefore leads to ensuring that traditional uses are limited to the maximum. Thus usage rights legally recognized to peasants translate this will to minimize their interference with those of foresters (Aubert & Sabir, 2014). The degradation of forest spaces would be, among other things, a consequence of this dissociation of the cultural from the natural, knowledge from identity in a perspective of abstraction of the human and natural objects of the possible world and the new natural contract, hence the loss of the object of participation would be the title of this chapter in contemporary forestry.

FOREST GOVERNANCE PUT TO THE TEST OF PARTICIPATION

The forest, long apprehended from the angle of essentially technico-legal management (Boutefeu, 2009), is today at the heart of profound normative, institutional and social recompositions. The rise of the sustainable development paradigm, enshrined since the end of the 20th century as a global reference framework for public action, has profondly reconfigured intervention frameworks in the forestry sector. Henceforth, forest management can no longer be thought of solely in terms of wood production or biological conservation, but must integrate in a systemic manner the ecological, economic and social dimensions, in a perspective that looks not only at the effectiveness of public action but essentially its legitimacy which refers to sustainability in its sense which echoes social acceptability, responsibility and participation.

Thus, if forest governance claims principles such as participation, decentralization and plurality of actors, their effective implementation comes up against persistent structural constraints: power asymmetries between actors (Ribot & Peluso, 2003), weak recognition of local uses, institutional fragmentation, dependence on international instruments and tensions between global environmental objectives and local socio-economic realities. The forest thus becomes a space of permanent negotiation and socio-political transactions, even latent conflictuality, where conservation logics, development imperatives and social demands confront each other.

Thus examining forest policy, its strategic and programmatic frameworks brings us back to noting that forest governance denotes not only the limits of legitimacy of technicist public action, its effectiveness is increasingly questioned. All kinds of more or less precise observations indicate that it is becoming increasingly difficult to practice consensual management. Current forest strategies testify to the difficulty of designing and implementing a sustainable management model. Participatory structures have difficulty establishing themselves and operating according to software that looks at the diversity of subjects and practices and recognizes incommensurable formations.

Certainly the new participation instruments put in place by forest administrations in the southern Mediterranean have their value, especially now that representative organizations (Association and cooperatives...) are trying to support politically representative structures (local authorities) and create a dynamic of territorial development in favor of forest development. New instruments, certainly, but which continue to carefully keep traditional structures aside. In light of these questions, one has the right to be surprised by such agoraphobia expressed by traditional structures whose incommensurability struggles to be recognized and represented as such. What system is able to observe this diversity and recognize its legitimacy to sit in its structures?

Reinventing Participation: Problem of Plurality of Subjects and Hybridization of Local Solidarities

Participation in environmental governance is a multidimensional problem. It will therefore be necessary to determine who participates and how? The nature and dimensions of degradation of nature and its spaces? The neutrality of participatory spaces is to be questioned. One of the manifestations of this participation is to establish a regime of membership and belonging and therefore in fact, exclusion. It is indeed by this that we can measure the advance of exclusion which goes hand in hand with the opportunities for action and inclusion that this participation offers (Menzou, 2024). Identifying the paradoxes of participation in the forest governance scheme requires particular attention that goes beyond political and institutional discursive analysis. Thus, can we speak of governance in a posture where administrative imposition and regulation is one of the main characteristics? Where social play is likely lacking both at the level of constitution and at the level of development of environmental policies...

At first glance, the simple meaning of the notion of participation could only attract the eye of researchers, like those of public authorities, towards the population and users, but it paradoxically obscures the participation of the State (institutions and public services) in fighting against forest degradation which makes the functioning of sustainable development conflictual (Farcy & Huybens, 2016).

If the objectivity of facts determines neither individual reactions nor collective habitus, it is important to question cognitive schemes and reference frameworks whose principles help understand socio-political facts and public action when it comes to participation and according to what perception it is elaborated (Goffman & Joseph, 2009). In this regard, population participation in forest resource management should take into account the incommensurability of subjectivities and look at the diversity of actors and stakeholders, not in relation to management imperatives, but from a paradigmatic point of view. In this sense, the approaches to organizing user populations in associations and other ESS structures would be a vertical approach rather than explaining it as reality in terms of participation (Menzou, 2020).

Thus, if the institutional setup in modern countries does not present a certain plurality of structures and relationships, it is difficult to perceive or recognize the autonomy of action of actors and stakeholders. Hence the interest in questioning the nature and scope of participation in a system that claims to be democratic - In its modern acceptance where it means the production of man, thus opposing the reduced sense where we only observe procedures for legitimizing power - (Nancy, 2009) or one in democratic transition. For countries where systems are hybrid, as illustrated by the experience of the southern Mediterranean, it is interesting to observe participation as ontological reality and/or social fact instead of constructing it as an object excluding any other form of participation transcending modern technical and political temporality. It is thus that participation could stand against democracy (Godbout, 2014a).

This critical reaction, facing the meaning of participation, is supposed to lead us not to implicitly avert our gaze from the will to exclude any definition of communities guaranteeing their self-determination and their links with the forest and recognizing their knowledge to avoid any drift of cognitive hierarchization (Farcy & Huybens, 2016). From this point of view, we see the dynamic link that exists between State and communities where we witness a likely imposition of a definition of participation excluding other definitions of rights holders with a view to supporting public power.

The democratic challenge, in its integrative and non-impolitical sense, comes up against, among other things, the conception and construction of participation through the prism of effectiveness and efficiency within the framework of modern governance. if 'The effectiveness of policy implementation requires stakeholder participation and commitment to good governance' (FAO, 2018), participation is designed and implemented as a means or mechanism of 'acting with' in the sense of J.L Nancy (Nancy, 2007) and no longer as an incommensurable paradigm calling into question the legitimacy of representations and powers which is historically opposed to the principle of division of labor that characterizes modern societies (Godbout, 2014b). In this perspective emerges a conception where legitimacy gives way to legality. This appears as a legal-political regulation mechanism more than social, favoring all the more socio-political conflictual possibilities and excluding any transactional eventuality.

Conflicts between rights holders and users of forest spaces testify to the limits of the governance model implemented in not yet modern countries. These conflicts translate the will to implement participation through the prism of 'acting with' where participants are considered as objects as soon as they are redefined as a legal category (cooperative or association), and because of their spatiotemporal proximity which is of a contingent nature when we exclude knowledge and bypass collective memory, which undoubtedly refer to an incommensurable sense of space and time looking at diversity as a principle of participation (Menzou, 2024).

Faced with this diversity, which escapes all analytical prospection and development perspective, and whose sociological observation is expressed not only by the proliferation of stakeholders but also by the multiplicity of complex situations loaded with uncertainties regarding public action, we can perceive how participation can constitute a privileged recourse. But what meaning to give to participation? what form can we profile for participation in order to look at the diversity of subjects without reducing them to legal categories?

The idea of involving local populations in the forest economy in 1976 in Morocco, while inscribing this idea in a deliberative framework, was innovative at the time. On the legal level, participation was a constructivist perspective resonating with territorial planning at that time when we witnessed communal, provincial and regional organization, marking the ramification of power at the territorial level (Mhirit & Benchekroun, 2006).

New stakeholders have already established themselves around the forest problem. Within this multiplicity of stakeholders, each stakeholder presents different structurations, sometimes divergent interests due to the incommensurability of representations that each stakeholder possesses of the forest. The problem thus posed by political bodies is that of participation spaces (Menzou, 2024). We are therefore accountable for the number of established bodies and financial receipts established for the benefit of local authorities by forestry projects. But, it is always degradation that constitutes the leitmotiv of official discourse (CESE, 2023), combined with several forms of social exclusion that denote the non-participation of the population in the fight against forest degradation.

Conditional participation' consists in constructing both the population and its relationship with the forest as a legal object, from which can certainly result achievements in terms of acceptance of forestry projects and awareness of forest conservation against degradation (Moukrim et al., 2019). However, it should be noted the foundations of this form of participation in other incommensurable registers. Even if this participation technically arouses the interest of researchers and would likely constitute a lever to fight against forest degradation (Aziz et al., 2014), it would only be a factor of degradation and exclusion of incommensurable forms of participation, particularly the *Jmaa*, where participation joins that of order and establishes the place of diversity. It does not simply mean taking part in, but is situated at the ontological level and consequently implies the relationship between incommensurable subjects (Modern and pre-modern...) within the framework of 'being-in-common' where the common ontologically precedes any individuality involving the participation of all subjects that modern participation contributes to their loss.

To understand this conditional participation through the prism of loss of subject, it is important to proceed with a change of perspective and focus on socio-legal reconfigurations in rural areas, in particular the forms and structures of forest solidarity that find their substantial reality in an extension of State power seeking its legitimation in this new formula of collectivization.

Forest Governance put to the Test of Administered Forest[2].

Analysis of participation in the forests of the southern Mediterranean shows that participatory dynamics are less a matter of institutionalized governance than a set of pragmatic, often informal and asymmetric compromises, aimed at maintaining an unstable social balance between actors with divergent interests. The use of conditional participation clearly illustrates the illusion of plurality of subjects to inscribe us in analytical perspectives towards non-hierarchical forms of social coordination, involving a plurality of public and private actors, aiming to resolve complex collective problems through a reconfiguration of modes of State intervention in a decentralization scheme (Jessop, 1998).

We thus witness substitution of governance for government, like that of regulation for reglementation, ethics for morality or norm for rule, signifies the suppression of the gap between being and ought to be (Supiot, 2015). This substitution relies on the univocity of technical language which ensures performance of action, seemingly allows to master not only spaces (Biodiversity, environment, city, forests...) but also subjects. It is in this sense that governance is a sort of depoliticization of citizenship (Rosanvallon, 2014) and a sort of dejuridicization of the person (Supiot, 2015) where the citizen is substituted by stakeholders, see a discourse of depoliticization where we witness a transformation of commons into resources to be managed, the replacement of collective uses by individual rights (Deneault, 2013).

Moreover, the reappropriation of forest spaces by users and rights holders would be a thesis that can prosper in view of the financial interests that are presented as levers of change in perception and social representation of forests as a source of income and local development (Lahssini & Vanuxem, 2016), however it obscures that governance of commons is incommensurable with the construction of forests as public goods and to law as an enemy of the common good

Ostrom's model governance of commons (Ostrom & Baechler, 2010), presupposes the existence of geographically delimited resources managed by a community. If forests provide ecosystem services, it is difficult to apprehend them as commons, according to this model, due to the absence of collective management by the community. State ownership, by its legal definition being an exclusive property, thus presents one of the substantial limits to inscribing forests in a governance posture on the one hand (Nadir, 2008), and on the other hand the notion of common does not seem particularly adapted to protect forests in the modern context (Herzenni, 2008).

Participation, as a sine qua non condition in the definition of governance, is to be defined not only as a process but essentially as a finality. It is by no means about coordination of competing interests but one where substantial normativity is substituted for axiological neutrality thus redefining authority as a relational and non-sovereign category where powers are asymmetric (Borrini & Jaireth, 2007). It is thus that governance falls within the framework of normativity and not description. The elaboration of a conceptual framework for forest governance would be a sketch that obscures not only this normative aspect but mainly the administered nature of forest spaces substituting appropriated nature.

The FAO conceptual framework on forest governance has progressively imposed itself as an international reference, by promoting principles such as participation, transparency, sustainability, equity and accountability. However, behind this apparent normative robustness emerge several conceptual, political and epistemological limits that reduce its analytical and operational scope, particularly in forest contexts of the Southern Mediterranean. Governance reduced to a set of 'good practices' where participation presents itself as procedure and not as a relationship that measures the acceptability of public action without any consideration of power asymmetry which generates a reproduction of existing hierarchies (State / communities, economic actors / local populations, experts / vernacular knowledge...). Indeed, the forest governance index applied to the case of the forest space of Mamora, the world's largest cork oak forest, located in the southern Mediterranean in Morocco, translates average forest governance with a score of 2.5/5 (Table 3).

Table 3. Result of the forest governance index of Mamora in Morocco (Menzou, 2024)

Dimension	Score (0-5)	Justification
1. Legal and political framework	3	National forest laws apply, but local implementation remains limited by insufficient financial and human resources.
2. Transparency and access to information	2	Forest data specific to the region (e.g., forest cover, forest status) are not very accessible to the public. There is a lack of communication tools to raise citizen awareness about forest management.
3. Stakeholder participation	2	Participatory initiatives remain marginal. Local communities are involved in some reforestation programs, but their influence on forest planning and management is limited.
4. Law enforcement and control	3	Although efforts exist (monitoring by forest agents and guards mobilized by forest cooperatives, silvopastoral associations, and Local Authorities), illegal logging and overgrazing continue to be major challenges, particularly in peri-urban and rural areas.
5. Sustainable resource management	1	Reforestation and conservation projects are underway (e.g., eucalyptus, acacia, and pine plantations and regenerations of Holm oak). However, anthropogenic pressures and lack of long-term monitoring hinder their effectiveness.
6. Social and economic equity	2	Benefits derived from forest exploitation benefit local communities little, who depend mainly on forests for firewood and grazing. Initiatives aimed at integrating these populations remain limited.

This result informs us, certainly of the effort to be deployed to increase the score and support forest governance, but it obscures the central problem of governance: govern what? The forest administered by the State or the forest appropriated by rights holders?

The assessment of management of the Mamora forest in Morocco informs us of public action efforts to plan Mamora with the aim of maximizing cork production to satisfy the international market. From then on, de-densification of cork oak stands continues to profile the posture of the Mamora forest. The area of cork oak forest has gone from 100,000 ha in 1951 to approximately 65,000 ha in 2012, an average regression rate of approximately 630 ha/year. Participation in sustainable forest management is also reduced to the number of associations and cooperatives established and having contracts with the administration, thus excluding any ontological conflictuality from quantification. The real was referred to the benefit of its mathematical representation (Menzou, 2024). In sum, the forest management plans that have succeeded at the level of the Mamora space bring out an observation according to which we can correlate project performance with perpetual degradation of Mamora and its native natural species for the benefit of introduced allocthonous species mainly for economic reasons.

CONCLUSION

This chapter aimed to call into question forest governance put to the test of participation in a perspective of socio-ecological sustainability. Dwelling both on the limits of the conceptual model of forest governance and on the syndrome of continuous degradation of forest spaces particularly in the southern Mediterranean, it thus questions the representativeness of participation structures to propose a bi-representative system where governance bodies should combine elections and sortition which undoubtedly makes use of what the direct democratic tradition has best (the horizontal dimension of participatory deliberation) and also integrates what the technocratic tradition has best by considering at their fair value the technical skills of the forest administration (the non-elected without however giving them the last word) and finally it revalues what classical representative democracy has best.

REFERENCES

Altman, I., & Low, S. M. (Eds.). (1992). *Place Attachment*. Springer US., DOI: 10.1007/978-1-4684-8753-4

Aspe, C., & Genin, D. (2014). À chacun sa biodiversité : Approches savantes et locales de la gestion des ressources et des milieux dans le Sud de la France et le Haut Atlas marocain. *Revue d'ethnoécologie*, (5).

Aubert, P.-M. (2014). Les évolutions de la politique forestière au Maroc : Entre réappropriation du modèle forestier français et idéalisation de la tribu. *Revue Forestière Française, 305*(4), 305. DOI: 10.4267/2042/53627

Aubert, P.-M., & Sabir, M. (2014). Du technique au politique et retour : État des lieux et enjeux dans la gestion des forêts au Maroc. Introduction au dossier. *Revue Forestière Française*, (4), 295. DOI: 10.4267/2042/53626

Aziz, L., Flouchi, S., & Msiah, F.-E. (2014). *Acquis et défis de la gestion participative des ressources forestières : Cas du Parc National d'Ifrane*. DOI: 10.60569/1-A4

Barrière, O. (2011). Pluralisme juridique et patrimonialisation : Entre paradigmes de l'« appropriation » et du « patrimoine commun ». In *La question du patrimoine en Guyane* (Illustrated, p. 453). Ibis Rouge.

Barrière, O., & Barrière, C. (Eds.). (1997). *Le foncier-environnement : Fondements juridico-institutionnels pour une gestion viable des ressources naturelles renouvelables au Sahel*. Organisation des Nations unies pour l'Alimentation et l'Agriculture.

Beau, R., & Larrère, C. (2018). *Penser l'anthropocène*. Sciences po, les presses.

Benabou, A., Moukrim, S., Laaribya, S., Aafi, A., Chkhichekh, A., Maadidi, T. E., & Aboudi, A. E. (2022b). Mapping Ecosystem Services of Forest Stands : Case Study of Maamora, Morocco. *Geography, Environment, Sustainability, 15*(1), 141–149. DOI: 10.24057/2071-9388-2021-047

Benabou, A., Moukrim, S., Lahssini, S., Aboudi, A. E., Menzou, K., Elmalki, M., Madihi, M. E., & Rhazi, L. (2022a). Impact of climate change on potential distribution of Quercus suber in the conditions of North Africa. *Biosystems Diversity, 30*(3), 289–294. DOI: 10.15421/012231

Borrini, G., & Jaireth, H. (Eds.). (2007). *Sharing power : Learning-by-doing in co-management of natural resources throughout the world*. Earthscan.

Boutefeu, B. (2005). L'aménagement forestier en France : À la recherche d'une gestion durable à travers l'histoire. *Vertigo, 6*(2), 6–2. DOI: 10.4000/vertigo.4446

Boutefeu, B. (2009). *La forêt mise en scène : Attentes des publics et scénarios de gestion de la forêt*. L'Harmattan.

CESE. (2023). *Les écosystèmes forestiers du Maroc : Risques, défis et opportunités*. Conseil Economique, Social et Environnementa. [Auto-saisine]

Charles, L., & Kalaora, B. (2001). La nature administrée. L'Europe, la France et la politique de la nature. [Cairn.info.]. *Le Debat, 116*(4), 47–63. DOI: 10.3917/deba.116.0047

Charles, L., & Kalaora, B. (2003). Sociologie et environnement en france. L'environnement introuvable ? *Écologie & politique, 27*(1), 31-57. Cairn.info. DOI: 10.3917/ecopo.027.0031

Chiasson, G., Boucher, J. L., & Thibault, M. (2005). La forêt plurielle : Nouveau mode de gestion et d'utilisation de la forêt, le cas de la Forêt de l'Aigle. *VertigO*, (Volume 6 Numéro 2). DOI: 10.4000/vertigo.4298

de Arano, I. M., Garavaglia, V., & Farcy, C. (2016). Chapitre 5—Les forêts face au changement global. In *Mediterra 2016 : Zéro gaspillage en Méditerranée* (p. 117-138). Presses de Sciences Po. DOI: 10.3917/scpo.cihea.2016.01.0117

Defarges, P. M. (2022). *La gouvernance* (6e éd. mise à jour). Que sais-je ? Deneault, A. (2013). *Gouvernance : Le management totalitaire*. Lux éd. [diff. Harmonia Mundi].

Descola, P. (2015). *Par-delà nature et culture*. Gallimard.

Etienne, L. R. (1997). Patrimonialité plutôt que propriété : Approches de politiques foncières en Afrique et dans l'océan indien dans une perspective de gestion patrimoniale et environnementale. In *Droits de propriété et environnement* (pp. 321–334). Dalloz.

FAO. (2013). *Etat des forêts méditerranéennes 2013*. l'Organisation des Nations Unies pour l'alimentation et l'agriculture.

FAO. (2018). *SOFO 2018—La situation des forêts du monde 2018*. www.fao.org. https://www.fao.org/state-of-forests/fr/

FAO. (2022). *La situation des forêts du monde (SOFO) 2022 : Des solutions forestieres pour une relance verte et... Des economies inclusives, resilientes et durables*. FOOD & AGRICULTURE ORG.

FAO. (2025). *The status of Mediterranean forests*. FAO., DOI: 10.4060/cd6921en

Farcy, C., & Huybens, N. (Eds.). (2016). *Forêts, savoirs et motivations*. L'Harmattan.

Fay, G. (1986). Désagrégation des collectivités et dégradation des milieux dans le Haut Atlas marocain. *Revue de l'Occident Musulman et de la Mediterranee*, *41*(1), 234–248. DOI: 10.3406/remmm.1986.2121

Foucault, M. (2004). *Sécurité, territoire, population : Cours au Collège de France, 1977-1978*. Seuil : Gallimard.

Godbout, J. (2014a). *La participation contre la démocratie : Suivi de L'engagement: une fidélité* (Deuxième édition). Liber.

Godbout, J. (2014b). *La participation contre la démocratie : Suivi de L'engagement: une fidélité* (Deuxième édition). Liber.

Goffman, E., & Joseph, I. (2009). *Les cadres de l'expérience*. de Minuit.

Goxe, A. (2010). Chapitre 5. L'évaluation des politiques territoriales au regard du développement durable. In B. Zuindeau (Éd.), *Développement durable et territoire : Nouvelle édition originale* (p. 71-81). Presses universitaires du Septentrion. DOI: 10.4000/books.septentrion.15389

Hermet, G. (2004). Un régime à pluralisme limité? À propos de la gouvernance démocratique. *Revue Française de Science Politique*, *54*(1), 159–178. DOI: 10.3917/rfsp.541.0159

Herzenni, A. (2008). *Aires forestières communautaires conservées dans le Haut Atlas central*. IUCN.

Hufty, M. (2005). La gouvernance internationale de la biodiversité. *Etudes Internationales*, *32*(1), 5–29. DOI: 10.7202/704254ar

Jessop, B. (1998). The rise of governance and the risks of failure : The case of economic development. *International Social Science Journal*, *50*(155), 29–45. DOI: 10.1111/1468-2451.00107

Kalaora, B. (1993). *Le musée vert : Radiographie du loisir en forêt*. L'Harmattan.

Kalaora, B., Poupardin, D., & Larrère, R. (2023). *Le corps forestier dans tous ses états : De la Restauration à la Belle Époque* (Reproduction en fac-similé). l'Harmattan.

Komarolya, D. V., & Maleknia, R. (2025). Identifying indicators of sustainable management of urban forests : A qualitative approach. *Trees, Forests and People*, *22*, 101084. DOI: 10.1016/j.tfp.2025.101084

Lahssini, S., & Vanuxem, S. (2016). Les Paiements pour Services Écologiques ou la reconquête de la Mâamora par ses habitants? *Développement durable et territoires. Économie, géographie, politique, droit, sociologie*, (Vol. 7, n°1). *Développement Durable & Territoires, 7*(1). Advance online publication. DOI: 10.4000/developpementdurable.11165

Lefebvre, H. (2000). *La production de l'espace* (4e éd). Anthropos.

Lionel, C., & Kalaora, B. (2001). La nature administrée. L'Europe, la France et la politique de la nature. *Le Débat, 116*(4), 47-63. Cairn.info. DOI: 10.3917/deba.116.0047

Luhmann, N. (2001). *La légitimation par la procédure.* Presses de l'Université Laval. DOI: 10.1515/9782763712598

Marghadi, S. (2009). Intégration des méthodes d'aide à la décision dans l'aménagement multifonctionnel des forêts au Maroc (Doctoral dissertation, UCL-Université Catholique de Louvain).

Menzou, K. (2020). Forêts marocaines et usages. La formation en économie sociale et solidaire des forestiers. In J.-L. Laville (Éd.), *Comment former à l'économie sociale et solidaire?* Presses universitaires de Rennes.

Menzou, K. (2024). *Dégradation des espaces forestiers au Maroc : Risques et action publique (Cas de l'espace forestier de la Mamora* [Thèse en Aménagement et Développement durable.]. l'Institut National d'Aménagement et d'Urbanisme (INAU) de Rabat.

Mhirit, O., & Benchekroun, F. (2006). *Les écosystèmes forestiers et périforestiers : Situation, enjeux et perspectives pour 2025* (Le maroc possible : une offre de débat pour une ambition collective, p. 393-483). http://www.rdh50.ma

Montouroy, Y. (2014). Le changement dans la gouvernance globale des forêts:La recentralisation de l'action européenne contre le bois illégal. *Gouvernement et action publique, 3*(1), 107-128. DOI: 10.3917/gap.141.0107

Moukrim, S., Lahssini, S., Naggar, M., Lahlaoi, H., Rifai, N., Arahou, M., & Rhazi, L. (2019). Local community involvement in forest rangeland management : Case study of compensation on forest area closed to grazing in Morocco. *The Rangeland Journal, 41*(1), 43–53. DOI: 10.1071/RJ17119

Moukrim, S., Lahssini, S., Rhazi, M., Menzou, K., El Madihi, M., Rifai, N., Bouziani, Y., Azedou, A., Boukhris, I., & Rhazi, L. (2022). Climate Change Impact on Potential Distribution of an Endemic Species *Abies marocana* Trabut. *Ekologia (Bratislava), 41*(4), 329–339. DOI: 10.2478/eko-2022-0034

Nadir, B. (2008). *Domanialité et environnement : Cas des eaux et forêts* (1ère éd). Editions IDGL.

Nancy, J.-L. (2007). L'être-avec de l'être-là. *Cahiers philosophiques, 111*(3), 66-78. Cairn.info. DOI: 10.3917/caph.111.0066

Nancy, J.-L. (2009). Démocratie finie et infinie. In *Démocratie, dans quel état?* (p. 77-94). La Fabrique Éditions. Cairn.info. DOI: 10.3917/lafab.colle.2009.01.0077

Ost, F. (1995). *La nature hors la loi : L'écologie à l'épreuve du droit*. Découverte.

Ost, F. (2003). *La nature hors la loi. L'écologie à l'épreuve du droit*. La Découverte. Cairn.info. https://www.cairn.info/la-nature-hors-la-loi--9782707139368.htm

Ostrom, E., & Baechler, L. (2010). *Gouvernance des biens communs : Pour une nouvelle approche des ressources naturelles*. De Boeck.

Peluso, N. L., & Vandergeest, P. (2001). Genealogies of the Political Forest and Customary Rights in Indonesia, Malaysia, and Thailand. *The Journal of Asian Studies, 60*(3), 761–812. DOI: 10.2307/2700109

Ribot, J. C., & Peluso, N. L. (2003). A Theory of Access*. *Rural Sociology*, *68*(2), 153–181. DOI: 10.1111/j.1549-0831.2003.tb00133.x

Romagny, B., Aderghal, M., Auclair, L., Ilbert, H., & Lemeilleur, S. (2018). Communs en crise. Agdals, terres collectives, forêts et terroirs au Maroc. *Revue internationale des études du développement, 233*(1), 53-73. Cairn.info. DOI: 10.3917/ried.233.0053

Romagny, B., Auclair, L., & Elgueroua, A. (2008). La gestion des ressources naturelles dans la vallée des Aït Bouguemez (Haut Atlas) : La montagne marocaine à la recherche d'innovations institutionnelles. *Mondes en développement, 141*(1), 63-80. Cairn.info. DOI: 10.3917/med.141.0063

Rosanvallon, P. (2014). *La contre-démocratie : La politique à l'âge de la défiance*. Éd. Points.

Schmutz, J. (2001). La Doctrine mediévale des causes et la théologie de la nature pure (XIIIe-XVIIe siècles). *Revue Thomiste*, *101*(1-2), 217–264.

Serres, M. (1990). *Le contrat naturel*. Editions F. Bourin.

Simenel, R., Mohammed, A., Sabir, M., & Auclair, L. (2016). Cairn, borne ou belvédère? Quand le naturalisme et l'analogisme négocient la limite entre espace cultivé et forêt au Maroc. *Anthropologica*, *58*(1), 60–76. DOI: 10.3138/anth.581.A04

Stone, C. D. (1974). *Should trees have standing? Toward legal rights for natural objects*. Kaufmann. DOI: 10.2307/1228012

Supiot, A. (2015). *La gouvernance par les nombres*. Inst. d'Études Avancées de Nantes. [u.a.]

Tuan, Y.-F. (2011). *Space and place : The perspective of experience* (7. print). Univ. of Minnesota Press.

UN. (2017). *Adoption du cs forets*. https://press.un.org/fr/201/ag11977.doc.htm

Vidalou, J.-B. (2017). *Être forêts : Habiter des territoires en lutte*. Zones.

ENDNOTES

1 The concept of "conditional participation" refers to a specific mode of integrating local populations into public forestry projects. It denotes a form of participation contingent upon the prior acceptance of a set of institutional and normative transformations, through which social solidarities inherited from customary systems are gradually reconfigured according to the categories of positive law and modern management. In other words, access to forestry development schemes is conditioned by the enrollment of communities in standardized legal structures—such as silvopastoral associations or forest cooperatives—that translate the local social universe into the administrative, contractual, and quantifiable language of public action.

This dynamic is part of a broader movement in which contemporary governance tends to replace regulations rooted in social and symbolic relations with technical mechanisms based on evaluation, contractualization, and legal normalization. In this context, participation is no longer merely a recognition of communities' political capacity to define the aims of collective action; rather, it becomes a procedure of alignment with predetermined institutional frameworks. Traditional forms of mutual aid, collective management, and shared responsibility—often grounded in kinship, neighborhood ties, or customary law—are thus reconfigured to meet the requirements of administrative legibility and indicator-based governance.

The anthropological consequence of this process is a transformation in the status of social actors. The community member, bearer of historicity, collective memory, and deliberative capacity, tends to be redefined as a functional agent within a programmatic apparatus. Conditional participation thus operates a shift from the social subject, capable of initiative and judgment, to a reactive object, adjusted to institutional incentives and constraints. This displacement is not merely organizational; it is epistemological and political, as it reshapes the criteria of legitimacy for local voices and the modes of recognition of vernacular knowledge

2 The administered forest appears as the object of a configuration of heterogeneous actors — environmentalists, experts, public authorities, and landowners — whose interactions contribute to structuring it according to specific sectoral logics. It is thus repeatedly confined within distinct policy frameworks and forms of compartmentalization that are both theoretical and institutional in nature. As a result of these processes of specialization and framing, the forest no longer emerges as a common good subject to collective deliberation, but rather as an object of partial appropriation by distinct groups, most often endowed with specialized expertise. For more details (Lionel & Kalaora, 2001)

Chapter 3
Silvopastoral Management and Perspectives for Participatory Forest Management in North Africa

Mustapha Naggar
http://orcid.org/0009-0009-9336-2249
Ecole Nationale Forestière d'Ingénieurs de Salé, Morocco

ABSTRACT

In North African countries (Morocco, Algeria, Tunisia), silvopastoral ecosystems display high diversity shaped by biotic and abiotic interactions, yet face critical challenges from overgrazing pressure by millions of small ruminants on declining forest areas, resulting in forage deficits and widespread desertification. Silvopastoral management has evolved from traditional community-based systems, dominant until the colonial period, through a productivist model introduced during the French Protectorate, toward contemporary integrated participatory frameworks. This multifunctional approach combines technical prescriptions, rotational grazing, controlled stocking rates, fodder shrub plantations, and assisted natural regeneration, with organizational measures involving local communities through pastoral associations. This approach integrates climate adaptation, ecosystem services valuation, and emerging technologies for real-time pasture monitoring, offering promising pathways toward sustainable silvopastoral management facing global change challenges in North Africa.

DOI: 10.4018/979-8-3373-6239-7.ch003

INTRODUCTION

Historically, the Mediterranean basin is the birthplace of cereal domestication, small ruminant domestication, and water management. Pastoral areas cover 120 million hectares, excluding desert zones (Bourbouze & Donadieu, 1987). Mediterranean rangelands are characterized by vegetation periods of varying lengths and consequently by significant variations in the quantity and quality of available fodder resources.

Most of the Mediterranean region produces less than one ton of dry matter per hectare per year. In their current state, rangelands can only be exploited extensively, requiring high herd mobility so that livestock can benefit from the complementarity between different grazing areas in forests and non-forest zones. Silvopastoral systems, integrating woody vegetation and livestock farming, are increasingly recognized as a sustainable land use practice that promotes biodiversity and the provision of ecosystem services in the Mediterranean (Peri et al., 2024). Silvopastoral areas constitute a major component of the Mediterranean space, having shaped the landscape for several centuries. However, both in their structures and operational modalities, they are characterized by great diversity resulting from ecological situations as well as the cultural, social, and economic contexts of the human societies that shape them.

Pastoral practices are governed by the availability of pastoral resources according to vegetation zonation, livestock management systems, climatic conditions, and silvopastoral practices specific to each territory (Bourbouze, 2003). It should be noted that wooded areas, whose woody and non-woody products are subject to other exploitation modalities, are also multi-use, and most often, several types of stakeholders interact there. Silvopastoral systems present complex impacts on a diverse range of outcomes, requiring an integrative approach to maximize their positive contributions to agricultural and natural production systems (Peri et al., 2024).

Indeed, the forest offers a privileged pastoral space due to the nature, diversity, and richness of its floristic species. These species, consisting of trees, shrubs, and evergreen bushes, provide through their branches, young shoots, and accessible foliage phytomass, fodder availability throughout the year in all seasons. These plant formations also constitute potential mitigation strategies to improve microclimatic conditions and reduce thermal stress risks in the context of climate change (Lemes et al., 2021). This particular characteristic of forest rangelands differentiates them from other non-forest rangelands whose fodder production remains highly dependent on climatic hazards. Therefore, as soon as drought persists, a limitation of non-forest pasture production is observed, and it is always the forest that suffers, through additional overload, the backlash of climatic hazards. In the current context of global changes, North African silvopastoral systems offer opportunities to create synergies between climate change mitigation and adaptation by maintaining mosaic landscapes that strengthen the provision of ecosystem services and increase the resilience of these areas (Hernández-Morcillo et al., 2018).

FORESTRY AND SILVOPASTORAL CONTEXT IN NORTH AFRICA

Common Elements and Features of the Forestry and Silvopastoral Sector in North Africa

The forests of North African countries play a strategic role on socio-economic, pastoral, and environmental levels. They constitute a heritage through the diversity of ecological systems and biodiversity they integrate and through the importance of their extent over approximately 14.2 million hectares, of which: (i) 9 million hectares in Morocco (holm oak, cork oak, thuya, cedar, argan, cypress, esparto grass stands and various secondary species), (ii) 4.1 million hectares in Algeria (cork oak, holm oak and pines...) and (iii) 1.1 million hectares in Tunisia (Aleppo pine, cork oak, zeen oak, maritime pine and others...).

Silvopastoral ecosystems constitute an important food resource for rural population livestock. Indeed, they are called upon to produce fuelwood, timber, secondary products (mushrooms, aromatic and medicinal plants, beekeeping, acorns...) and to protect soils against various forms of erosion. Thus, the North African forest offers a multi-use space where pastoral and social vocation predominates.

As an essential component of natural resources, North African forest ecosystems are subject to Mediterranean climatic conditions, characterized by water scarcity during a good part of the year, which considerably limits the growth of forest species. Indeed, aridity affecting more than 90% of the territory, coupled with excessive resource use, contributes to the fragility of forest ecosystems. As a result, almost all lands are threatened or affected, to varying degrees, by desertification. In Africa, approximately 45% of the continental area is affected by desertification; the five Mediterranean countries of North Africa (Egypt, Algeria, Libya, Morocco and Tunisia) are also exposed to the risk of this phenomenon and present zones of high to very high severity (UNCCD, 2022).

Disruptions of ecological and environmental balances, under the effect of human action and developments related to global changes, position the forestry sector at the heart of sustainable development. Indeed, forest areas that play a primary protective role generate a multitude of goods and services for the community, often little perceptible to different stakeholders. Facing climate change, the North African region has experienced during the last decade some of the most intense droughts ever recorded, with significant drops in biomass production compared to usual annual averages.

However, despite their determining roles in soil and water protection and the fight against desertification, North African forests remain subject to strong energy and pastoral pressures. These constraints are essentially linked to:

i. Population growth and the poverty conditions it generates, imposing strong dependence of these rural populations on forest resources;
ii. Low incomes in rural areas: Production systems remain characterized by rainfed agriculture and extensive livestock farming. Difficulties in accessing socio-economic services and supervision have placed the rural population of forest areas in a situation of poverty and precariousness;
iii. Isolation and lack of access to information: The low involvement of rural populations in the design and implementation of development projects is linked to their weak organization, but above all to insufficient support in terms of supervision, training, and information. Infrastructure insufficiency is also considered a limiting factor for the valorization of agricultural production and access to services, particularly health and education.

Given the constraints mentioned and the magnitude of global changes, it becomes imperative to continuously adapt silvopastoral management and planning approaches to establish the foundations for sustainable management of North African forest ecosystems.

Silvopastoral Sector in Morocco

In Morocco, wooded and esparto grass areas cover nearly 9 million hectares and have an important social function mainly devoted to extensive silvopastoral livestock farming activity. These areas constitute a fodder reserve for national livestock and produce, in normal years, the equivalent of 1.943 billion fodder units per year, representing a pastoral production level of about 207 FU/ha/year. Compared to the level of pastoral withdrawals (4.392 billion FU), or a withdrawal level of 467 FU/ha/year, this reveals a pastoral overload of about 2.26 times the production potential (FAO, 2020).

The foliage phytomass capital of the tree and arborescent stratum can play a strategic fodder reserve role during lean periods and scarcity for local livestock and herders from other areas more affected by drought. Indeed, livestock grazing in forests amounts to more than 10 million heads of small ruminants (sheep and goats), representing 50% of national livestock. The fodder needs of this livestock exceed 3.5 billion FU, of which the equivalent of 42% comes from silvopastoral areas (FAO, 2020).

As a result, silvopastoral ecosystems remain subject to increasingly strong social pressure due to several factors, the most important of which are: (i) The transformations and upheavals experienced by agricultural production systems and land development in general, and the management of extensive livestock farming systems in particular, and (ii) The fragility of ecological motivations in favor of productivist arguments in most livestock and rangeland development projects and programs in arid and semi-arid zones (Naggar, 2021).

Moreover, silvopastoral livestock farming contributes significantly to social security in forest and periforest areas and remains a basis of the economic and social organization of rural communities. Goat farming in northern Morocco illustrates this economic importance well, with a contribution to household income varying from 68% for meat producers up to 100% for milk producers who do not have agricultural land (Chebli et al., 2021). These elements suggest that sustainable development initiatives for silvopastoral ecosystems require appropriate governance mechanisms in favor of the planning, management, and exploitation of silvopastoral resources and which integrate the evolution of pastoral practices in relation to forest areas.

In terms of silvopastoral resource governance in Morocco, it should be specified that silvopastoral space is part of the State's private domain, and that pastoral practices are considered use rights granted to local populations. The main stakeholders concerned are the forestry administration, agricultural services, user herders, pastoral associations and herder groups, territorial and ethnic communities, etc. These stakeholders are consulted during the preparation of participatory silvopastoral management plans and participate in the implementation of the prescriptions of said plans and in the organization of pastoral practices in forests, particularly in the management and exploitation of silvopastoral improvement perimeters.

Silvopastoral Sector in Tunisia

Tunisia has rich and diversified forest ecosystems that have always played determining environmental and socioeconomic roles in local and regional development. The total area of forests and rangelands amounts to 5.642 million ha (including esparto grass stands), representing 34% of the national territory. These ecosystems constitute an essential bulwark against desertification, particularly in a context where the majority of lands are threatened or affected by this phenomenon to different degrees (Bedoui, 2020).

Indeed, silvopastoral areas remain generally located in the North and Center-West. These ecosystems are diversified, ranging from maquis, garrigue to high-altitude forests globally dominated by formations based on Aleppo pine, cork oak, and zeen oak. These silvopastoral areas cover nearly 1,141,628 ha (Source: 2nd IFPN, 2010), in a context where aridity affects more than three-quarters of the national territory, coupled with excessive resource use, contributes to the fragility of silvopastoral ecosystems. As a result, approximately 65% of the Tunisian territory, mainly representing arid and semi-arid zones, is affected by desertification, particularly threatening productive agricultural lands.

The rural population of forest regions is about 800,000 to 1 million inhabitants, or 8 to 10% of the total population, drawing their essential subsistence from forest and silvopastoral areas and practicing livestock farming activities and collection of woody and non-woody products. For this population, these areas are considered a pastoral and energy reserve usable free of charge throughout the year.

The pastoral production of silvopastoral areas amounts to 400 million fodder units, equivalent to 40% of the national fodder balance in normal years (AINI, 2004). Moreover, the livestock using silvopastoral areas is estimated at 5 million small ruminants, including 4.5 million sheep and 500,000 goats (FAO-FOSA, 2001).

Indeed, Tunisia faces numerous challenges, mainly the preservation of silvopastoral resources and biodiversity conservation, sustainable valorization of woody and non-woody resources, restoration of natural rangelands threatened by desertification, socio-economic progress of forest and pastoral populations, and respect for commitments made under international conventions. Within the framework of the National Strategy for Sustainable Development and Management of Forests and Rangelands 2015-2024, Tunisia has implemented soil and water conservation programs, notably through the construction of benches and slopes to limit runoff and promote water infiltration into the soil (Saadani, 2015). With a view to improving silvopastoral territory management practices and the organization of pastoral practices, the global approach adopted by the General Directorate of Forests (DGF) in Tunisia is based on a participatory and partnership approach. Indeed, it involves promoting eco-development actions developed with and for user rural communities: valorization of woody and non-woody forest products, infrastructure, village water systems, silvopastoral development, income-generating activities, organization of populations around the promotion of sectors, etc., which condition, in return, the success and effectiveness of silvopastoral management programs. It goes without saying that this evolution must be accompanied by an adaptation of the missions and attributions of field forestry structures and, moreover, by a strengthening of the required capacities of forest managers and stakeholders concerned.

The new pastoral code in Tunisia established in 2023 structures the management, securitization, and improvement of pastoral systems. This code organizes pastoral areas, regulates transhumance, creates management bodies (Commissions). It defines rules for the development and exploitation of rangelands and silvopastoral areas through the establishment of management bodies (national and regional) and the recognition of sector stakeholders. The pastoral sector in Tunisia involves a multitude of public, technical, private, and community stakeholders, under the aegis of the Ministry of Agriculture (MARHP). Key stakeholders include the forestry administration (DGF), the institute of arid regions (IRA), the state land office (OTD), the Tunisian union of agriculture and fisheries (UTAP), and organized local communities such as mutual agricultural services societies (SMSA) focused on the sustainable management of pastoral and silvopastoral areas.

Forestry Sector in Algeria

In Algeria, silvopastoral areas consisting mainly of forests and maquis cover 4.1 million hectares, representing a afforestation rate of 16.4% for northern Algeria and only 1.7% if arid Saharan regions are also taken into consideration. The main natural forest formations are dominated by Aleppo pine, cork oak, holm oak, junipers, and thuya. Added to these forest areas are the esparto grass stands totaling 2.7 million hectares (FAO-FOSA, 2000). These areas remain subject to progressive degradation characterized by the replacement of forest species by maquis and scrubland, a phenomenon exacerbated by pastoral pressure and arid climatic conditions (Benabderrahmane & Chenchouni, 2010).

The practice of forest grazing in Algeria constitutes an important factor in the evolution of forest and esparto grass areas. The livestock grazing on these environments is estimated at nearly 11 million heads of small ruminants dominated by more than 80% by sheep. Considering agricultural activities, which often exclude herds for a long period of the year, the largest part of the current herd, except cattle, feeds in forests and maquis. The pastoral pressure on silvopastoral areas is about 3.5 small livestock units per hectare (SLU/ha), against a possibility barely exceeding 0.5 SLU/ha, while noting that the pastoral load has almost quadrupled in 30 years (World Bank, 2023).

Facing these challenges, Algeria launched in the 1970s the Green Dam Project (Green Dam Project), extending over approximately 3 million hectares between the Moroccan border to the west and the Tunisian border to the east, in the pre-Saharan zone located between the 200 and 300 mm isohyets, with the objective of controlling the progression of desertification in arid and semi-arid regions (Benhizia et al., 2021). This Project was successful for a few years, but then the plantations, mainly of pines, experienced severe dieback due to forest exploitation pressures, livestock overgrazing, and the proliferation of the pine processionary caterpillar. Also, in terms of desertification, the work of Mihi et al. (2022) highlighted the magnitude of the desertification risk which can reach up to 59% of the total area, with the aridity index and soil granulometry as the most critical conditioning factors.

Stakeholders in the silvopastoral domain in Algeria mainly include herders from forest and steppe zones, forestry administration services (DGF), Ministry of Agriculture services and research institutes for sustainable development, herder groups, which are structured around specific sectors that manifest themselves through integrated enterprises such as Production and Logistics Groups (AGROLOG), or family village groups, and entities specialized in genetics or feed.

PROBLEMATIC

In North Africa, the livestock farming system practiced in forests is of the extensive silvopastoral type represented essentially by small ruminants (26 million heads) of rustic local breeds well adapted to their environment and which are majority breeding stock. This livestock farming remains held by a majority of small herders, poorly organized and poorly supervised, and not favored by the current marketing system (World Bank, 1995).

In these countries, the pre-colonial period is characterized by tribal order and the predominance of various forms of community appropriation of rangelands and forests. With the colonial episode, the forest code crossed the Mediterranean and radically transformed the land ownership map and forest exploitation. The State became the owner and manager of almost all forests, both conservator of a threatened natural space and guarantor of productive silviculture. In Algeria, the State domain extends over rangeland lands, which partly received in Morocco and Tunisia the status of collective lands.

In the past, the practice of forest grazing referred to traditional concepts of extensive livestock farming and collective use of pastoral resources. Also, these practices were based on the complementarity of different grazing areas in forests and non-forest areas (transhumance; nomadism...) and communities imposed temporary closures (Agdal) to regenerate rangelands and extend the grazing period (Mahdi 2002).

Since the promulgation of the forest law of the beginning of last century, forest grazing has been recognized as a use right devolved to local populations and obeys administrative exploitation rules (grazing registration, payment of fees, defensibility of regeneration quarter rangelands, etc.).

Confronted with the social realities of pastoral populations neighboring forests, who are generally small herders with very limited productivity performance and whose fodder system is based on the free nature of pastoral resources, the application of these new pastoral rules could only be observed partially. And also, the organization of forest grazing practice remains confronted with serious challenges of silvopastoral ecosystem restoration and their adaptation to global changes (FAO, 1989).

Generally, silvopastoral ecosystems of North African countries present great diversity, and their production level does not reflect the real potentialities of the environment. This situation is explained by overexploitation of silvopastoral resources: overgrazing, topping and branch stripping of the tree and arborescent stratum during drought periods and even clearings (DGF-Algeria, 2007). A recent study reveals that in North Africa, overgrazing is concentrated in a particularly intense manner, contributing to the rapid degradation of silvopastoral systems, particularly in Tunisia where the loss of natural vegetation cover and biodiversity has accelerated in recent decades under the combined effect of overgrazing, cultivation, and unpredictable weather conditions including frequent droughts (Louhaichi et al., 2023). The main features characterizing this problematic are summarized as follows:

- *Silvopastoral areas threatened by desertification and irregular silvopastoral resources:* Given the magnitude of population growth in forest and periforest zones, the sedentarization of pastoralists, the progression of the market economy, and the recurrence of droughts, the modes and practices of forest rangeland use have undergone profound mutations inducing major imbalances between supply and pastoral demand at the forest area level. Pastoral activity has become very dependent on free silvopastoral resources regardless of climatic conditions and practically throughout the year. In case of climatic hazards, drought or cold, pressure intensifies on the silvopastoral capital through livestock concentration in forests. Climate changes amplify these imbalances, with more intense droughts and record temperatures that negatively affect the capacity of ecosystems to regenerate, even the distribution of several species (Benabou et al. 2022; Moukrim et al., 2019a, 2022).
- *A fodder deficit due to excessive withdrawals of pastoral resources:* Various silvopastoral management studies reveal a fodder deficit exceeding 30%. It emerges that these rangelands cannot by themselves satisfy the needs of livestock that is maintained by free fodder resources offered by the silvopastoral space (Naggar, 2000).
- *Use conflicts in silvopastoral areas:* Since the beginning of last century, silvopastoral space has passed from the state of property of ethnic communities to a state domain burdened with regulated use rights (grazing, fuelwood and timber collection for domestic use, collection of aromatic and medicinal plants, argan nut collection...). Thus, the use rights recognized by the legislator only reflect the access and use rights of goods and services provided by forests since always. But the exercise of grazing rights free of charge has resulted in rental livestock farming practiced without any concern for silvopastoral resource restoration.
- *The multiplication of pastoral offenses:* Through their green, persistent, and mobilizable phytomass throughout the year, forest stands, particularly deciduous trees, remain subject to branch stripping and topping whose products will serve as supplementary fodder for livestock in forest and periforest zones

during seasons with low fodder supply (summer, autumn), snow periods in mountain areas, and/or drought years.

- *Limited knowledge of silvopastoral systems and insufficient valorization of experiences:* Research work in the silvopastoral domain is generally insufficient. Indeed, the implementation of silvopastoral improvement programs suffers from the insufficiency or even absence of scientific and technical references for the choice and diversification of plant material (native species) adapted to the context of areas to be managed and to climatic constraints. Identified challenges include high demand for a limited range of plant species, insufficient nursery capacity, and incomplete understanding of best practices for promoting the planting of certain native species (Louhaichi et al., 2023).
- *A multi-stakeholder (multidimensional) problematic and lack of common vision:* Livestock farming activity is multi-sectoral, multi-stakeholder, and operates in the three components of territories: forest, collective, and agriculture. Indeed, pastoral practices in forest areas of state domain status, state forests, obey rules and customs specific to different human groups marked by strong rurality. The coordination and convergence of different sectoral strategies in this matter are essential to slow down the degradation level of silvopastoral ecosystems in a context marked by global changes.

PARTICIPATORY SILVOPASTORAL MANAGEMENT GUIDELINES

The Participatory Silvopastoral Management Approach

The elaboration of a participatory silvopastoral management plan requires aligning the possibilities of valorization and exploitation of all silvopastoral resources with the involvement of local stakeholders concerned, particularly user herders belonging to human groups surrounding these silvopastoral territories. This approach aims to favor participation and partnership in the concerted treatment of the pastoral question in forests (Naggar & Lahssini, 2017). Participatory approaches for natural resource management are essential to ensure that community members living in and depending on ecosystems can influence planning and decision-making, thus promoting transformative and inclusive governance towards sustainable outcomes (Fry et al., 2024).

The establishment of the participatory silvopastoral management plan is based on the following considerations: i) knowledge of the study environment, namely, the development of basic studies relating to the silvopastoral, forestry, geographical, and socio-economic context of the environment; ii) identification of the vocations of silvopastoral space in relation to multi-objective (or multi-functional) management; iii) identification of forest territories that represent a user and livestock movement space at the scale of a well-defined forest area; iv) diagnosis and participatory management with the involvement of community user groups (pastoral associations or cooperatives...) in the management of certain forest areas or the exploitation and development of their resources; v) integration of rangeland, biodiversity, energy biomass, and valorization of woody and non-woody forest products aspects into the proposed management.

It should also be noted that the three North African countries (Morocco, Algeria, Tunisia) have ratified and adhered to the three (3) United Nations conventions on the conservation of biological diversity (CBD), the fight against desertification (UNCCD), and the fight against climate change (UNFCCC), and have developed national adaptation and mitigation measures in this regard (FAO, 2010).

With a view to sustainable silvopastoral management that takes into account global changes (climatic, socio-economic), FAO has identified seven strategic axes for the MENA region that North African countries adapt to their specific contexts (FAO-MENA, 2022): i) development of forest legal frameworks to support silvopastoralism and adoption of sustainable pastoral practices; ii) strengthening of social capital and capacities of local communities; iii) promotion of good governance through stakeholder dialogue and the establishment of participatory management mechanisms; iv) co-production and mobilization of silvopastoral knowledge integrating traditional and scientific knowledge; v) consideration of the multifunctionality of silvopastoral systems including production, conservation, and ecosystem services; vi) diversification of value chains for woody and non-woody forest products, notably aromatic and medicinal plants, beekeeping, and ecotourism; vii) strengthening of monitoring and information systems for continuous evaluation of management programs.

These strategic orientations translate into priority actions on the ground: the development of participatory management plans, the reconstitution and restoration of degraded silvopastoral ecosystems to combat soil erosion, the development of alternative energy sources to reduce excessive fuelwood collection, the conservation of genetic resources and the use of drought-resistant ecotypes, the promotion of agroforestry as a flexible land use system, the valorization of forest products and identification of new economic opportunities, as well as the integrated management of rangeland areas in forests and non-forest areas with the restoration of community management pastoral practices.

Technical Prescriptions for Participatory Silvopastoral Management

The implementation of the participatory silvopastoral management plan on the ground consists, in each rangeland park exploited by the same ethnic group, of delimiting and materializing two zones of unequal areas. The smaller (in Morocco, at most 20% of the forest area) called "regeneration quarter" is regenerated during a period, during which it remains closed to direct livestock grazing, but grass mowing may be permitted. The rest of the forest, while continuing to receive the silvicultural treatments provided for by forest management (thinning, clearing...), is open to grazing while respecting as much as possible the usual principles of grazing registration and payment of fees...

The Regeneration Quarter (Area Reserved for Forest Rejuvenation)

No sustainable management of a forest can be guaranteed if, for a determined time, part of this ecosystem, "the regeneration quarter", is not protected and regenerated. Conditioning the sustainability and renewal of resources, this measure applies to all. However, it is often the subject of contestation from users who reproach it for: (i) diverting from grazing a significant area of their silvopastoral park and (ii) protecting, for too long a period, the silvopastoral space.

Regarding the regeneration quarter area, the practice in North Africa reserves at most one-fifth of the silvopastoral park. At this level, consultation with users or their representatives will focus on the size and spatial distribution of parcels to be regenerated, the choice of which is to be made consensually.

The duration of regeneration quarter protection depends essentially on the forest species and the ease with which it fructifies and regenerates. From 15 to 20 years for conifers (Cedrus atlantica), the theoretical duration of protection is about 10 years for broadleaves and conifers that regenerate by stump shoots (*Quercus suber, Q. rotundifolia, Tetraclinis articulata...*). The opening of closures can, of course, occur earlier if regeneration is achieved before the deadlines set by the planner.

Pastoral Interventions Outside the Regeneration Quarter

Outside the regeneration quarter which remains a sanctuary closed to grazing throughout the period planned by the planner to rejuvenate it, all the rest of the forest (at least 80% of the silvopastoral park area) is open to grazing and is subject to interventions aimed at improving the production of herbaceous cover and that of forest trees and shrubs. The main techniques for quantitative and qualitative improvement of silvopastoral resources are summarized as follows:

a) Treatment in Pastoral High Forest (Clearing-Thinning): Case of Holm Oak

The treatment in pastoral high forest of holm oak coppices, a species common to North African countries, has as its main objective the opening of stands, raising the canopy, and soil illumination necessary for the development of herbaceous cover and the increase of fodder production. This treatment also allows the improvement of the growth of conserved stems in order to subsequently obtain the best results in terms of fuelwood and timber production necessary for the local economy.

The intensity of interventions is dosed to obtain the compromise that allows optimizing pastoral and woody productions. Experiments conducted in the holm oak forest of the Middle Atlas have led to the adoption of a technical itinerary to be applied for holm oak coppices (Qarro, 2006) defined as follows: i) 20 years after the regeneration and coppicing cut, clearing with removal of 1/3 of the material among poorly performing shoots; ii) canopy control will continue every 10 to 15 years depending on the development state of holm oak so as to maintain an average cover of 45 to 60%; iii) the closure to be applied after each silvicultural treatment will last about two years. This to allow on the one hand the herbaceous cover to develop and on the other hand any new shoots not to take significant development and, for this, to allow their consumption by animals.

b) Rotational Pastoral Closure

Annual rest allows the reconstitution of plant reserves and seed production. This technique is gradually entering silvopastoral management and the ambition of the technician is to do everything possible for this practice to gain ground by implementing a simple movement scheme in the different rangeland territories that takes into account their level of fodder production.

This objective will be all the more easily achieved as popularization succeeds in making herders understand the advantages that rotation presents to rationalize the use of pastoral space with a view to extending the grazing period in forests and improving its contribution to the fodder balance.

c) Planting of Fodder Shrubs

The range of shrub species used in Maghreb countries remains limited to a few species of the genus *Cactus*, *Acacia*, *Atriplex*, *Chamaecytisus*..., the objective of these plantations is to increase fodder production in the short and medium term in order to relieve pressure on rangelands, by providing livestock feed less dependent on rainfall fluctuations, essential during lean periods or drought. These plantations must be integrated into the participatory management and valorization system of silvopastoral resources. Challenges remain, however, notably high demand for a limited range of plant species, insufficient nursery capacity, and lack of national coordination for the collection and multiplication of native species seeds (Louhaichi et al., 2023).

Organizational and Eco-Development Measures of the Participatory Silvopastoral Management Plan

The main measures to be retained are summarized as follows:

- Organizational measures and improvement of silvopastoral resource governance: the proposed activities focus on the mapping of users and rights holders of silvopastoral areas, support and accompaniment for the creation of pastoral organizations, rehabilitation of good silvopastoral practices, capacity building of pastoral associations, contractualization of commitments with pastoral organizations, and possibly the establishment of pastoral conflict resolution mechanisms. The experience of participatory rangeland management has demonstrated that this type of approach improves rangeland governance and the inclusive participation of herders in decisions relating to their lands, while improving rangeland productivity.
- Governance of silvopastoral resources: In terms of governance and based on a common and shared vision of local stakeholders of the silvopastoral space subject to the participatory management study, it is recommended to establish a silvopastoral management charter with the possibility open to stakeholders to take charge of part of the activities financed by the participatory silvopastoral management plan. Transformative and inclusive governance requires the presence of local actors with appropriate capacities, including the ability to participate significantly in decision-making processes and occupy leadership positions (Fry et al., 2024).
- Eco-development measures and income-generating activities: it involves proposing modalities for improving livestock system management adapted to the silvopastoral context of the study area, contributing to the valorization of livestock sectors with herder groups (meat, milk, wool, goat cheese...), to the development of valorization activities for non-woody forest products, such as aromatic and medicinal plants, beekeeping, ecotourism..., and to strengthening user capacities in terms of sustainable valorization of these products.
- The participatory silvopastoral management plan must also possibly propose research axes to progressively improve technical itineraries for restoring silvopastoral areas and anticipate new silvopastoral management technologies in a context of global changes and establish interactive databases be-

tween researchers and managers and the different organs of the participatory management system for silvopastoral ecosystems.

Compensation for Forest Closures: A Tool for Participatory Silvopastoral Space Management

With a view to initiating participatory management of silvopastoral areas, and in order to support silvopastoral ecosystem restoration efforts, the Moroccan Water and Forests Department strengthened the regulatory framework governing forest grazing by promulgating a legal instrument instituting the granting of compensation for forest closures in 2002.

This measure is part of a dynamic aimed at evolving legislative texts governing use rights in the direction of a progressive transfer of responsibilities to organized local populations, and is likely to reduce the tension of conflicts between the Forestry Administration and user populations. This participatory approach marks a transition towards social forestry where community involvement has evolved towards a contribution to forest resource management. Indeed, this measure is similar to the temporary buyback of use rights through the adoption of a regulatory mechanism for managing and compensating forest closures within the framework of participatory forest space management.

The regulatory provisions of compensation are stipulated in decree no. 1855-01 of March 21, 2002 (SGG, 2002), setting the limits, conditions, and modalities for requesting and granting compensation for closures of forest domain to be exploited or regenerated. This mechanism is based on a financial incentive based on the temporary buyback of grazing use rights through compensation for areas withdrawn from grazing, through the granting of a financial subsidy of 250 dh/ha/year of closure with a minimum eligible area for compensation of 300 ha for all types of silvopastoral ecosystems, except for argan where this subsidy is 350 dh/ha/year of closure with a minimum eligible area for compensation of 100 ha (dh: dirham is the currency of Morocco; 1 MAD is equivalent to 0.1 Euro). The granting of compensation for forest closures is subject to the presentation of a file including a technical part and an administrative one.

This instrument has enabled the emergence of an associative dynamic of rural user communities in favor of participatory management of silvopastoral resources. Indeed, this approach consecrates the creation of pastoral associations with the objective of rethinking the exercise of grazing rights and making it compatible with the sustainable management of silvopastoral ecosystems (Moukrim et al., 2019b).

The pastoral associations created for this purpose constitute the appropriate form of user organization for effective participation in the resolution of local problems of conservation and sustainable development of forest and silvopastoral ecosystems.

The experience capitalized in Morocco has made it possible to realize that the supervision of these associations consists of pedagogy and social engineering work that requires time and know-how. Proximity work with these associations has greatly contributed to the treatment of the pastoral question in forests.

It emerges from this experience that the compensation mechanism has constituted, since its entry into force in 2005, an incentive lever aimed at strengthening the organization and participation of the user population in a perspective of participatory management of silvopastoral resources. Concretely, on the ground, seven (7) associations were created in 2005 over an area of 6,150 ha. The implementation pace of this component has progressed in a very variable manner according to regions, reaching in 2024, 198 associations grouping 15,634 users, and which participate in the participatory management of forest closures over an area of 76,135 ha (ANEF, 2024).

This compensation instrument has globally enabled establishing the conditions for better respect of silvopastoral areas under closure and considerably reducing the oppositions raised by users against reforestation and forest space restoration programs on the eve of each campaign.

On the socio-economic level, this compensation has enabled the creation of a consultation and mutual interest space within the concerned user communities and the initiation of income-generating activities and the realization of community socio-economic projects (track maintenance, electrification, well drilling and equipment, grouped drinking water supply...), sometimes in partnership with other programs of the Social Development Agency (ADS) and those of the National Human Development Initiative (INDH).

It should be noted in passing that the forest closure compensation mechanism (CMD) is similar to the notion of Payment for Ecological Service (PES) which emerged in the 2000s and whose approach involves appropriate forms of contractualization that may involve private actors or community organizations and public actors. In our case, it consists of an agreement by which a “closure” service is remunerated by public authorities to organized forest users who consent to it and commit to providing said service and securing it. As such, the granting of compensation for closure is assimilated to an explicit payment for environmental service in which silvopastoral associations will know that they offer an environmental service when they accept and ensure compliance with the closure of forest parcels and that they will be remunerated for this practice (conditionality of compensation).

By capitalizing on the achievements of this experience of participatory management of silvopastoral resources built around compensation and management of forest closures, the new forest strategy called “Forests of Morocco 2020-2030”, aims to reconcile users with wooded and silvopastoral areas according to a sustainable, participatory, and inclusive management model. This new strategy has retained, among its axes, "Reinventing and structuring the participatory approach, with the population" and whose planned activities consist of: (i) Bringing about territorial governance bodies through the creation of more than 200 forest development organizations (FDO), whose mission is to contribute to supporting participatory management programs for forest and silvopastoral areas at the local level, (ii) creating a new corps of more than 500 territorial facilitators to ensure close support for planned activities, and (iii) Contractualizing participatory management with community organizations established according to the specificities and vocations of forest and silvopastoral areas.

Also, within the framework of the Forests of Morocco 2020-2030 strategy, the Water and Forests Department, elevated since 2020 to a National Water and Forests Agency, proceeded in 2021 to update the provisions of the legal text on forest closure compensation, with the objective of strengthening the participatory management mechanism for silvopastoral resources. As such, the amendments focused on a clear improvement through the revaluation of the compensation amount to 1,000 dh/ha/year for different forest and silvopastoral ecosystems and 1,100 dh/ha/year for the case of argan forest. As for the minimum eligible area for forest closure compensation, it is reduced to 50 ha for different forest and silvopastoral ecosystems and to 40 ha for the case of argan forest (SGG, 2021). This new initiative is likely to enable better integration of users and local actors in taking charge of sustainable social forestry activities for which the forestry service is only one actor among others whose main mission is limited to technical supervision of forestry and silvopastoral activities and support for community organizations of forest and silvopastoral areas.

By way of conclusion to this part of the article, it emerges that the forest closure compensation tool has opened new perspectives in the direction of: (i) Involving organized user populations in the participatory management of forest resources, (ii) Consolidating achievements in terms of user organization for the emergence of a social economy in the periforest area and an associative fabric working in favor of sustainable development, (iii) Making the created pastoral associations a basis for developing rural economy likely to mitigate pastoral pressure on forest ecosystems, and (iv) Improving the living conditions of user populations and initiating income-generating activities within the framework of community projects through good placement of funds from forest closure compensation. However, forest governance in North African countries requires long-term multidimensional cooperation, aligned objectives, and a sustainable vision in order to meet the country's international climate commitments and ensure the sustainability of forest and silvopastoral ecosystems in the face of global change challenges.

PERSPECTIVES FOR PARTICIPATORY SILVOPASTORAL MANAGEMENT

The perspectives for improving the management of the participatory silvopastoral management process in the context of global changes are based on an integrated and multidimensional approach that articulates participatory governance, technological innovation, scientific research, and capacity building.

Participatory Governance and Institutional Strengthening

The adaptation of participatory and integration approaches to the specificities of silvopastoral ecosystems and the social context of their landscapes constitutes a fundamental prerequisite. Silvopastoral systems are complex and require holistic management that recognizes that the viewpoints, values, and experiences of local populations are deeply connected to the design and management of these systems. This approach involves strengthening consultation and formalizing the involvement of concerned stakeholders (professionals, consulting firms, elected officials, users, NGOs, concerned services, etc.) in the process of developing, validating, implementing, monitoring, and evaluating management plans. Participatory governance constitutes a prerequisite to ensure exchanges between stakeholders, thus enabling inclusive and sustainable forest management.

Strengthening the capacities of local actors, particularly pastoral communities, in multifunctional management of silvopastoral areas proves essential to ensure local ownership of management plans. This approach requires innovating by creating new forms of partnership with user organizations likely to facilitate the implementation of forest management plans, as well as the integration into basic forestry training of issues relating to the valorization and conservation of different ecosystem resources: multiple resource management, communication and negotiation techniques, etc.

The revision of the regulatory framework for silvopastoral management to make it binding for actors also constitutes an important lever to ensure compliance with contractual commitments and the sustainability of restoration and conservation actions.

Strategic Planning and Integrated Spatial Approach

The development of a national strategic plan that defines guidelines and forestry management procedures for development and implementation by forest ecosystem is imperative as a priority. This planning must ensure a clear and long-term vision of the urgent and priority needs of silvopastoral resource management and development programs, based on a bottom-up participatory approach.

It is appropriate to promote and consolidate the approach favoring a spatial entry to ensure the internal integration of silvopastoral management plans and identify opportunities for complementarity and integration with programs of related sectors, which could be the subject of contractual partnerships. This approach requires having a bank of territorialized silvopastoral projects that are technically evaluated and relevant, economically efficient, and socially acceptable, thus enabling rapid implementation adapted to local contexts.

Adaptation to Global Changes and Ecosystem Services

The integration of new dimensions of sustainable development into participatory silvopastoral management approaches, in terms of adaptation to global changes and evaluation of goods and services provided by different natural ecosystems, constitutes a major strategic axis. Silvopastoral systems play a key role in resilience to drought by improving water cycle management, while offering a range of other benefits including carbon sequestration, biodiversity conservation, and food security (FAO, 2022). The integration of multifunctionality of silvopastoral systems into management allows maximizing positive impacts on herders while minimizing potential negative effects on fodder production (Mackay-Smith et al., 2025).

In this perspective, it is necessary to improve knowledge on the relevance of silvopastoral area restoration measures in a context characterized by global changes (climatic and socio-economic). Silvopastoral systems constitute a promising strategy to strengthen resilience to drought, with environmental benefits including improved water quality, soil conservation, carbon sequestration, biodiversity conservation, and reduction of forest fuels, while generating financial benefits for local communities (Ghaffariyan, 2025).

The development of tools for integrating non-market goods and services and NWFPs into the establishment of management plans, as well as the development of models for integrating social aspects into management decisions, will ensure a holistic evaluation of silvopastoral ecosystems and their contributions to human well-being. Providing foresters and managers with the tools and arguments of environmental economics necessary to defend management proposals also proves essential.

Technological Innovation and Decision Support Tools

The modernization of management techniques through the development of pastoral inventory models, decision support tools, and new technologies for resource assessment, data compilation, and mapping constitutes an essential lever to improve the efficiency of silvopastoral management. The integration of remote sensing technologies, coupled with artificial intelligence and machine learning, now allows near real-time monitoring of pasture quality, fodder biomass, and degradation conditions at large scale (Shahi et al., 2025).

The use of IoT (Internet of Things) sensor networks also enables continuous and automated monitoring of key factors such as soil moisture, temperature, plant health, and environmental conditions, thus facilitating data-driven decision-making in real time to optimize any intervention (Arion et al., 2025). The implementation of a popularization program for managers to popularize the use of these low-cost and freely accessible technological tools as a basis for grazing management constitutes a priority.

Scientific Research and Knowledge Development

The coordination of research programs on themes related to silvopastoral ecosystem restoration, improvement of extensive livestock management, and adaptation of pastoral practices to social changes affecting silvopastoral zones is essential. The spatial heterogeneity of environmental and social conditions requires site-specific decisions based on solid scientific knowledge and capitalization of local experiences.

The development of multidisciplinary research involving ecologists, agronomists, social scientists, and economists will enable a comprehensive understanding of silvopastoral systems and their dynamics in the context of global changes. Strengthening collaboration between research institutions, forest administrations, and local communities through participatory research-action approaches will promote the co-construction of locally adapted and scientifically validated management solutions.

The establishment of long-term monitoring systems for silvopastoral ecosystems, using standardized methodologies and integrating the collection of ecological, social, and economic data, will enable the evaluation of management interventions' effectiveness and adaptive adjustment of management strategies in the face of evolving environmental and social conditions.

Economic Valorization and Sustainable Finance

The diversification of income sources for local communities through the sustainable valorization of forest products (timber, fuelwood, NWFPs) and ecosystem services (carbon, water, biodiversity) constitutes an essential pillar for reducing pressure on silvopastoral resources. The development of quality value chains for pastoral products (meat, dairy products, wool) and forest products (aromatic and medicinal plants, honey, ecotourism) will create economic incentives for sustainable resource management.

The exploration of innovative financing mechanisms, including payments for ecosystem services (PES), green bonds, climate funds, and public-private partnerships, will mobilize the resources necessary for implementing ambitious silvopastoral restoration and management programs. The integration of silvopastoral systems into national and international carbon offset mechanisms could generate significant resources for local communities while contributing to global climate change mitigation efforts.

Regional Cooperation and Knowledge Exchange

Strengthening regional cooperation between North African countries in the field of silvopastoral management will promote the exchange of experiences, good practices, and technological innovations. The establishment of regional platforms for knowledge exchange and coordination of research and development efforts will avoid duplication and optimize the use of limited resources.

Collaboration with international organizations (FAO, UNCCD, CBD) and participation in regional and global networks on sustainable land management will enable access to technical and financial resources and integration of local experiences into global knowledge on silvopastoral systems and their sustainable management.

CONCLUSION

Silvopastoral ecosystems in North Africa constitute a vital heritage for rural populations and play a crucial role in maintaining ecological balances and combating desertification in a region characterized by aridity and vulnerability to climate change. Despite their socio-economic and environmental importance, these ecosystems face multiple pressures including overgrazing, unsustainable resource exploitation, population growth, and intensifying climate change effects.

The analysis of forest and silvopastoral contexts in Morocco, Algeria, and Tunisia reveals common challenges but also specificities related to each country's historical, institutional, and socio-economic contexts. The three countries have developed regulatory frameworks and management strategies for their forest and pastoral resources, but implementation of sustainable management approaches remains confronted with institutional, technical, financial, and social constraints.

The participatory silvopastoral management approach, based on the involvement of local communities in planning, decision-making, and implementation of management actions, emerges as a promising pathway to reconcile the objectives of resource conservation and improvement of rural populations' living conditions. Experiences of participatory management, particularly the forest closure compensation mechanism in Morocco, demonstrate the potential of financial incentive approaches to mobilize local communities in favor of sustainable ecosystem management.

However, the transition towards truly sustainable silvopastoral management requires a profound transformation of governance systems, strengthening of local actors' capacities, integration of scientific and technological innovations, and mobilization of adequate financial resources. The perspectives outlined in this article emphasize the need for a holistic and integrated approach that combines participatory governance, strategic planning, adaptation to global changes, technological innovation, scientific research, economic valorization, and regional cooperation.

In a context marked by accelerating climate change and intensifying pressures on natural resources, the sustainability of North African silvopastoral ecosystems will depend on our collective capacity to implement ambitious and innovative management models that recognize the central role of local communities, value multifunctional ecosystem services, and integrate the best available scientific and traditional knowledge.

The challenge is considerable, but the stakes justify the mobilization of all stakeholders—states, local communities, researchers, NGOs, development partners—around a common vision of sustainable and equitable management of forest and pastoral resources for present and future generations.

REFERENCES

ANEF. (2024). Bilan des activités de l'Agence Nationale des Eaux et Forêts (ANEF) au titre de l'année 2024 (Rapport interne), 65 P.

Arion, I. D., Morar, I. M., Truta, A. M., Chereches, I. A., Isarie, V. I., & Arion, F. H. (2025). Internet of Things (IoT)-Based Applications in Smart Forestry: A Conceptual and Technological Analysis. *Forests*, *17*(1), 44. DOI: 10.3390/f17010044

Bedoui, C. (2020). Study of desertification sensitivity in Talh region (Central Tunisia) using remote sensing, GIS and the MEDALUS approach. *Geoenvironmental Disasters*, *7*(1), 16. DOI: 10.1186/s40677-020-00148-w

Benabderrahmane, M. C., & Chenchouni, H. (2010). Assessing environmental sensitivity areas to desertification in Eastern Algeria using Mediterranean desertification and land use "MEDALUS" model. *Int J Sustain Water Environ Syst*, *1*(1), 5–10. DOI: 10.5383/swes.01.01.002

Benabou, A., Moukrim, S., Lahssini, S., El Aboudi, A., Menzou, K., Elmalki, M., & Rhazi, L. (2022). Impact of climate change on potential distribution of Quercus suber in the conditions of North Africa. *Biosystems Diversity*, *30*(3), 289–294. DOI: 10.15421/012231

Benhizia, R., Kouba, Y., Szabó, G., Négyesi, G., & Ata, B. (2021). Monitoring the spatiotemporal evolution of the Green Dam in Djelfa Province, Algeria. *Sustainability (Basel)*, *13*(14), 7953. DOI: 10.3390/su13147953

Bourbouze, A. (2003). Enjeux et débats sur la réorganisation des espaces pastoraux du sud méditerranéen. Actes du colloque international « Organisation spatiale et gestion des ressources et des territoires ruraux», 25-27 février 2003, Montpellier (France). 7 P.

Bourbouze, A., & Donadieu, R. (1987). L'Élevage sur parcours en régions méditerranéennes. Montpellier: CIHEAM/ IAM. *Options méditerranéennes*. 102 p.

Chebli, Y., El Otmani, S., Elame, F., Moula, N., Chentouf, M., Hornick, J. L., & Cabaraux, J. F. (2021). Silvopastoral system in Morocco: Focus on their importance, strategic functions, and recent changes in the Mediterranean side. *Sustainability (Basel)*, *13*(19), 10744. DOI: 10.3390/su131910744

FAO. (1989). *Bilan de trente ans de développement pastoral dans le bassin méditerranéen. Organisation des Nations Unies pour l'Alimentation et l'Agriculture*. FAO.

FAO. (2010). *Les forêts et le changement climatique, Document de travail 9. Les forêts et le changement climatique dans la région du Proche-Orient. Organisation des Nations Unies pour l'Alimentation et l'Agriculture*. FAO.

FAO. (2020). *Détermination des valeurs de productions des essences forestières et des prélèvements de bois de feu et pastoraux au Maroc dans le cadre du Projet IKI (réalisé par le consultant national Naggar mustapha). Organisation des Nations Unies pour l'Alimentation et l'Agriculture*. FAO.

FAO. (2022). *Grazing with trees: A silvopastoral approach to managing and restoring drylands* (Haddad, F. F., Herrera, P. M., & Besbes, B., Eds.). Vol. 187). Food & Agriculture Organisation.

FAO-FOSA. (2000). *Etude prospective du secteur forestier en Afrique (FOSA)-Algérie Organisation des Nations Unies pour l'Alimentation et l'Agriculture*. FAO.

FAO-FOSA. (2001). *Etude prospective du secteur forestier en Afrique (FOSA)-Tunisie Organisation des Nations Unies pour l'Alimentation et l'Agriculture*. FAO.

FAO-NENA. (2022). *A turning point for drought management: Upscaling the silvopastoral approach*. Food and Agriculture Organization, Regional Office for the Near East and North Africa.

Fry, A. S., Weaver, M. J., Libala, N., Ralekhetla, M. M., Gusha, B., Mtati, N., & Palmer, C. G. (2024). Stepping stones to transformative participatory governance capabilities in a rural landscape restoration project. *Ecosystems and People (Abingdon, England)*, *20*(1), 2425378. DOI: 10.1080/26395916.2024.2425378

Ghaffariyan, M. R. (2025). Silvopastoral systems as a strategy for drought resilience: A short international review. *Silva Balcanica*, *26*(1), 29–37. DOI: 10.3897/silvabalcanica.26.e127074

Hernández-Morcillo, M., Burgess, P., Mirck, J., Pantera, A., & Plieninger, T. (2018). Scanning agroforestry-based solutions for climate change mitigation and adaptation in Europe. *Environmental Science & Policy*, *80*, 44–52. DOI: 10.1016/j.envsci.2017.11.013

Lemes, A. P., Garcia, A. R., Pezzopane, J. R. M., Brandão, F. Z., Watanabe, Y. F., Cooke, R. F., Sponchiado, M., de Paz, C. C. P., Camplesi, A. C., Binelli, M., & Gimenes, L. U. (2021). Silvopastoral system is an alternative to improve animal welfare and productive performance in meat production systems. *Scientific Reports*, *11*(1), 14092. DOI: 10.1038/s41598-021-93609-7 PMID: 34238990

Louhaichi, M., Hassan, S., Jamel, K., & Ouled Belgacem, A. (2023). Challenges and opportunities for promoting use of native species for the rehabilitation of degraded silvopastoral systems in Tunisia. International Center for Agricultural Research in the Dry Areas (ICARDA). Beirut, Lebanon. https://hdl.handle.net/20.500.11766/68814

Mackay-Smith, T. H., Spiekermann, R. I., Richards, D. R., Harcourt, N., & Burkitt, L. L. (2025). An integrative approach to silvopastoral system design: Perspectives, potentials and principles. *New Zealand Journal of Agricultural Research*, *68*(2), 218–258. DOI: 10.1080/00288233.2023.2298922

Mahdi, M. (2002). *Mutations sociales et réorganisation des espaces steppiques.* Imp. Najah El Jadida.

Mihi, A., Ghazela, R., & Wissal, D. (2022). Mapping potential desertification-prone areas in North-Eastern Algeria using logistic regression model, GIS, and remote sensing techniques. *Environmental Earth Sciences*, *81*(15), 385. DOI: 10.1007/s12665-022-10513-7 PMID: 35891927

Moukrim, S., Benabou, A., Lahssini, S., Aafi, A., Chkhichekh, A., Moudden, F., Ben Bammou, M., El Aboudi, A., & Laaribya, S. (2022b). Spatio-temporal analysis of North African forest cover dynamics using time series of vegetation indices–case of the Maamora forest (Morocco). *Biosystems Diversity*, *30*(4), 372–379. DOI: 10.15421/012236

Moukrim, S., Lahssini, S., Naggar, M., Lahlaoi, H., Rifai, N., Arahou, M., & Rhazi, L. (2019b). Local community involvement in forest rangeland management: Case study of compensation on forest area closed to grazing in Morocco. *The Rangeland Journal*, *41*(1), 43–53. DOI: 10.1071/RJ17119

Moukrim, S., Lahssini, S., Rhazi, M., Alaoui, H. M., Benabou, A., Wahby, I., El Madihi, M., Arahou, M., & Rhazi, L. (2019a). Climate change impacts on potential distribution of multipurpose agro-forestry species: Argania spinosa (L.) Skeels as case study. *Agroforestry Systems*, *93*(4), 1209–1219. DOI: 10.1007/s10457-018-0232-8

Moukrim, S., Lahssini, S., Rhazi, M., Menzou, K., El Madihi, M., Rifai, N., Bouziani, Y., Azedou, A., Boukhris, I., & Rhazi, L. (2022a). Climate change impact on potential distribution of an endemic species Abies marocana Trabut. *Ekologia (Bratislava)*, *41*(4), 329–339. DOI: 10.2478/eko-2022-0034

Naggar M., (2000). Éléments de base d'une stratégie de sylvopastoralisme en Afrique du Nord. *Options méditerranéennes*, série A, n° 39, pp. 191-202.

Naggar, M. ((2021). Renouveau des pratiques pastorales en forêt au Maroc: De la tradition à l'aménagement. *Éditions universitaires européennes*. 87 p.

Naggar M., Lahssni S., (2017). L'aménagement forestier et les enjeux de restauration des écosystèmes forestiers au Maroc. *Revue Forêt Méditerranéenne* t. XXXVIII, n° 1. PP: 77-82.

Peri, P. L., Chará, J., Viñoles, C., Dube, F., Sotomayor, A., Kremer, K., Bahamonde, H., Varella, A., Paniagua, P., Enriquez, A., Teglia, C., Bussoni, A., & Lencinas, M. V. (2024). Current trends in silvopastoral systems. *Agroforestry Systems*, *98*(7), 1945–1953. DOI: 10.1007/s10457-024-01093-5

Qarro, M. (2006). Le sylvopastoralisme et la gestion durable des écosystèmes naturels arides. 14th International Soil Conservation Organisation - Conférence Water Management and Soil Conservation in Semi-Arid Environments, Marrakech, Morocco, May 14-19, 2006, 12 p.

Qarro M., Berkat O., M'Hamedi M., Yessef M., (2000). L'aménagement sylvopastoral: contraintes et développement durable, 2000. Rupture: Revue Options Méditerranéennes: Série A. *Séminaires Méditerranéens*; n. 39. PP: 171-174.

Saadani, Y. (2015). La stratégie de développement durable des forêts et parcours en Tunisie 2015-2024 - Pour une transition vers l'économie verte. Revue Foret Méditerranéenne, T. XXXVI, n°2, 2015. PP: 235-240.SGG, 2002. Arrêté du ministre chargé des eaux et forêts fixant les limites, conditions et modalités de demande et d'octroi de la compensation pour mise en défens du domaine forestier à exploiter ou à régénérer. Bulletin officiel n°5000 publié par le Secrétariat Général du Gouvernement (SGG), PP:456-457.

SGG. (2021). Arrêté du ministre de l'agriculture, de la pêche maritime, du développement rural et des eaux et forêts, modifiant et complétant l'arrêté fixant les limites, conditions et modalités de demande et d'octroi de la compensation pour mise en défens du domaine forestier à exploiter ou à régénérer. Bulletin officiel n°7080 publié par le Secrétariat Général du Gouvernement (SGG), PP:564-565.

Shahi, T. B., Balasubramaniam, T., Sabir, K., & Nayak, R. (2025). Pasture monitoring using remote sensing and machine learning: A review of methods and applications. *Remote Sensing Applications: Society and Environment*, *37*, 101459. DOI: 10.1016/j.rsase.2025.101459

UNCCD. (2022). Global Land Outlook: Second edition. United Nations Convention to Combat Desertification. Bonn, Germany.

World Bank. (1995). Une stratégie de développement des parcours en zones arides et semi-arides. Rapport technique Maroc. 82 p.

World Bank. (2023). Notes sur les forêts algériennes- Gestion durable des forêts pour lutter contre les incendies de forêts. 112 P.

Section 2
Ecosystem Services, Biodiversity, and Climate Mitigation

Chapter 4
Temporary Ponds in a Southern Mediterranean Forest Context:
Biodiversity, Functioning, and Threats

Mohammed El Madihi
http://orcid.org/0000-0002-2425-5396
Faculty of Sciences, Chouaib-Doukkali University, El-Jadida, Morocco

Abdessadeq Boudjaj
Research Center of Plant and Microbial Biotechnologies, Biodiversity, and Environment, Laboratory of Botany and Valorization of Plant and Fungal Resources, Department of Biology, Faculty of Sciences, Mohammed V University in Rabat, Morocco

Mouhssine Rhazi
Independent Researcher, Morocco

Mohamed Ben Bammou
Research Center of Plant and Microbial Biotechnologies, Biodiversity, and Environment, Laboratory of Botany and Valorization of Plant and Fungal Resources, Faculty of Sciences, Mohammed V University in Rabat, Morocco

Loubna Tifarouine
Faculty of Sciences, Chouaib-Doukkali University, El-Jadida, Morocco

Rachid Fghire
Faculty of Sciences, Chouaib-Doukkali University, El-Jadida, Morocco

Said Moukrim
http://orcid.org/0000-0002-3172-3776
Faculty of Sciences, Mohammed V University in Rabat, Morocco

Lahssini Said
http://orcid.org/0000-0003-1986-8240
National School of Forest Engineers, Salé, Morocco

Saber Er-Riyahi
Faculty of Letters and Human Sciences, Department of Geography, Moulay Ismail University, Morocco

DOI: 10.4018/979-8-3373-6239-7.ch004

Laila Rhazi
Research Center of Plant and Microbial Biotechnologies, Biodiversity, and Environment, Laboratory of Botany and Valorization of Plant and Fungal Resources, Faculty of Sciences, Mohammed V University in Rabat, Morocco

ABSTRACT

Mediterranean temporary ponds in Morocco are essential components of forest ecosystems, characterized by a seasonal alternation between flooded and dry phases driven by Mediterranean climate. Their hydrological functioning strongly influences the structure, composition, and dynamics of biological communities, including numerous rare, endemic, and threatened species. These ponds maintain close functional relationships with surrounding forest formations, jointly contributing to hydrological regulation, carbon storage, soil protection, and biodiversity maintenance. Their distribution covers a wide bioclimatic gradient, from the Rif to the High Atlas, with high concentrations in coastal forest areas. However, temporary ponds and adjacent forests are increasingly exposed to converging threats related to climate change and human activities. Their conservation requires integrated management that considers pond–forest interactions, supported by ecological monitoring, community awareness, and integration into conservation frameworks, to ensure the resilience of Mediterranean forest landscapes

INTRODUCTION

Temporary ponds are wetlands of remarkable ecological singularity, defined by the seasonal alternation of inundated and dry phases throughout the annual cycle. This cyclical character, which is closely dependent on precipitation regimes and temperature patterns, profoundly influences their ecological functioning and the dynamics of their biological communities (Grillas et al., 2004). These small ecosystems, generally occupying closed depressions, are mainly filled by autumn and winter rainfall and then gradually dry out during the summer period. The alternation between aquatic and terrestrial conditions confers a unique ecological identity on temporary ponds, making them particularly dynamic and heterogeneous habitats.

Temporary ponds, also referred to as ephemeral ponds, vernal pools (in California), or *dayas* (in Morocco), belong to the broader category of temporary wetlands, that is, freshwater habitats subject to alternating dry and wet phases. These ecosystems vary widely in size, shape, depth, and duration of inundation (hydroperiod), parameters that strongly determine the biological diversity they support (Brendonck & Williams, 2000; Blaustein & Schwartz, 2001; Williams, 2006; Boix et al., 2016). Despite this, their ecological and hydrological value is often underestimated, even though they represent an important water resource and provide refuge for numerous rare or endemic species (Schwartz & Jenkins, 2000; Beja & Alcazar, 2003). Limited recognition of their ecological role has led to their disappearance at an alarming rate, with losses recorded over the last century ranging from 60% to 97% depending on the region of the world (Brendonck & Williams, 2000; Nicolet et al., 2004; Rhazi et al., 2012).

In the Mediterranean context, Mediterranean temporary ponds (MTPs) represent a distinct subtype of temporary wetlands. They occur primarily within the Mediterranean Basin (Grillas et al., 2004), but also in other regions characterized by a Mediterranean-type climate, with mild, wet winters and hot, dry summers, such as south-western Australia (Jacobs & Brock, 1993), California (Barbour & Major, 1977; Zedler, 1987), central Chile (Bliss et al., 1998), and South Africa. These habitats develop within a wide range of landscape contexts, notably in agricultural and forested environments (El Madihi et al., 2017; Van den Broeck, 2016).

Mediterranean temporary ponds exhibit high morphological and ecological diversity, which is closely related to substrate characteristics, hydrological regimes, and water physico-chemical properties (Grillas et al., 2004). Although they share common features, most notably winter flooding followed by summer drying, they constitute a highly heterogeneous set of habitats in which contrasting ecological conditions coexist. Under such unstable conditions, flora and fauna have evolved remarkable adaptive strategies, including rapid seed germination, resistance to desiccation, short life cycles, and egg or larval diapause in invertebrates (Cucherousset, 2006).

Despite their limited surface area, these habitats support exceptional levels of biodiversity, both floristic (Médail et al., 1998; Quézel, 1998; Deil, 2005) and faunistic (Boutin et al., 1982; Thiéry, 1987; Grillas et al., 2004; Van den Broeck, 2016). They function as reservoirs of genetic diversity and refuges for endemic or threatened species, thereby playing a crucial role in maintaining regional ecological balance. However, their high vulnerability to land-use change, agricultural drainage, and climate warming makes them among the most threatened wetland types in the Mediterranean region (Moukrim et al., 2025).

In forested landscapes, these ecosystems play a strategic role as biodiversity reservoirs and as key elements of ecological connectivity, facilitating species dispersal and gene flow among habitats (Rhazi et al., 2012). They also contribute to the maintenance of local microclimates by buffering the strong thermal and hydrological fluctuations typical of Mediterranean environments. Within the Mediterranean Basin, and particularly in Morocco, temporary ponds are of considerable ecological and socio-economic interest, providing both high-heritage-value habitats and a wide range of ecosystem services (Bouahim et al., 2015). Nevertheless, they are currently facing multiple threats, including climate change, intensification of human activities, and degradation of forest habitats (El Madihi, 2020; Bouahim, 2010). This chapter aims to analyze the specificity of temporary ponds associated with Moroccan forested landscapes, with a particular focus on their ecology, functions, biodiversity, the pressures they face, and the functional relationships they maintain with forest ecosystems.

Figure 1. Forest temporary pond located within the cork oak (Quercus suber) woodland of the Maâmora Forest (Morocco)

(© El Madihi, M. & Rhazi L.)

ECOLOGY AND FUNCTIONING OF TEMPORARY PONDS

Temporary ponds are small, shallow freshwater wetlands occupying depressions that are often endorheic. They are characterized by a fluctuating hydroperiod, cyclically alternating between phases of inundation and desiccation, which represents a fundamental ecological feature (Van den Broeck, 2016; El Madihi, 2020) (Fig. 2). Inundation is mainly driven by autumn and winter precipitation, and may occasionally be supplemented by groundwater rise or snowmelt. The duration of flooding generally ranges from three to eight months, followed by a phase of complete drying during the warm season.

Figure 2. Alternation between flooded and dry phases of a temporary pond within the Benslimane Forest (Morocco)

(© El Madihi, M.& Rhazi L.)

The typology of temporary ponds is primarily based on the dynamics of their hydrological regime, which determine ecological structure and the composition of biological communities. Two complementary approaches are commonly used. Grillas et al. (2004) proposed a Mediterranean classification based on the duration, frequency, and regularity of inundation, collectively referred to as the hydroperiod. According to these criteria, ponds may be classified as ephemeral, with short flooding periods ranging from a few weeks to two months and high unpredictability, supporting communities dominated by species with rapid life cycles and high dormancy capacity, such as resistant seeds or resting eggs; seasonal, characterized by predictable annual inundation lasting on average three to six months, hosting diverse and relatively stable communities with high biological productivity; or intermittent to quasi-permanent, with prolonged flooding phases interrupted by occasional drying events, and supporting larger-bodied or long-lived species as well as stable biological assemblages.

At a broader scale, Boulton and Brock (1999) proposed a conceptual framework applicable to all temporary aquatic systems, based on inundation frequency, predictability, and duration. This framework distinguishes ephemeral systems, which are rarely and irregularly flooded; seasonal systems, which experience predictable annual inundation; and intermittent or quasi-permanent systems, characterized by long hydroperiods punctuated by periods of drying. The integration of these two approaches allows for a comprehensive understanding of the ecological continuum of temporary ponds, ranging from highly unstable systems strongly constrained by climatic variability to more regular and persistent habitats offering enhanced ecological stability. This classification is essential for understanding species distribution patterns according to degrees of hydrological specialization, assessing community resilience to climatic disturbances, and providing a robust framework for the conservation and sustainable management of these sensitive habitats (Table 1).

Table 1. Simplified typology of temporary ponds based on the duration and regularity of inundation and associated implications for biological communities

Type	Hydrological Criteria	Ecological Examples	Characteristic Biodiversity
Ephemeral	Short and irregular inundation, low predictability; long periods of desiccation	Temporary depressions, salt marshes, residual water bodies	Dominance of species with rapid life cycles and high dormancy capacity (resistant seeds or resting eggs)
Seasonal	Predictable annual inundation; intermediate duration (3–6 months)	Temporary ponds, seasonal marshes, small ponds	Diverse and relatively stable communities; high biological productivity
Intermittent / Quasi-permanent	Prolonged inundation with occasional drying events	Large marshes, temporary lakes, deep forested wetlands	Presence of large-bodied or long-lived species; increased stability of biological assemblages

The hydrological variability of temporary ponds, accentuated by the unpredictability of Mediterranean climates, characterized by mild, rainy winters and hot, dry summers, imposes strong ecological constraints on biological communities, which must adapt to alternating aquatic and terrestrial environments across seasons (Grillas et al., 2004). Flora and fauna have therefore developed exceptional adaptive strategies, including desiccation tolerance, prolonged dormancy, rapid reproduction during the wet phase, and the formation of survival reserves in the sediment in the form of seeds, resting eggs, or spores, ensuring species persistence and community reassembly with the return of rainfall (Brock et al., 2003; Bonis et al., 1995).

Vegetation is structured along a hydrological gradient: aquatic hydrophytes dominate during inundation, amphiphytes appear as water recedes, and opportunistic terrestrial species, often originating from surrounding forested or agricultural areas, colonize temporarily exposed substrates (Rhazi et al., 2006; Fennane et al., 2014). Hydrological parameters such as maximum water depth, duration, and timing of inundation directly control the structure and composition of plant communities (Keeley & Zedler, 1998; Bouahim et al., 2014), while soil characteristics, including texture, pH, and nutrient content, influence germination, growth, and survival according to species' tolerance to physico-chemical constraints (Bonis, 1993). Climate also plays a central role, as variations in temperature and precipitation determine not only the duration of inundation but also productivity and biodiversity (Pyke, 2005; Brendonck et al., 2014).

Hydroperiod stability is crucial for the persistence of characteristic species, several of which are rare or endemic. Studies conducted in Morocco (Rhazi et al., 2012; El Madihi et al., 2017) have shown that floristic richness decreases when inundation duration shortens or when soils become more eutrophic. In forested contexts, such as the Benslimane Forest, ponds benefit from specific microclimatic conditions, including shading, litter presence, and reduced evaporation. These factors prolong the inundation phase, maintain cooler and more humid conditions, provide ecological refuges for species sensitive to rapid drying, and contribute to overall ecological diversity in the forest landscape (Bouahim, 2010).

The fauna of Moroccan temporary ponds is equallyunique, encompassing a diversity of taxa adapted to the discontinuity of water. Amphibians such as *Discoglossus scovazzi* and *Pleurodeles waltl*, branchiopod crustaceans like *Branchipus schaefferi* and *Triops mauritanicus*, as well as copepods, aquatic insects, mollusks, and zooplankton, all exhibit exceptional biological adaptations. These include the production of resting eggs capable of surviving for several years in dry sediment and life cycles synchronized with hydroperiod duration (Van den Broeck, 2016).

Ecological interactions within these ponds are complex and dynamic. Primary production, although limited, is essential and is provided by macrophytes, algae, and periphyton, forming the basis of food webs for herbivorous invertebrates and zooplankton, which in turn support amphibians and migratory waterbirds. During inundation, these ecosystems play key roles in hydrological and biogeochemical regulation, retaining water and recycling nutrients, whereas during dry phases, they function as temporary terrestrial habitats, enhancing ecological connectivity between wetlands and surrounding drylands.

However, these habitats are now under significant threat. Human activities,including urbanization, drainage, agriculture, and intensive grazing, alter hydrological and physico-chemical parameters, disrupting ecological functioning. Eutrophication, invasive species introduction, and gradual infilling of depressions further accelerate the regression of these unique habitats (Grillas et al., 2004; Rhazi et al., 2012). Climate change exacerbates these pressures, potentially prolonging dry phases and reducing water surface areas, thereby compromising the survival of many specialized species (IPCC, 2014; Rosset et al., 2010). Despite their small size and apparent fragility, temporary ponds play a major ecological role: they serve as biodiversity reservoirs, refuges for rare and endemic species, regulators of hydrological processes, and providers of traditional ecosystem services such as grazing, medicinal plant harvesting, and livestock watering. Their conservation is therefore essential for maintaining Mediterranean biodiversity and ensuring the sustainability of Moroccan rural landscapes.

GEOGRAPHICAL DISTRIBUTION AND MEDITERRANEAN CONTEXT

Temporary ponds are small wetland ecosystems whose global distribution is closely linked to regions experiencing strong climatic seasonality. They occur on all continents; however, their distribution is discontinuous and mainly concentrated in areas where the alternation between wet and dry periods promotes the formation of temporary water bodies (Grillas et al., 2004). Such habitats are found in temperate, subtropical, and semi-arid regions, with a particularly high representation in Mediterranean-type climates. At the global scale, the Mediterranean Basin represents one of the major hotspots for temporary ponds, both in terms of their abundance and their wide spatial distribution.

This biogeographical region is well known for the density and diversity of its wetlands, including a large number of temporary ponds distributed throughout all Mediterranean countries (Grillas et al., 2004). These ponds occur across southern Europe (Spain, France, Italy, Greece), North Africa (Morocco, Algeria, Tunisia), and the Near East, where they occupy a wide range of natural and semi-natural landscapes (Thiéry, 1997). The strong representation of temporary ponds in the Mediterranean Basin has led to their recognition as habitats of major conservation interest, particularly in Europe, where they are classified among priority habitats. This recognition is largely based on their spatial importance and regional concentration, making the Mediterranean Basin a key global area for the study and conservation of temporary ponds (Grillas et al., 2004).

Within the Mediterranean Basin, Morocco stands out for its particularly high abundance of temporary ponds, locally referred to as *dayas*. Several studies have emphasized that Morocco ranks among the richest Mediterranean countries in terms of temporary pond density, especially within the Atlantic plains, where these habitats are exceptionally numerous (Ramdani, 1986; Thiéry, 1987). This spatial richness can be explained by the country's climatic, geological, and geomorphological diversity, which favors the formation of numerous endorheic depressions. The distribution of temporary ponds in Morocco extends across much of the national territory, although pronounced regional contrasts are observed. Their frequency is particularly high along the Atlantic coastal zone, from Tangier to Agadir, and decreases towards the east, south, and higher-altitude regions (Thiéry, 1997). Moreover, the duration of inundation follows a spatial gradient, generally decreasing from north to south and from west to east across the country (Thiéry, 1997).

The pioneering studies of Ramdani (1986) and Thiéry (1987) identified several major regional assemblages of temporary ponds in Morocco, reflecting this spatial heterogeneity. These assemblages include ponds of the eastern high plateaus and Saharan zones, those of arid inland plains, ponds of the Atlantic coastal plains, including those of the Maâmora cork oak forest and the Benslimane region, as well as ponds located in the mountainous massifs of the Rif, Middle Atlas, and High Atlas (Fig. 4). Some regions are characterized by an exceptionally high concentration of temporary ponds, notably the Benslimane area and the Maâmora cork oak forest, which represent some of the richest sectors in the country in terms of temporary pond density (El Madihi, 2020). The origin of many dayas in these areas dates back to the Early Pleistocene (Muller et al. 2022), when sedimentary infilling processes led to the formation of closed depressions favorable to the temporary accumulation of water (Amami et al., 2013).

At the scale of the Mediterranean Basin, Morocco thus emerges as a key territory for the study of temporary pond distribution, owing to their abundance, wide geographical extent, and the diversity of regional contexts in which they occur. Although the assertion that Morocco represents the leading Mediterranean country in terms of temporary pond density is primarily based on regional inventories, the available literature consistently supports the idea of an exceptional national representation of these habitats (Ramdani, 1986; Thiéry, 1987; El Madihi, 2020).

Figure 3. Distribution of the main temporary pond systems in Morocco: 1 = Western Rif; 2 = Rabat area; 3 = Chaouia and Doukkala; 4 = Haouz and Souss; 5 = High Atlas; 6 = Middle Atlas; 7 = Eastern region; 8 = Moroccan Sahara.

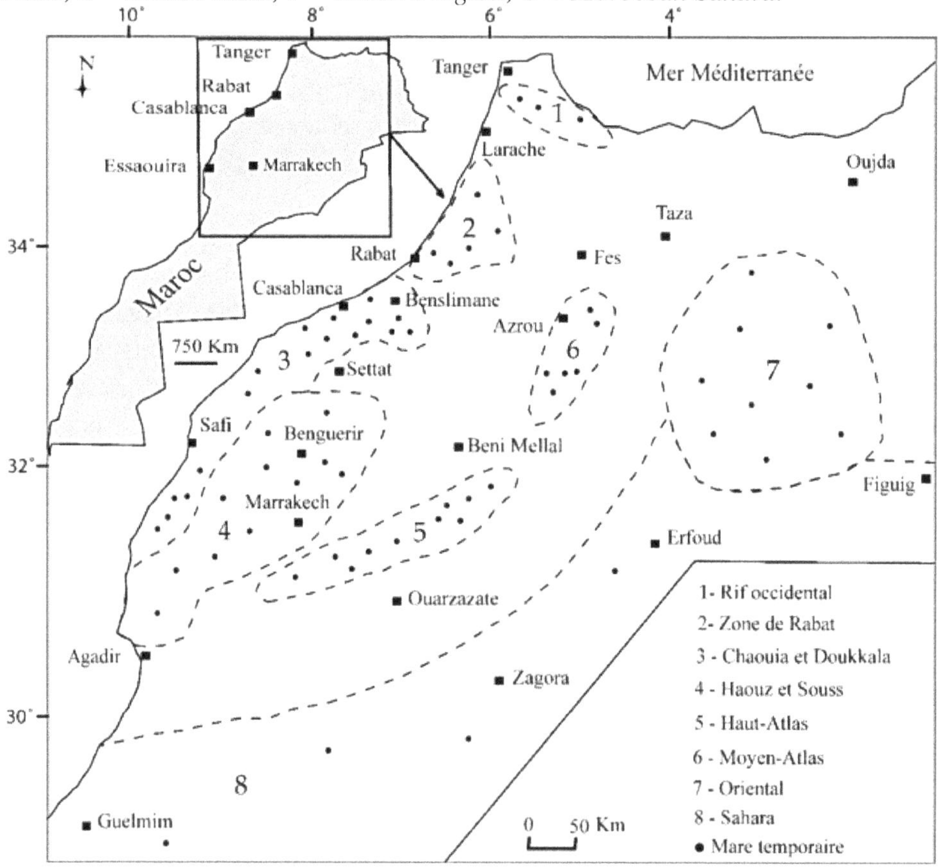

(After Thiéry, 1987; modified by Saber, 2006)

ECOLOGICAL IMPORTANCE AND ECOSYSTEM SERVICES

Mediterranean temporary ponds represent ecosystems of major ecological importance within Mediterranean landscapes, despite their often-discreet nature and limited surface area. These habitats provide a wide range of ecosystem services whose ecological and socio-economic value remains largely underestimated in many environmental management policies. According to the Millennium Ecosystem Assessment (2005) classification, these services can be grouped into four main categories: provisioning, regulating, supporting, and cultural services.

These wetlands primarily function as localized reservoirs of resources, particularly water and plant biomass. During wet periods, they supply water for livestock, irrigation of surrounding agricultural fields, and maintain soil moisture conditions favorable for the development of herbaceous vegetation used as grazing resources (Bouahim et al., 2015) (Fig. 3). In forested areas, they complement the role of tree cover by serving as watering points for both wild and domestic fauna. In addition, plant communities associated with these seasonal wetlands include species traditionally used in local pharmacopoeia or artisanal harvesting, while sediment deposits in these depressions are sometimes exploited for sand and clay extraction. Thus, although their direct economic contribution is relatively limited compared to other aquatic ecosystems, these habitats locally support sustainable rural activities and contribute to the socio-ecological resilience of Mediterranean territories.

Beyond provisioning services, temporary ponds play a crucial role in hydrological regulation and ecological stability of Mediterranean landscapes, which are often characterized by high interannual rainfall variability and increased vulnerability to drought. Their capacity to temporarily store rainwater allows for gradual infiltration into groundwater, contributing to aquifer recharge and downstream flood attenuation (Van den Broeck, 2016). They therefore function as natural sponges that slow surface runoff, reduce water erosion, and promote percolation. These hydrological functions are particularly important in Mediterranean catchments subjected to intense anthropogenic pressures, such as urbanization, agricultural intensification, and deforestation. By absorbing excess water and slowly redistributing moisture into the soil, these aquatic ecosystems help maintain soil structure and fertility in surrounding areas. They also influence the local microclimate by regulating air temperature and humidity in their vicinity, thereby indirectly contributing to the stability of surrounding vegetation cover.

Temporary ponds further provide essential supporting functions for global ecological processes. The alternation of wet and dry phases promotes soil formation through the activation of mineralization and redox cycles in surface horizons. These processes sustain primary productivity and nutrient regeneration necessary for plant growth (Bouahim et al., 2015). Moreover, these ecosystems ensure biological connectivity between terrestrial and aquatic environments by serving as transitional habitats for a wide diversity of plant and animal species adapted to hydrological fluctuations. Their habitat mosaic, ranging from permanently inundated zones to temporarily dry margins, offers a variety of ecological niches and promotes the coexistence of species with contrasting ecophysiological strategies.

Figure 4. Some ecosystem services provided by Moroccan temporary ponds. (A) Provisioning service: watering and grazing resources for livestock; (B) Provisioning service: harvesting of medicinal plants; (C) Cultural service: educational value; (D) Cultural service: recreational value.

The exceptional biodiversity of MTPs, often characterized by endemic or rare species, confers upon these ecosystems a level of importance that is disproportionate to their limited surface area. Numerous species of amphibians, aquatic invertebrates, and hygrophilous plants complete their entire life cycle within these habitats, relying directly on the temporal dynamics of flooding. Mediterranean temporary ponds therefore act as reservoirs of genetic and functional diversity, contributing to the maintenance of ecological resilience at the landscape scale. As highlighted by Cardinale et al. (2012), biodiversity directly influences ecosystem functioning by enhancing primary production, nutrient recycling, and resilience to disturbances. In the Mediterranean context, where climatic variability intensifies hydrological constraints, this biological diversity plays a critical stabilizing role by enabling ecosystems to continue delivering services even under environmental stress.

Temporary ponds also possess a strong cultural and educational dimension that enhances their heritage value. As distinctive elements of the landscape, they offer valuable opportunities for environmental education, awareness-raising regarding sustainable water management, and biodiversity conservation (Bouahim, 2010). Direct observation of these systems allows for a concrete understanding of the interactions between climate, hydrology, soil, and vegetation, making them particularly valuable natural laboratories for teaching ecological sciences. From a landscape perspective, temporary ponds contribute to visual diversity and the aesthetic quality of rural and forested environments. Their often-seasonal presence marks the rhythm of the seasons and reflects the cyclical functioning of Mediterranean ecosystems. These intangible values contribute significantly to the ecological and cultural identity of the regions in which they occur.

Finally, it is important to emphasize that the ecological functions and ecosystem services provided by temporary ponds are tightly interconnected and interdependent. Biodiversity, hydrological regulation, and soil fertility form a network of positive feedbacks that underpin the overall stability of these systems. The degradation or loss of temporary ponds through drainage, urbanization, or pollution inevitably results in the loss of multiple services, including reduced groundwater recharge, decreased primary productivity, the disappearance of key species, and the disruption of biogeochemical cycles. In forested environments, such losses are particularly detrimental, as they disrupt the functional complementarity between tree cover and temporary wetlands, thereby weakening the resilience of the entire ecosystem. Mediterranean temporary ponds thus emerge as vital components of ecological and hydrological functioning within the landscapes they occupy. Their role extends far beyond that of simple seasonal water bodies: they regulate essential processes that sustain life and, through their biological and functional complexity, support a wide range of human and ecological uses. In a context of climate change and increasing anthropogenic pressures, their conservation represents a strategic challenge for the sustainability of Mediterranean socio-ecosystems. Recognition of their multiple ecosystem services should therefore lead to their explicit integration into land-use planning, water resource management, and biodiversity conservation policies, in order to ensure the long-term persistence of these unique systems, which serve as true ecological archives of Mediterranean natural functioning.

BIODIVERSITY AND INTERACTIONS WITH FOREST COVER

Moroccan temporary ponds constitute ecosystems of major ecological importance, characterized by high floristic and faunal biodiversity that is strongly structured by seasonal hydrological dynamics and by close interactions with the surrounding forest cover. These habitats are distinguished by the coexistence of highly specialized strictly aquatic and amphibious species, and opportunistic terrestrial species that temporarily colonize ponds during drying phases. The flora characteristic of inundated phases is dominated by families typical of Mediterranean temporary wetlands, such as Isoetaceae, represented by taxa of the genus *Isoetes* (e.g. *Isoetes velata*), which are indicators of oligotrophic ponds with short hydroperiods; Alismataceae, with amphibious species such as *Baldellia ranunculoides* and *Damasonium bourgaei*; Marsileaceae and Pilulariaceae, including rare pteridophytes of high conservation value such as *Marsilea strigosa* and *Pilularia minuta*; and Elatinaceae, with several species of the genus *Elatine* that are strictly associated with temporarily flooded habitats (Fig. 5). These species exhibit adaptive strategies tightly linked to the alternation between submersion and desiccation, relying on short life cycles, predominantly sexual reproduction, and high seed production, which allow them to complete their biological cycle within often very narrow hydrological windows (El Madihi, 2020).

During dry phases, ponds are colonized by assemblages of opportunistic terrestrial species, largely originating from the surrounding forest matrix and frequently belonging to the families Poaceae, Asteraceae, Fabaceae, and Caryophyllaceae. This opportunistic floristic component reflects the direct influence of forest cover on species composition within ponds and contributes to a temporary increase in observed floristic richness, without challenging the functional dominance of species characteristic of temporary wetlands. More than half of the plant species recorded in these habitats are considered rare, endemic, or threatened at the national scale, conferring an exceptional conservation value on forested temporary ponds. This visible diversity is further enhanced by the presence of a hidden floristic heritage, represented by the sediment seed bank, which contains viable seeds of aquatic, amphibious, and terrestrial forest species. This seed bank plays a central role in the resilience of plant communities by ensuring population persistence despite interannual hydrological fluctuations, prolonged drying events, and frequent reproductive failures at the pond surface, and by enabling rapid recolonization once favorable hydrological conditions return (El Madihi, 2020).

Figure 5. Some remarkable plant species of Moroccan temporary ponds. (A) Damasonium bourgaei; (B) Pilularia minuta; (C) Marsilea strigosa; (D) Crassula vaillantii; (E) Ranunculus peltatus; (F) Baldellia ranunculoides.

The fauna associated with forest temporary ponds also exhibits remarkable diversity and includes groups that are tightly dependent on hydrological temporality. Aquatic macroinvertebrate communities are dominated by Crustacea, particularly large branchiopods belonging to the families Branchipodidae, Chirocephalidae, Thamnocephalidae, and Notostracidae, with emblematic species such as *Branchipus schaefferi*, *Tanymastix affinis*, and *Triops mauritanicus*. These organisms display life cycles synchronized with pond inundation and produce resistant eggs capable of surviving prolonged dry periods within the sediments. Microcrustaceans, especially Cladocera (families Daphniidae and Chydoridae) and Copepoda (families Diaptomidae and Cyclopidae), play a fundamental role in the trophic functioning of temporary ponds by ensuring energy transfer between primary producers and higher trophic levels.

Aquatic insects represent another major faunal group, dominated by Coleoptera (families Dytiscidae and Hydrophilidae) and Diptera (families Chironomidae and Culicidae), which use these habitats as sites for reproduction, larval development, and feeding before recolonizing adjacent forest habitats as adults (Van den Broeck, 2016) (Fig. 6).

Forest temporary ponds also play a crucial role for amphibians, particularly those belonging to the families Salamandridae and Hylidae, with species such as *Pleurodeles waltl* and *Hyla meridionalis*, which use these habitats as breeding sites and for larval development. The absence of predatory fish, the availability of trophic resources, and the relative stability of hydrological conditions under forest cover promote reproductive success and make these ponds true nurseries for regional amphibian populations. Beyond their importance for strictly aquatic or semi-aquatic fauna, forest temporary ponds constitute complementary habitats for overall forest biodiversity. They serve as hydric refuges during dry periods, feeding areas for numerous insects, birds, reptiles, and mammals, and as essential water resources for terrestrial fauna, particularly in Mediterranean environments subject to strong climatic constraints.

Figure 6. Examples of typical fauna of Moroccan temporary ponds. (A) Pleurodeles waltl; (B) Triops mauritanicus; (C) Branchipus schaefferi; (D) Cyzicus bucheti; (E) Hemidiaptomus maroccanus; (F) Tanymastix affinis

(Van den Broeck, 2016)

Forest cover exerts a decisive influence on the ecological functioning of temporary ponds. By reducing evaporation, buffering temperature fluctuations, and prolonging the duration of inundation, it promotes the persistence of species sensitive to rapid desiccation and contributes to the stability of biological communities. Forest wetlands also play a role in water regulation and quality improvement by filtering nutrients, retaining water, and creating ecologically rich interfaces with diverse microhabitats. Positioned at the interface between aquatic and terrestrial environments, temporary ponds enhance ecological connectivity within forest landscapes, functioning as true biological corridors linking different compartments of the territory. Thus, the close interaction between temporary ponds and forest cover confers a key role on these systems in maintaining forest biodiversity, ecosystem resilience to climatic fluctuations, and the conservation of both visible and hidden biological heritage. Despite their exceptional biological richness and ecological importance, forest temporary ponds remain highly vulnerable ecosystems. The biodiversity they support is increasingly exposed to multiple natural and anthropogenic pressures that threaten their ecological integrity and long-term persistence.

THREATS AND PRESSURES

Temporary ponds are highly vulnerable ecosystems due to their shallow depth, often small size, and isolation within the landscape. The species that inhabit them, although inconspicuous and sometimes poorly known, play a crucial ecological role in maintaining local biodiversity (Grillas et al., 2004). Historically, the existence of temporary ponds has been compatible with agricultural activities for several millennia, particularly in Mediterranean regions (Grillas et al., 2004). However, these ecosystems are now strongly threatened by a combination of natural and anthropogenic pressures, leading to rapid and widespread habitat loss across the Mediterranean and globally (King, 1998; Gallego-Fernandez et al., 1999; Rhazi et al., 2012).

In the southern Mediterranean, particularly in North Africa, temporary ponds face intense anthropogenic pressures. Urbanization, intensive agriculture, drainage, infilling, and unregulated recreational activities contribute to their degradation or complete disappearance (Rhazi et al., 2006; Saber, 2006; Bouahim et al., 2014). A study conducted in the province of Benslimane (Western Meseta) revealed a 23% decline in the number of ponds and a 68% decrease in their area between 1955 and 2001, with the number of ponds falling from 871 to 670 (Saber, 2006). This decline is largely linked to the negative perception of ponds by local communities and land managers: their small size and location within highly anthropized agricultural landscapes, combined with fertile soils (particularly in Benslimane and Chaouia), favor their conversion into arable land (Rhazi et al., 2012).

Direct pressures identified by Bouahim et al. (2014) can be grouped into nine main categories: livestock grazing, medicinal plant harvesting, drying (pumping, drainage), recreational activities, sediment extraction, domestic activities (e.g., laundry), agriculture (at the margins or within the pond), infilling, and urbanization. Globally, temporary ponds also face widespread threats such as alterations of the hydroperiod, habitat loss and fragmentation, eutrophication, pollution, sediment disturbance, fires, increased livestock density, introduction of invasive species, and climate change (Grillas et al., 2004; Zacharias & Zamparas, 2010; Boix et al., 2016). These pressures have led to a rapid decline of these habitats, with estimated loss rates between 23% and 97% over the last century.

In addition to direct threats, degradation of surrounding forests amplifies the impacts on temporary ponds. These forests play a key role in microclimate regulation, water availability, and protection against erosion. Their degradation leads to reduced water retention, increased soil erosion, and progressive infilling of ponds, compromising their ecological functionality.

In response to these threats, the international recognition of temporary ponds as priority habitats, particularly under the Ramsar Convention and the code 3170 for habitats of community interest in southern Europe, highlights their ecological value and the need for their conservation. However, despite recently proposed monitoring methods (Quintana et al., 2015; Lumbreras et al., 2016), there remains an urgent need for simple, scalable tools to assess the ecological status of these habitats. According to Hutchinson's theory (1957), understanding the impact of natural and anthropogenic factors on target ecosystems requires observing changes in biodiversity and community structure, which serve as reliable indicators of ecological disturbances (Legendre & Gauthier, 2014). Therefore, the conservation and sustainable management of Mediterranean temporary ponds rely on improved knowledge of biodiversity and community structure. The implementation of appropriate monitoring tools and ecological quality indicators is essential to identify priority habitats, guide conservation actions, and ensure the long-term persistence of these fragile ecosystems (Pinto-Cruz et al., 2009; Zacharias & Zamparas, 2010; Boix et al., 2016).

Figure 7. Examples of threats to Moroccan temporary ponds. (A) Drainage; (B) Water pumping; (C) Agriculture within the pond; (D) Urbanization; (E) Agriculture along the pond edges; (F) Infilling.

Beyond climatic and anthropogenic pressures, biological invasions are increasingly considered an additional factor capable of altering the ecological functioning and biodiversity of Mediterranean temporary ponds. Biological invasions represent an emerging threat to Mediterranean temporary ponds, particularly in landscapes affected by agricultural intensification, forestry activities, and urban expansion. Invasive plant species may colonize pond margins and shallow areas, competing with specialized hydrophilous vegetation adapted to seasonal inundation cycles

(Boix et al., 2016; Grillas et al., 2021). Species such as *Paspalum distichum*, *Azolla filiculoides*, can form dense stands that modify habitat structure, alter nutrient dynamics, and limit the regeneration of characteristic plant communities including *Isoetes*, *Marsilea*, or *Pilularia*.

Faunal invasions may also disrupt ecological processes. Although invasive species remain relatively limited in many Moroccan seasonal wetlands due to their temporary hydrology, increasing human activities and hydrological alterations may facilitate their spread, highlighting the importance of monitoring and early detection programs.

FUNCTIONAL LINK BETWEEN PONDS AND FORESTS

Temporary ponds and forests form intimately connected ecosystems, sharing and reinforcing essential ecosystem services for the overall functioning of the landscape. These services include hydrological regulation, biodiversity maintenance, erosion control, and carbon storage (Costanza et al., 1997; Joyce, 2012; Van den Broeck, 2016). Forests play a key role in water retention and microclimate regulation around ponds, helping to stabilize their hydroperiod and reduce water stress for aquatic and semi-aquatic species. Conversely, temporary ponds provide vital resources for forest communities, serving as water points and refuges for numerous species (Oertli et al., 2010; Grillas et al., 2021).

This interconnection is also reflected in species sharing both habitats at different life stages. Amphibians, for instance, reproduce in temporary ponds but use surrounding forests for foraging and shelter during their terrestrial phase (Jacobs & Wilson, 1996; Jenkins & Buikema, 1998; Bouahim, 2010; Van den Broeck, 2016). Similarly, some aquatic and semi-aquatic insects depend on forest vegetation for nesting, feeding, and dispersal, highlighting the importance of complementary habitat networks for biodiversity maintenance.

However, these environments share common vulnerabilities to natural and anthropogenic pressures. Climate change, in particular, alters precipitation patterns and inundation periods, disrupting species life cycles and community dynamics (El Madihi et al., 2017; Grillas et al., 2021; Moukrim et al., 2022). Habitat fragmentation due to urbanization, intensive agriculture, and forestry reduces ecological connectivity and increases population isolation, raising the risk of local extinctions (Van den Broeck, 2016). Moreover, direct anthropogenic pressures such as drainage, infilling, overgrazing, and waste deposition exacerbate the degradation of these interdependent ecosystems.

SUSTAINABLE MANAGEMENT OF FOREST TEMPORARY PONDS

Sustainable conservation of forest temporary ponds relies on a multidimensional approach combining ecological monitoring, integrated management, local community involvement, and regulatory frameworks. These fragile ecosystems, highly interconnected with their forest environment, require adapted strategies to maintain their ecological functions and the services they provide.

In Morocco, several forest landscapes illustrate the practical application of these management principles. In the Maâmora cork oak forest, temporary ponds have been the focus of ecological monitoring programs aimed at documenting floristic diversity, hydroperiod variability, and amphibian populations, providing valuable data for conservation planning (Bouahim et al., 2015; El Madihi, 2020). In the Benslimane forest region, management efforts have emphasized the protection of ponds from agricultural encroachment and waste deposition, which represent major threats to wetland integrity. In northern Morocco, particularly in the Rif Mountains, temporary ponds embedded within forest ecosystems support endemic and rare plant species and have been integrated into biodiversity assessment and conservation initiatives (Rhazi et al., 2012; Grillas et al., 2021).

Similarly, in the Middle Atlas, several ponds located in cedar and oak forest landscapes have been studied to understand the effects of grazing pressure and hydrological fluctuations on aquatic communities, leading to recommendations for controlled grazing regimes and long-term ecological monitoring (Moukrim et al., 2022). These examples highlight the importance of site-specific management strategies that consider both ecological processes and local land-use practices in order to ensure the long-term conservation of Moroccan forest temporary ponds.

Ecological Monitoring and Quality Indicators

Rigorous ecological monitoring is the cornerstone of sustainable management. It includes implementing biodiversity indicators (amphibians, aquatic invertebrates, macrofauna, hydrophilous plants) and water quality parameters (physico-chemical variables, turbidity, nutrients) to assess the ecological status of ponds (MedWet, 2018; Oertli et al., 2010). Mapping sites and tracking temporal changes in pond surface area, depth, and connectivity provide essential data to detect the impacts of anthropogenic and climatic pressures and guide restoration interventions (Grillas et al., 2021; Moukrim et al., 2025).

Integrated Pond–Forest Management

Managing ponds and surrounding forests in an integrated manner is crucial to maintain ecosystem services. This involves including ponds in forest management plans and, conversely, considering forest condition in wetland conservation strategies (HCEFLCD, 2015; Van den Broeck, 2016). Such an approach maintains microclimate, hydrological regulation, and ecological corridors essential for biodiversity, while minimizing fragmentation and population isolation.

Community Awareness and Education

Social acceptance of temporary ponds is a key factor in their conservation. Awareness campaigns and environmental education targeting local communities promote understanding of the ecological and socio-economic value of these habitats (Bouahim et al., 2015; MedWet, 2018). Encouraging agricultural and recreational practices compatible with pond preservation helps reduce direct anthropogenic pressures such as drainage, infilling, and overgrazing.

Regulatory Frameworks and Protection Instruments

Integrating temporary ponds into national and international regulatory frameworks is essential for effective protection. International conventions, such as the Ramsar Convention on Wetlands and international forestry agreements, provide standards and tools for ecosystem conservation (HCEFLCD, 2015; MedWet, 2018). At the national level, including ponds and their watersheds in conservation policies, land-use planning, and forest regulations prevents their destruction and regulates human use.

By combining these four pillars, ecological monitoring, integrated management, community awareness, and regulatory frameworks, it is possible to implement sustainable conservation of forest temporary ponds, ensuring biodiversity protection, the maintenance of ecosystem services, and resilience to anthropogenic and climatic pressures (Boix et al., 2016; Grillas et al., 2021).

In addition to providing habitat complementarities, the functional link between temporary ponds and forests is expressed through several key ecological processes operating at the landscape scale. Forest vegetation surrounding ponds contributes organic matter inputs in the form of leaf litter and woody debris, which represent an important source of nutrients and energy for aquatic food webs. These inputs stimulate microbial activity and primary production, thereby supporting the development of aquatic invertebrate communities that constitute the base of the trophic network (Batzer & Wissinger, 1996; Williams, 2006). In return, ponds participate in nutrient cycling by acting as temporary sinks for sediments and nutrients transported by runoff, thus limiting their transfer to downstream ecosystems and contributing to water purification processes (Oertli et al., 2010; Boix et al., 2016).

Moreover, these habitats and forests together form ecological networks that enhance landscape connectivity for many organisms. These small wetlands function as stepping stones facilitating dispersal and gene flow among populations, particularly for amphibians, aquatic insects, and semi-aquatic plants whose life cycles depend on both aquatic and terrestrial environments (Semlitsch & Bodie, 2003; Oertli et al., 2010). The maintenance of forest buffers around ponds therefore plays a crucial role in preserving these ecological flows, ensuring habitat continuity, and increasing the resilience of species to environmental changes. Through these combined processes of nutrient exchange, trophic interactions, and biological connectivity, forest temporary ponds contribute significantly to the ecological integrity and functioning of Mediterranean forest landscapes.

CONCLUSION AND PERSPECTIVES

Mediterranean temporary ponds are highly unique ecological ecosystems, characterized by a seasonal alternation between flooded and dry phases, primarily driven by precipitation regimes. This hydrological pattern closely determines the structure, composition, and dynamics of the biological communities they host. Despite their often-small size, these wetlands play a major ecological role, providing essential habitats for distinctive biodiversity, including many rare, endemic, or threatened species specifically adapted to water fluctuations through strategies such as dormancy, rapid germination, or short life cycles.

Within southern Mediterranean forest landscapes, temporary ponds are structuring elements of the habitat mosaic, significantly contributing to overall biodiversity and enhancing ecological connectivity. Functional interactions between these wetlands and forest ecosystems are close and interdependent: forests help regulate the water cycle feeding the ponds, while ponds contribute to the hydric resilience of forest stands under climatic stress. Hydroperiod variability, ranging from a few months to nearly a year, shapes complex and dynamic trophic networks involving primary producers, aquatic invertebrates, amphibians, and birds, strengthening biodiversity at the landscape scale.

In Morocco, temporary ponds are distributed along a broad bioclimatic gradient, with notable concentrations in Atlantic coastal forests and some large cork oak groves. However, these ecosystems are now subject to multiple and interdependent pressures, including urbanization, intensified agricultural practices, eutrophication, and climate change impacts. These threats jointly affect ponds and surrounding forests, highlighting their shared vulnerability to global change.

Sustainable conservation of Mediterranean temporary ponds (MTPs) thus requires an integrated and holistic approach, combining: (i) rigorous ecological monitoring based on biological and physico-chemical indicators, (ii) explicit integration of these ecosystems into forest management plans at the landscape scale and restoration of degraded sites, (iii) raising awareness among local communities about their ecological and socio-economic value, and (iv) inclusion in national and international regulatory frameworks. Within this perspective, truly sustainable forest management must recognize the preservation of temporary ponds as a priority objective, an essential condition for ensuring the resilience, functionality, and long-term sustainability of Mediterranean forest landscapes.

REFERENCES

Amami, B., Rhazi, L., Chaibi, M., Fauquette, S., Ougougdal, M. A., Charif, A., Ridaoui, M., Bouahim, S., Carré, M., & Daoud-Bouattour, A. (2013). Late Quaternary history of a Mediterranean temporary pool from western Morocco, based on sedimentological and palynological evidence. *Palaeogeography, Palaeoclimatology, Palaeoecology, 392*, 281–292. DOI: 10.1016/j.palaeo.2013.09.024

Barbour, M. G., & Major, J. (1977). *Terrestrial vegetation of California*. University of California Davis.

Batzer, D. P., & Wissinger, S. A. (1996). Ecology of insect communities in nontidal wetlands. *Annual Review of Entomology, 41*(1), 75–100. DOI: 10.1146/annurev.en.41.010196.000451 PMID: 15012325

Beja, P., & Alcazar, R. (2003). Conservation of Mediterranean temporary ponds under agricultural intensification: An evaluation using amphibians. *Biological Conservation, 114*(3), 317–643. DOI: 10.1016/S0006-3207(03)00051-X

Blaustein, L., & Schwartz, S. S. (2001). Why study ecology in temporary pools? *Israel Journal of Zoology, 47*(4), 303–312. DOI: 10.1560/CKMU-Q2PM-HTGC-P9C8

Bliss, L. C., Marticorena, C., Matthei, O., & Quezada, M. (1998). Plant diversity in Chile: a comparative ecology of Mediterranean ecosystems. In Rundel, P. W. (Eds.), *Landscape disturbance and biodiversity in Mediterranean-type ecosystems*. Springer.

Boix, D.. (2016). Pond research and management in Europe: “Small is beautiful”. *Hydrobiologia, 793*, 1–16.

Boix, D., Sabater, S., Gascón, S., Sala, J., Quintana, X. D., & Badosa, A. (2016) Ecology of Mediterranean temporary ponds. In Batzer, D., & Boix, D. (Eds.), *Invertebrates in Freshwater Wetlands* (pp. 229–255). Springer. DOI: 10.1007/978-3-319-24978-0_5

Bonis, A., Grillas, P., Van Wijck, C., & Lepart, J. (1993). The Effect Of Salinity On The Reproduction Of Coastal Submerged Macrophytes From Experimental Communities. *Journal of Vegetation Science, 4*(4), 461–468. DOI: 10.2307/3236073

Bonis, A., Lepart, J., & Grillas, P. (1995). Seed Bank Dynamics And Coexistence Of Annual MacrophytesIn A Temporary And Variable Habitat. *Oikos, 74*(1), 81–92. DOI: 10.2307/3545677

Bouahim, S. (2010). Usage et conservation des mares temporaires méditerranéennes: cas des mares temporaires de la région de Benslimane (Maroc occidental). Thèse de doctorat, Université Hassan II, Casablanca & Université Montpellier II.

Bouahim, S., Rhazi, L., Amami, B., Waterkeyn, A., Rhazi, M., Saber, E. R., Zouahri, A., Van Den Broeck, M., Muller, S. D., Brendonck, L., & Grillas, P. (2014). Unravelling the impact of anthropogenic pressureon plant communities in Mediterranean temporary ponds. *Marine and Freshwater Research*, *65*(10), 918–929. DOI: 10.1071/MF13194

Bouahim, S., Rhazi, L., Ernoul, L., Mathevet, R., Amami, B., Er-Riyahi, S., & Grillas, P. (2015). Combining vulnerability analysis and perceptions of ecosystem services in sensitive landscapes: A case from western Moroccan temporary wetlands. *Journal for Nature Conservation*, *27*, 1–9. DOI: 10.1016/j.jnc.2015.05.003

Boulton, A. J., & Brock, M. A. (1999). *Australian freshwater ecology: processes and management*. Gleneagles Publishing.

Boutin, C., Jobin, B., & Bélanger, L. (1982). Écologie des mares temporaires méditerranéennes. *Ecologia Mediterranea*, *8*, 47–58.

Brendonck, L., & Williams, W. D. (2000). Biodiversity in wet-lands of dry regions (drylands). In Gopal, B., Junk, W. J., & Davis, J. A. (Eds.), *Biodiversity in Wetlands: assessment,Function and Conservation* (pp. 181–194). Backhuys.

Brock, M. A., Nielsen, D. L., Shiel, R. J., Green, J. D., & Langley, J. D. (2003). Drought and Aquatic Community Resilience: The Role Of Eggs And Seeds In Sediments Of Temporary Wetlands. *Freshwater Biology*, *48*(7), 1207–1218. DOI: 10.1046/j.1365-2427.2003.01083.x

Cardinale, B. J., Duffy, J. E., Gonzalez, A., Hooper, D. U., Perrings, C., Venail, P., & Naeem, S. (2012). Biodiversity loss and its impact on humanity. *Nature*, *486*(7401), 59–67. DOI: 10.1038/nature11148 PMID: 22678280

Costanza, R., d'Arge, R., de Groot, R., Farber, S., Grasso, M., Hannon, B., & van den Belt, M. (1997). The value of the world's ecosystem services and natural capital. *Nature*, *387*(6630), 253–260. DOI: 10.1038/387253a0

Cucherousset, J. (2006). Adaptations des organismes aux milieux temporaires. *Annales de Limnologie*, *42*, 1–12.

Deil, U. (2005). A review on habitats, plant traits and vegetation of ephemeral wetlands – a global perspective. *Phytocoenologia*, *35*(2-3), 533–705. DOI: 10.1127/0340-269X/2005/0035-0533

El Madihi, M. (2020). Étude de la biodiversité végétale et des traits fonctionnels des plantes des zones humides temporaires du Maroc le long d'un gradient de latitude et de pression anthropique. Thèse de doctorat, Université Mohammed V, Rabat.

El Madihi, M., Rhazi, M., Bouahim, S., Amami, B., Waterkeyn, A., & Grillas, P. (2017). Functional traits of plant communities in Mediterranean temporary pools under different disturbance regimes. *Plant Ecology & Diversity*, *10*(4), 291–303. DOI: 10.1080/17550874.2017.1346716

Fennane, M., Ibn Tattou, M., & El Oualidi, J. (2014). Flore pratique du Maroc.: Vol. 3. *Série Botanique 40*. Travaux de l'Institut Scientifique.

Gallego-Fernandez, J. B., Garcia-Mora, M. R., & Garcia-Novo, F. (1999). Small wetlands lost: A biological conservation hazard in Mediterranean landscapes. *Environmental Conservation*, *26*(3), 190–199. DOI: 10.1017/S0376892999000272

Grillas, P., Gauthier, P., Yavercovski, N. & Perennou, C. (2004). *Mediterranean Temporary Pools. Volume 1: Issues relating to conservation, functioning and management*. Tour du Valat.

Grillas, P., Rhazi, L., Lefebvre, G., El Madihi, M., & Poulin, B. (2021). Foreseen impact of climate change on temporary ponds located along a latitudinal gradient in Morocco. *Inland Waters*, *11*(4), 492–507. Advance online publication. DOI: 10.1080/20442041.2021.1962688

HCEFLCD. (2015). *Stratégie nationale des zones humides du Maroc 2015–2024*. Haut-Commissariat aux Eaux et Forêts et à la Lutte Contre la Désertification, Rabat.

Hutchinson, G. E. (1957). Concluding remarks. *Cold Spring Harbor Symposia on Quantitative Biology*, *22*(0), 415–427. DOI: 10.1101/SQB.1957.022.01.039

Jacobs, S. W. L., & Brock, M. A. (1993). Wetlands of Australia: southern (temperate) Australia. In Whigham, D. F., Dykyjov, D., & Hejny, S. (Eds.), *Wetlands of the World: Inventory, Ecology and Management* (Vol. 1, pp. 244–304). Kluwer Academic Publishers.

Jacobs, S. W. L., & Wilson, K. L. (1996). A biogeographical analysis of the freshwater plants of Australasia. *Australian Systematic Botany*, *9*(2), 169–183. DOI: 10.1071/SB9960169

Jenkins, D. G., & Buikema, A. L.Jr. (1998). Do similar communities develop in similar sites? *Ecological Monographs*, *68*(3), 421–443. DOI: 10.1890/0012-9615(1998)068[0421:DSCDIS]2.0.CO;2

Joyce, C. (2012). Preface: Wetland services and management. *Hydrobiologia*, *692*(1), 1–3. DOI: 10.1007/s10750-012-1179-9

Keeley, J. E., & Zedler, P. H. (1998). Characterization and global distribution of vernal pools. In Witham, C. C. (Ed.), *Vernal pool ecosystems* (pp. 1–14). California Native Plant Society.

King, J. L. (1998). Loss of diversity as a consequence of habitat destruction in California vernal pools. In *Ecology, conservation, and management of vernal pool ecosystems—Proceedings from a 1996 conference*. California Native Plant Society.

Legendre, P., & Gauthier, O. (2014). Statistical methods for temporal and space–time analysis of community composition data. *Proceedings. Biological Sciences*, *281*(1778), 20132728. DOI: 10.1098/rspb.2013.2728 PMID: 24430848

Lumbreras, A., Marques, J., Belo, A., Cristo, M., Fernandes, M., Galioto, D., Machado, M., Mira, A., Sá-Sousa, P., Silva, R., Sousa, L. G., & Pinto-Cruz, C. (2016). Assessing the conservation status of Mediterranean temporary ponds using biodiversity: A new tool for practitioners. *Hydrobiologia*, *782*(1), 187–199. DOI: 10.1007/s10750-016-2697-7

Médail, F., Michaud, H., Molina, J., Paradis, G., & Loisel, R. (1998). Conservation de la flore et de la végétation des mares temporaires dulçaquicoles et oligotrophes de France méditerranéenne. *Ecologia Mediterranea*, *24*(2), 119–134. DOI: 10.3406/ecmed.1998.1856

MedWet. (2018). *MedWet Framework for Action 2019–2030: Wetlands for sustainable development in the Mediterranean region*. Tour du Valat.

Millennium Ecosystem Assessment. (2005). *Ecosystems and human well-being: Synthesis*. Island Press.

Moukrim, S., Lahssini, S., Rhazi, M., Menzou, K., El Madihi, M., Rifai, N., Bouziani, Y., Azedou, A., Boukhris, I., & Rhazi, L. (2022). Climate change impact onpotential distribution of an endemic species Abies marocana Trabut. *Ekologia (Bratislava)*, *41*(4), 329–339. DOI: 10.2478/eko-2022-0034

Moukrim, S., Tellal, J., Lahssini, S., Boudjaj, A., Rhazi, M., El Madihi, M., Ben Bammou, M., Moukrim, M., Saber, E.-R., Wahby, I., Rhazi, L., & Grillas, P. (2025). Long-time dynamics in a North African coastal Ramsar site, the Sidi Boughaba Lake. *Ekologia (Bratislava)*, *44*(2), 147–159. DOI: 10.2478/eko-2025-0017

Nicolet, P., Biggs, J., Fox, G., Hodson, M. J., Reynolds, C., Whitfield, M., & Williams, P. (2004). The wetland plant and macroinvertebrate assemblages of temporary ponds in England and Wales. *Biological Conservation*, *120*(2), 261–278. DOI: 10.1016/j.biocon.2004.03.010

Oertli, B., Céréghino, R., Hull, A., & Miracle, R. (2010). Pond conservation in Europe. *Hydrobiologia*, *634*(1), 1–9. DOI: 10.1007/s10750-009-9891-9

Oertli, B., Joye, D. A., Castella, E., Juge, R., Cambin, D., & Lachavanne, J.-B. (2010). Ponds conservation: From science to practice. *Hydrobiologia*, *634*(1), 1–9. DOI: 10.1007/s10750-009-9891-9

Pinto-Cruz, C., Molina, J. A., Loidi, J., & Costa, J. C. (2009). Plant communities as a tool in temporary pond conservation. *Hydrobiologia*, *634*(1), 11–24. DOI: 10.1007/s10750-009-9885-7

Pyke, C. P. (2005). Assessing climate change impacts on vernal pool ecosystems and endemic branchiopods. *Ecosystems (New York, N.Y.)*, *8*(1), 95–105. DOI: 10.1007/s10021-004-0086-y

Quézel, P. (1998). La végétation des mares transitoires à isoetes en région méditerranéenne, intérêt patrimonial et conservation. *Ecologia Mediterranea*, *24*(2), 111–117. DOI: 10.3406/ecmed.1998.1855

Quintana, X. D., Boix, D., Gascón, S., & Sala, J. (2015). Monitoring temporary ponds. *Limnetica*, *34*, 1–20.

Ramdani, M. (1986). Contribution à l'étude écologique des mares temporaires du Maroc. Thèse de doctorat, Université de Rabat.

Rhazi, L., Grillas, P., Saber, E., Rhazi, M., Brendonck, L., & Waterkeyn, A. (2012). Vegetation of Mediterranean temporary pools: A fading jewel? *Hydrobiologia*, *689*(1), 23–36. DOI: 10.1007/s10750-011-0679-3

Rhazi, L., Grillas, P., Tan Ham, L., & El Khyari, D. (2001). The seed bank and the between-years dynamics of the vegetation of a Mediterranean temporary pool (NW Morocco). *Ecologia Mediterranea*, *27*(1), 69–88. DOI: 10.3406/ecmed.2001.1907

Rhazi, L., Rhazi, M., Grillas, P., & Khyari, D. (2006). Richness and structure of plant communities in temporary pools from western Morocco: Influence of human activities. *Hydrobiologia*, *570*(1), 197–203. DOI: 10.1007/s10750-006-0180-6

Rosset, V., Lehmann, A., & Oertli, B. (2010). Warmer and richer? Predicting the impact of climate warming on species richness in small temperate water bodies. *Global Change Biology*, *16*(8), 2376–2387. DOI: 10.1111/j.1365-2486.2010.02206.x

Saber, E. (2006). *Approche de la dynamique spatio-temporelle des géosystèmes forêt et mares temporaires de la province de Benslimane (Maroc occidental)* (Doctoral dissertation, Université de Provence, Aix-Marseille I).

Serge Muller. (2022). *Amina Daoud-Bouattour, Séverine Fauquette, Marion Bottollier-Curtet, NabilRifai, et al. Holocene history of peatland communities of central Rif(Northern Morocco). Geobios.* Elsevier., DOI: 10.1016/j.geobios.2021.12.001

Schwartz, S. S., & Jenkins, D. G. (2000). Temporary aquatic habitats: Constraints and opportunities. *Aquatic Ecology*, *34*(1), 3–8. DOI: 10.1023/A:1009944918152

Semlitsch, R. D., & Bodie, J. R. (2003). Biological criteria for buffer zones around wetlands and riparian habitats. *Conservation Biology*, *17*(5), 1219–1228. DOI: 10.1046/j.1523-1739.2003.02177.x

Thiéry, A. (1987). Les mares temporaires: Écologie et biogéographie. *Annales de Limnologie*, *23*, 91–105.

Thiéry, A. (1997). Temporary pools: Ecological and biogeographical aspects. *Hydrobiologia*, *345*, 1–9.

Van den Broeck, M. (2016). *Monitoring and conservation of Mediterranean temporary ponds.* PhD thesis, Catholic University of Leuven, Belgium, 181 p.

Williams, D. D. (2006). *The Biology of temporary waters.* Oxford Biology.

Zacharias, I., & Zamparas, M. (2010). Mediterranean temporary ponds: A disappearing ecosystem. *Biodiversity and Conservation*, *19*(14), 3827–3834. DOI: 10.1007/s10531-010-9933-7

Zedler, P. H. (1987). The ecology of Southern California vernal ponds: A community profile. *US Fish and Wildlife Service Biological Report.*, *85*, 7–11.

Chapter 5

Economic Evaluation of Forest Management Options for Generating Cultural and Environmental Services:
Example of South and East Mediterranean Countries

Hamed Daly Hassen
http://orcid.org/0000-0001-8712-6252
University of Carthage, Tunisia

Nelly Bourlion
Independent Researcher, France

Magali Maire
INRAE, France

ABSTRACT

In a context of an increasing demand for recreational services and protection of water resources in the Mediterranean region, it is important to carry out forest management options for generating these services. The objective of this chapter is to evaluate the social profitability of these options. Four pilot sites, notably in national parks and biosphere reserves, were selected in Algeria, Lebanon, Tunisia, and Turkey. The Cost-Benefit Analysis (CBA) method was used to determine the net present value of the different options, and to determine the gains and losses for the

DOI: 10.4018/979-8-3373-6239-7.ch005

different stakeholders. Several evaluation methods were used to estimate the values of ecosystem services, such as the travel cost method and the contingent valuation method for recreation, as well as the avoided damage cost method for the protection of water resources. The results show that at least one management option is socially profitable for each site, and it must be sufficiently attractive for all stakeholders through economic incentives to be implemented.

1. INTRODUCTION

Mediterranean forests had a surface area of 85 million hectares in 2010, unequally distributed between the North (76.3%), the East (14.5%) and the South (9.2%) of the Mediterranean region (FAO, 2013). They represent ecosystems that are often similar, but different socio-economic contexts induce differences both in the importance and in the use of goods and services provided by wooded areas (FAO, 2013). Forests in Southern and Eastern Mediterranean Countries (SEMCs) are often characterized by their public land (State, Municipal, religious-owned) and by the diversity of goods and services used (firewood, fruit, pastureland, etc.) (Palahi et al., 2008). In Maghreb countries, the State manages forests, but the local populations often have use rights. However, there is considerable use of forest resources (pastureland, fruit and mushroom picking, honey production) that is not always legal (overuse, unregulated use, etc.), which often leads to stronger pressures than forest or sylvo-pastoral ecosystems are able to withstand in the context of sustainable management. Because resources are shared, local households seek to meet their subsistence needs and increase their short-term income with little interest in using resources sustainably or in the long-term effects of over-use. There is therefore a need to create a balance between uses, pastureland, and forest conservation. Benefit distribution varies depending on use rights and the type of products provided by forest ecosystems. Over the last two decades, several attempts to evaluate the market and non-market benefits of Mediterranean forests have been carried out, including environmental and cultural services. The first exhaustive estimation of the economic value of goods and services provided by Mediterranean forests was carried out by Merlo and Croitoru (2005) for the year 2001. Estimations used different evaluation techniques ranging from the use of market prices for the evaluation of provisioning goods, to damage costs for regulation services and cost-based approaches for biodiversity conservation (Khalfaoui et al., 2020; Corona et al, 2018). In general, the degree of confidence becomes lower for these latter values.

However, these values have often been underestimated because of the scarcity or lack of estimation of several benefits (Croitoru, 2007). Among the values of provisioning goods, the one linked to fodder production dominates in most countries, notably in Maghreb countries. Values relating to other non-wood forest products are also quite high in the different countries. On the other hand, the value of wood constitutes a low value in most countries (Merlo & Croitoru, 2005). For regulation services, watershed protection constitutes the first or second benefit of forests in the three Maghreb countries (Merlo & Croitoru, 2005).

Indeed, South and East Mediterranean countries are characterized by water scarcity and a high sensitivity to erosion—critical issues that are increasingly exacerbated by global change. Current advancements integrate multi-site comparative frameworks (Karimi et al., 2025) and more sophisticated valuation techniques—ranging from damage cost assessments for regulation services (Robertson et al., 2024) to refined cost-based approaches for biodiversity conservation (Acharya et al., 2020)—to better reflect the socio-economic context of forest management. Specifically, recent studies have highlighted the importance of large-scale economic evaluations of carbon sequestration as a critical component of forest ecosystem services (Russo et al., 2023).

While traditional Cost-Benefit Analysis (CBA) has been the standard for comparing management options against a "no intervention" baseline, recent global literature emphasizes the need for more dynamic and exhaustive estimations of both market and non-market benefits. Recent developments in Cost-Benefit Analysis (CBA) emphasize the "relative price" effect (Heckenhahn & Drupp, 2024), acknowledging that as natural resources become scarcer and demand for environmental services rises, their value increases faster than that of market goods. While international government recommendations have historically lagged in this regard, recent studies suggest that applying a relative price increase (e.g., 1% to 2% per year) for ecosystem services can significantly alter Net Present Value (NPV) results, making conservation projects more socially profitable over long-term horizons. This approach aligns with recent assessments of forest ecosystem services that call for more dynamic modeling of value over time.

In the Mediterranean Basin, climate change is projected to have predominantly negative impacts on forest ecosystem services, including water regulation and fodder production (Gauquelin et al., 2018). Specifically, warming above the 2°C threshold is expected to compromise the provision of regulating services by reducing freshwater quantity and quality. To account for these risks, alternative scenarios incorporate potential losses in service provision; for instance, projections for North African watersheds suggest that monetary losses due to climate change could range from moderate to substantial by the mid-21st century depending on the RCP scenario adopted. The increasing demand for forest services in the Mediterranean must be evaluated against the backdrop of climate threats, which challenge the assumption of stationary environmental conditions.

2. METHODOLOGY

2.1. Study Sites and Proposed Forest Management Options

The identification of pilot sites and management options (see Table 1) and the estimation of their social profitability relied on a participatory framework involving local and national officials, as well as community-level actors. Data collection involved direct interviews and surveys of visitors (e.g., in Chrea National Park) to elicit values for recreational services and identify potential impacts on local livelihoods. In Jabal Moussa, management options were specifically designed to generate additional revenue and create jobs for local populations, thereby compensating for income losses resulting from conservation restrictions. This distribution analysis ensures that the proposed options contribute to improving the incomes of local communities and are socially attractive enough for implementation.

Figure 1. Localization map of four pilot sites

1. **Chrea National Park (PNC)-Algeria**: Located 50 km South-West of Algiers, classified as a biosphere reserve. The forested land covers 22,673 hectares and is publicly owned. The natural vegetation consists of pure and mixed forest stands of Atlas cedar, cork oak, holm oak, and other species. This park plays an important role in biodiversity conservation and recreation. However, the forestry administration faces high management costs and risks of degradation and pollution due

to visitor overcrowding. To mitigate the effects of tourism pressure with limited and sporadic intervention by park rangers, the park manager has proposed two management alternatives: - OPNC1: Measures to support and supervise visitors by nature guides (guided tours, monitoring, creation of marked trails and signposts); - OPNC2: Operation of a new recreation area to channel some of the visitors present on the existing site while maintaining the same total number of visitors (Ouadah, 2016).

2. **Jabal Moussa Biosphere Reserve (RJM)-Lebanon**: The Jabal Moussa Biosphere Reserve (JMR) in Lebanon is located 45 km from Beirut. It is owned by two municipalities (Ebre and Yahchouch), as well as religious institutions. The management of the wooded areas includes the municipalities, the Jabal Moussa Protection Association (JMRA), the Ministry of Agriculture, and the Ministry of the Environment. The forest covers 5,500 hectares, comprising 1,250 hectares of protected area, 1,700 hectares of buffer zone, and 2,550 hectares of development zone. The local population is approximately 15,000, of whom 1,500 use the wooded areas for grazing and thyme harvesting. The protected area designation prohibits resource exploitation, particularly within the protected zone, resulting in a loss of income for the local population. In contrast, recreational activities, particularly hiking, are developed and managed by the APJM association, and therefore have a positive impact on the sale of forest products. The identified management option (ORJM) is the opening of new hiking trails and the creation of accommodation infrastructure (guesthouses) to increase the number of visitors. The aim is to generate additional revenue and create jobs for local people. Other income-generating activities (development of thyme cultivation, development of beekeeping) are planned to compensate the loss of income for local households following regulated resource exploitation (Karam, 2010).
3. **Oued Barbara Watershed (BVB)-Tunisia**: The Oued Barbara (BVB) watershed in Tunisia covers an area of approximately 16,400 hectares (DGF, 2010), of which 5,065 hectares (31% of the forest cover) are primarily cork oak and state-owned. Other land uses are mainly annual crops (39%) and agroforestry areas (12%) (Daly et al., 2012). The main functions of the forests are cork and fodder production and protection against dam sedimentation. The population living in the watershed area was approximately 25,000 in 2004. The inhabitants are generally poor and practice subsistence farming and livestock breeding. In connection with the construction of a dam downstream in 1998, several forestry interventions were implemented by the administration over the period 1994-2008 in order to reduce sedimentation from the dam downstream, reduce overgrazing and restore degraded forests, including: - OBVB1: Acacia plantation on the

edges of ravines (678 ha) and- OBVB2: Artificial regeneration of cork oak over an area of 110 ha (World Bank, 2010)

4. **Düzlerçamı Pilot Site (DZ) (Turkey)**: Düzlerçamı Pilot Site (DZ) (Turkey) is located very close to Antalya Province. The total surface area of the pilot site is 29168 ha. In the pilot site, there exist 11188 ha of productive and 6500 ha of unproductive forests. Furthermore, privately owned agricultural lands and settlements of 11480 ha exist inside the pilot site. Forests are state owned and managed by the General Directorate of Forestry. Düzlerçamı Pilot Site host maquis and Turkish pine (*Pinus brutia*). This site is influenced by recreation (mainly in the form of picnicking) and tourism activities with the presence of a couple of recreation spots rented to public (Municipalities) or private bodies by the forest administration. There exists an increasing demand for recreation activities by the local community and this was identified as the management option (O_{DZ}) leading to the opening of a new recreation spot (Balkiz, 2016).

2.2. Cost-Benefit Analysis (CBA) Method

The CBA method is a tool that allows to compare different proposed management options with the absence of intervention, in terms of costs and benefits (Hanley and Barbier, 2009). CBA thus makes it possible to estimate the impact of the management option in monetary terms through the net present value (NPV), and to target interventions with the greatest social impact. It involves several steps: (1) identifying the additional costs and benefits of each option compared to no intervention; (2) estimating the additional annual costs of each option. (3) the estimation of additional annual benefits, the benefits taken into consideration including all benefits: wood, cork, fodder, recreation, water protection, and carbon sequestration, and (4) the estimation of the net present value (Daly-Hassen et al., 2019). Each management option affects managers and beneficiaries differently. The distribution of costs and benefits per stakeholder is necessary to show the economic impacts, i.e., gains and losses, that will be incurred by all stakeholders following changes in forest management. This is particularly necessary to show the extent to which management options contribute to improving incomes of local communities.

The analysis steps are presented below:

2.2.1. Estimation of Additional Costs

The financial costs of the proposed management options, including initial investments and annual operating expenses, were provided by the respective organizations responsible for forest management at each site. These estimations are based on official investment budgets, technical reports from the World Bank, and national forest inventories, which reflect actual market prices for labor and materials in the region. To ensure reliability, all costs were updated to 2024 values using World Bank consumer price indices and exchange rates, providing a verified and consistent baseline for the social profitability analysis.

2.2.2. Evaluation of Additional Benefits

Environmental economic literature (Bishop (Editor), 1999; Brahic & Terreaux, 2009 ; Daly-Hassen et al., 2012; Jumiyati, 2025; Mavsar et al., 2014; Merlo & Croitoru, 2005 ; Plan Bleu, 2014 ; TEEB, 2010 ; World Bank, 2005) provides a wide selection of methods and techniques for valuing ecosystem services. This section presents the methods and techniques used for estimating the value of goods and services in the different sites and discusses their specific implementation for the valuation of forest goods and services. The valuation methods vary between approaches based on revealed preferences (observations of the current behavior of consumers in goods and services of interest such as market price, or substitute markets) and approaches based on expressed preferences (willingness of consumers to pay for a benefit) (Table 1). Estimates based on observed behavior are generally preferred over those based on assumed behavior (CAS, 2009), and direct measurements are better than indirect measurements (Merlo & Croitoru, 2005). However, the method selection depends on the characteristics of the case study and available data. The benefit transfer method must also be used with caution when sites have similar characteristics.

Table 1. Valuation methods used for different types of goods and services

Country	Option	Type of goods and services	Valuation Method
Algeria	O_{PNC1}	Recreation	Contingent Valuation Method
	O_{PNC2}	Recreation	Travel Cost Method
Lebanon	O_{RJM}	Recreation	Market Price (Entry Fees)
Tunisia	O_{BVB}	Wood, cork	Market Price
		Animal fodder	Substitute Price
		Water resource protection	Avoided Damage Cost Method
		Carbon	Market Price
Turkey	O_{DZ}	Wood	Benefit Transfer Method
		Recreation	Market Price
		Carbon	Benefit Transfer Method

The methods are discussed below in detail for each good or service.

Wood and Non-Wood Forest Products (NWFP)

The market price method is used to assess the value of these products. The price should be the local price and not include operating costs, transport, market or processing costs.

Fodder

The substitution price method was used by all the countries. When the market price for a product does not exist the value of a product can be estimated using substitution products available on existing markets. In this case, the price of barley was used as a substitution product for fodder. One kg of barley is equivalent in energy to one fodder unit. Fodder production depends on climate conditions and grazing practices. In Tunisia, productivity was based on data provided by a pastoral inventory.

Recreation

The Travel Cost Method is typically used to assess the value of recreation and cultural services in Algeria. This method considers the cost of travel and the number of visitors to the site observed in order to estimate the value of recreational amenities. The higher the cost of visit to reach the site, the fewer the number of visitors (function of demand). The recreational value is determined based on the consumer's surplus, i.e. the difference between the maximum price that a consumer is willing to pay and the actual price paid. This method requires a survey of visitors, the development of a function related to the number of visits on demand depending on the cost of the visit and deduction of the surplus that visitors gain from visiting the site, corresponding to the value of the visit. The cost of visit often used is the total cost: the sum of transport cost, the entry cost and the opportunity cost for the time spent on the site.

The benefit transfer method was used in Lebanon and Turkey. The method involves applying the values used for one site (willingness to pay) to another site if the site has similar characteristics. Although adjustments can be made, this method has weaknesses that are inherent to the approach. The quality of the results also depends on the scientific validity of the original study, the similarities between the original site and the studied site, and the method used to transfer the values. In addition to the margins of error related to the method used, the number of visitors is often estimated based on the number of vehicles having visited the site.

Water Resource Protection

The cost-of-damage method was applied to estimate the value of the reservoir protection against sedimentation at the BVB site (Tunisia) (Mastrorilli et al., 2018). It assumes that the loss of ecosystem services will incur costs of damage (degradation) to society, meaning that these services must be worth at least these costs. Applying this method requires data on the degradation costs that would be incurred without the forest. This leads to the use of models to estimate dam siltation with and without the forest, and to assessments of the dam's construction cost. The avoided cost method is also used to evaluate water purification.

Carbon Sequestration

For the Barbara watershed, the monetary valuation is based on the average price of carbon on the international market in the reference year. For Düzlerçamı, a "Shadow Price" of carbon was used to account for the failure of the carbon market (Russo et al., 2023). This price reflects the marginal opportunity cost of carbon in the absence of a market price (Mavsar et al., 2014).

Biodiversity Conservation

While biodiversity is a primary objective for the management of sites such as the Chréa and Jabal Moussa Biosphere Reserves, it is treated in this study as a foundational support service rather than a direct provisioning or cultural service. Biodiversity serves as the 'infrastructure' of the ecosystem, providing the necessary resilience and stability that allow for the sustained delivery of other services such as recreation, water regulation and fodder production (Mace et al., 2012; TEEB, 2010). Economically, biodiversity represents an 'insurance value'—the capacity of the forest to maintain its functions in the face of environmental shocks like climate change or pest outbreaks (Baumgärtner, 2007). By characterizing biodiversity as a supporting service, the risk of 'double-counting', which occurs when the underlying biological diversity and the final services it generates are both monetized and summed in a single Cost-Benefit Analysis, is avoided (Plan Bleu, 2014; TEEB, 2010). Consequently, while its value is not explicitly aggregated into the final Net Present Value, it is recognized as the essential qualitative condition upon which all other socio-economic benefits depend.

2.2.3. Net Present Value (NPV) Estimation

NPV is the difference between additional benefits and additional costs. Since costs and benefits are realized at different time intervals, a discount rate is used to translate future values into their present value.

Net benefit (Nb) = $\sum$ Additional Benefits of goods and services – $\sum$ Additional Costs (1)

The Net Present Value (NPV) corresponds to the sum of present values (PV) of net benefit flows.

$NPV = \sum Nbi/(1+r)^i$ for the year i ; r: Discount rate (2).

The management option is worthwhile if the NPV is positive when discounted at a suitable discount rate. Generally, an option with a higher NPV would be preferred. If the NPV is positive from the social perspective but negative for private stakeholders, this means that the management option is beneficial for the society because of its high environmental and non-marketed benefit. In this case, an economic incentive might be justified.

Analysis Period. The value associated with the different management options is calculated over a 20-year period. When the lifespan of certain forest species (such as cork oak) exceeds this period, we include their residual value in the analysis, i.e., the present value of the net benefits for the period beyond 20 years.

Discount Rate. This rate represents the opportunity cost of the capital employed by the different countries (8.5% in Algeria, 10% in Tunisia, 7.3% in Lebanon, and 5% in Turkey) in various analyses of natural resource management projects. Other rates are also used (2%, 5% and 10%) to analyze the sensitivity of the net present value.

Because data were obtained in different years, an adjustment of values to 2024 was done using the price index of consumers and the Euro change rate, using the World Bank data.

3. RESULTS

3.1. Estimation of Additional Costs

After identifying the additional costs of the forest management options, their values were estimated by the managers (see Table 2). The additional costs are quite high except for option OPNC1, for which the involvement of nature guides in raising visitor awareness will lead to a decrease in monitoring costs for the State. The costs are often concentrated during the first year (investment), while the costs are much lower in subsequent years. For Düzlerçami, mature forest stands are chosen as recreation areas. A bidding process is carried out whereby management of the site is leased for 29 years to the selected public or private entity. In the leased forest lands, wood harvesting by the State is no longer continued. In this case, a lease of a 20-hectare area in the pilot site as a recreation area is assumed.

Table 2. Estimation of the value of additional costs

Option		Costs – Value 2024 (€)	Stakeholders
O_{PNC1}	Salaries Monitoring costs (reduction)	5564 per year -2623 per year	Visitors Government
O_{PNC2}	Management costs Guarding and cleaning costs	135209 - year 1 8346 per year	Government Municipality
O_{BVB1}	Plantation cost	560469 (827 /ha) - year 1	State
O_{BVB2}	Plantation cost	109016 (991 /ha) - year 1	State
O_{RJM}	Investissement cost Maintenance cost Salaries	100902 - year 1 3027 per year 8476 - 12713/year	Association
O_{DZ}	Investissement cost Maintenance cost Renting cost	57000 - year 1 27034 - 1135171 /year 6365 /year	Municipality

Sources: Ouadah, 2016 ; Banque mondiale, 2010 ; Karam, 2016; Balkiz, 2016

Estimation of Additional Benefits

3.2.1. Wood, Forage and Non-Timber Forest Products

The values of forest products were determined only for the two sites in Tunisia and Turkey, as they are affected by the management options indicated.

However, the availability of and access to data on quantities traded and the market price prevent it from being strictly applied. Calculation assumptions were developed: (i) for Tunisia, the volume of firewood was estimated based on previous studies; (ii) for Turkey, it is considered no firewood harvesting. Furthermore, data concerning the quantity of wood harvested for a given year (Turkey) does not always reflect the wood supplied by forests because actual harvesting depends on the age of the forest and the forest exploitation plan.

There is even less available data on NWFP than on wood. Although data on cork quantities and prices are easily accessible, annual quantity and price fluctuations lead to use various assumptions. For Tunisia, the average production from cork harvesting (over a 12-year rotation) was used. For other NWFP, data based on the expert's knowledge, such as average productivity was used, which provides an estimate close to reality.

For the option OBVB1 (acacia plantation), the value of firewood production was estimated at €173,049 (year 15) based on 3.8 m^3/ha/year and a price of €4.5/m^3. Similarly, the value of forage production was estimated at €91,602/year based on 500 Forage Units (FU)/ha/year with a unit value of €0.27/FU. Option OBVB2 (cork oak plantation) considers cork production at 217 kg/ha/year (from age 31) with a unit price of €1.02/kg, representing a value of €32,803 per year, and a forage production value of €8,828/year (World Bank, 2010).

For the option DZ, the creation of a recreation area will result in the non-harvesting of timber on the recreation area (20 ha). Considering the average value of timber production in 2024, €12.27/ha, the reduction in timber harvesting on the site is €245/year.

3.2.2. Recreation

For the option PNC-Algeria, an examination of the survey data showed that the average frequency of visits per visitor (freqvisit) is 2.38 per year, with a standard deviation of 1.63. Visitors come from Algiers (46% of the total number of visitors), followed by Blida (22%). The most frequently cited reasons for visiting are nature viewing, landscape beauty, tranquility, and observing wildlife, particularly Barbary macaques.

The resulting demand function is as follows:

freqvisit = 1.1085 - 0.0035 couvisit (cost of visit); the significance level of the coefficients is high for the constant ($p \leq 0.01$), but low for the explanatory variable ($p \leq 0.1$).

Consumer surplus SCt = €3.31 per visit in 2024.

The recreational value is the result from multiplying the value of the visit obtained by the model and the number of visits for a given year. However, an inventory is not always kept of the number of visits to a site, especially when there is no entry fee. The number of people visiting one of the recreational sites was estimated at least 85,000. It is assumed that 10% of current visitors will opt for one of the two management options, representing 8,500 visitors per year, with a projected annual increase of 2% in visitor numbers.

The provision for visiting the Shouf Reserve site (Sattout, 2007) has been transferred to Jabal Moussa. The provision for payment (DAP) is US$52 per family, or US$10.40 per person (assuming a family of 5 people). This value is multiplied by the inflation rate to determine the recreational value at 2024 prices, which is €14.07 per visit (see Table 3). A gradual increase in the number of visitors of 1500 per year is assumed from the first year until year 10. The additional number of visitors is estimated at 1500 per year over a period of 10 years.

For Düzlerçami (Turkey), the recreation value assessment of the Kurşunlu Waterfall nature park was chosen (Ortaçeşme et al. 2002) for value transfer. Kurşunlu Waterfall was chosen as it is a recreation spot located in Antalya, close to the city center (25 km). In this study, the authors have used the Travel Cost Method to come up with the recreational use value and the consumer surplus was estimated as €0.19 in 1999. In order to generate corresponding value of the consumer surplus in 2024, an inflation rate was taken into account, hence, the consumer surplus for 2024 is €1.2 per visitor. Information existing in the existing recreation spot consists on the number of tickets for vehicles to get in the site. It was assumed that one ticket was cut for every 4 visitors entering the recreation spot. Secondly, information existing on the use of the Güver Cliff (one of the tourism hotspots in the pilot site) was incorporated into the case study, using the most up to date information from 2010.

Finally, the travel cost method (TCM) and the benefit transfer method (BTM) were used in national case studies. Table 3 shows a wide variation in the benefit values of recreation depending on the site. Even though reasoning per hectare of forest is not appropriate, the values obtained show that this service can be relatively very important, and moreover, it can be developed on very limited areas (23 ha in Turkey). Recreation is increasingly becoming a revenue-generating service through the payment of entrance fees (Lebanon, Turkey). When entrance fees are not applied, the benefits of recreation are threatened by overcrowding and damage caused. The income generated for the local population depends on the tourism activities developed.

Table 3. Values of recreation in the sites studied

	Method	Number of visits	Value per visit (€)	Value per ha (recerative forest)
PNC - Algeria	TCM	85000	3.3	172.0
RJM - Lebanon	TBM	5919	14.1	45.2
DZ - Turkey	TBM	206110	1.2	10753.5

Sources: Ouadah, 2016 ; Karam, 2016; Balkiz, 2016

3.2.3. Hydrological Services

For the Barbara watershed, Tunisia, sediment input was estimated using two semi-quantitative models: the PSIAC (Pacific Southwest Inter-Agency Committee) model and the FSM (Factorial Scoring Model). These models incorporate gully erosion processes through appropriate factors (World Bank, 2010). The impacts of vegetation cover on dam siltation were estimated according to proximity to water courses and slope. For example, in steep terrain and in the area near the reservoir and watercourses, this reduction is estimated at 7.4 t/ha/year compared to cereals, whereas in the area far from watercourses, this reduction is estimated at only 1.1 t/ha/year compared to degraded forest. Thus, forest plantations have significant impacts on reducing siltation for land near watercourses.

The costs of damage caused by dam siltation can be approached in two different ways:

- Additional costs to replace the storage capacity lost by siltation. The existing alternatives are raising the dams, dredging, and building replacement dams. Raising is not always technically possible, while dredging of deposits is considered an expensive solution, so the cost of building new dams is considered. The construction cost of the Barbara dam was 2.55 Tunisian Dinars (DT)/m3 in 1998 (DGBTH), which corresponds to €2.5 / m3 in 2014, after updating.
- Decrease in agricultural production caused by a reduced amount of water available for irrigation. This decrease would limit the current and future water availability for irrigation, especially during drought years (9 years observed over the past 30 years). The value of the avoided damage costs in terms of water capacity was thus estimated through the present value over the entire 30-year period with a discount rate of 2% to €3.8/m^3 of reservoir water capacity. For this reason, a conservative value of €2.7/m3 of reservoir capacity is considered, as the value of avoided damage related to soil protection against sedimentation. It is deduced that the value of sedimentation reduction is estimated at €22.4/ha for acacia planting and €3.3/ha for cork oak regeneration.

For the Chrea National Park (PNC)-Algeria, the economic value is obtained by multiplying the volume of water transferred from the study site for drinking water supply (4.927 million m^3) by the unit cost of avoided purification (€0.37/m^3), resulting in an economic value of €1.818 million, or €92.76 /ha/year. We can conclude that hydrological services such as water purification and protection of reservoirs from sedimentation provided by the forest are often considered as extremely important benefits in Southern Mediterranean countries. These countries are characterized by water scarcity and the sensitivity of soil to erosion and landslides. This is the case for the two sites studied (Algeria, Tunisia) (See Table 4).

Table 4. Value of hydrological services in the sites studied

	Service	**Quantity/ ha**	**Unit value**	**Value (€/ha for the entire site)**
PNC-Algeria	Water purification	251 m^3	€0.37 /m^3	92.76
BVB - Tunisia	Reduction of reservoir sedimentation	5.5 t	€4.59 /t	24.55

Sources: Ouadah, 2016 ; Banque mondiale, 2010.

3.2.4. Carbon Sequestration

For the Barbara watershed, Tunisia, the annual carbon flux was estimated based on the IPCC (Intergovernmental Panel on Climate Change) methodology and data concerning wood growth from the national forest inventory (DGF, 2010) and removals. Carbon sequestration for acacia is estimated at 3.67 t CO_2/ha for acacia in ravines, and at 1.82 t CO2/ha for cork oak. With the price being €4/t CO2, the value of carbon sequestration is estimated at €14.7/ha of acacia and €7.3/ha of cork oak.

For Düzlerçamı, Turkey: According to the National Inventory Report (NIR), annual carbon stock changes are calculated for all of the forests belonging to the state in Turkey. This is being carried out following the methodology of Intergovernmental Panel on Climate Change (IPCC). In this respect, site-specific data on annual gains (from old and young stands), and annual losses (from wood production and fires) were gathered together for the period between 2008 and 2013. Furthermore, national data on Harvested Wood Products (HWP) were extrapolated to Düzlerçamı Pilot Site to take into account the carbon stored. The average value of the carbon sequestration between 2008 and 2013 was calculated as 2.06 tCo2/year/ha. For the valuation of the carbon sequestration service, "Social Cost of Carbon" was used as 30 USD/tons of CO2 (World Bank Group, 2014), hence, the value of this service is estimated at €59.74 /ha/year.

The values of additional benefits per beneficiary are summarized in Table 5.

Table 5. Values of additional benefits

	Benefits	**Value 2024 (€)**	**Beneficiary**
O_{PNC1}	Recreation	6652 */year* 838 */year*	Guides nature Society
O_{PNC2}	*Shops Renting* *Recreation*	14330 */year* -22535 */year (+2% per year)*	Municipality Visitors
O_{BVB1}	*Wood* *Forage* *Protection against sedimentation* *Carbon sequestration*	*173049 (year 15)* *91602 /year (year 4 et +)* *15242 /year* *12359 /year*	*State* *Local Population* *Society* *Global community*
O_{BVB2}	*Cork* *Forage* *Protection against dam sedimentation* *Carbon sequestration*	*32803 /year (year 31 et +)* *8828 /year* *367 /year* *996 /year*	*State* *Local Population* *Society* *Global community*

continued on following page

Table 5. Continued

	Benefits	Value 2024 (€)	Beneficiary
O_{RJM}	*Entree fees of additional visitors* *Recreation*	*7989 (year 1) - 79890 (year 10 et +)* *13025 - 130250 (year1 ; >=10)*	*Manager association* *Society*
O_{DZ}	*Renting* *Wood (loss)* *Renting and entry fees* *Recreation* *Carbon sequestration*	*6365/year* *-245/year* *33171 - 165854 (year 2; 6 et +)* *50347 - 251733 (year 2; 6 et +)* *908/year*	*State* *Municipality* *Municipality* *Society* *Global Community*

Sources: Ouadah N. (2016) ; Banque mondiale. 2010 ; Karam B. (2016) ; Balkiz (2016)

3.3. Net Present Value of Management Options

For the Chréa National Park, the OPNC1 option, which involves providing visitor support, is economically attractive compared to the reference situation, which is not the case for the second option (Development of a relaxation area) (see Table 6). The first option benefits both the public manager (reduction of surveillance costs) as well as visitors and nature guides.

For the Barbara watershed, the integration of the value of non-market benefits (reduction of erosion, carbon sequestration) in the Cost-benefit Analysis of forest planting options shows that acacia planting is socially profitable, however, planting action results in a loss for the local population. Cork oak planting is not profitable considering a discount rate of 10%, mainly due to the long production cycle, including high initial costs and late benefits (Table 6).

For the Jabal Moussa Reserve, the benefits to society are far greater than the costs associated with opening new trails as well as operating costs. The results of the economic analysis show that the NPV is highly positive, which mainly benefits to visitors. The managing association also generates an additional net profit, despite the significant initial investment and management costs.

For Düzlerçami, the addition of a new recreation spot over 20 years has yielded profitable results for all of the relevant parties. The net benefit resulted to a high income of the private manager or the municipality generated from the management of one new recreation spot and an increase wellbeing for the visitors (Table 6)

Table 6. Net Present Values (NPV) of management options for sites studied over a 20-year period – year 2024

Options	NPV- Value 2024		Gains/losses per stakeholder - Value 2024 (€)			
Algeria	**Discount rate**	**Value (€)**	**Government**	**Municipality**	**Visitors**	**Nature Guides**
O_{PNC1}	8.5%	47828	26946	0	6269	14613
O_{PNC2}		-346818	-135209	55214	-266823	
Tunisia			**Government**	**Local Population**	**Society**	**Global Community**
O_{BVB1}	10%	133674	58010	-52125	107612	20177
O_{BVB2}		-203645	-163134	-45421	3274	1637
Lebanon			**Municipality**	**Manager**	**Visitors**	
O_{RJM}	7.3%	895285		1486	895285	
Turkey			**Government**	**Manager**	**Visitors**	**Global Community**
O_{DZ}	5%	2737711	82248	187192	2461982	6288

Sources: Ouadah, 2016 ; Banque mondiale, 2010 ; Karam, 2016; Balkiz, 2016

Sensitivity analysis to discount rates shows that the benefit linked to the improvement of the recreation service in Algeria does not cover the costs even when using a low discount rate of 2% (Table 7). On the other hand, for Tunisia, the analysis with a rate of 2% highlights the importance of cork oak as one of the most attractive species for society. Indeed, *The economics of ecosystems and Biodiversity* (TEEB) study recommends using low discount rates (from 1 to 4%) for projects affecting natural capital since the availability of this resource cannot be guaranteed in the future (TEEB, 2010).

Table 7. Variation of the net present value of management options among the discount rate for a period of 20 years –value 2024

	Options	NPV (€)		
		2%	**5%**	**10%**
PNC -Algeria	O_{PNC1} O_{PNC2}	81911 -499167	62650 -413044	43150 -325937
BVB - Tunisia	O_{BVB1} O_{BVB2}	285845 146084	215224 -157406	133674 -203645
RJM – Lebanon	O_{RJM}	1722821	1179469	657208
DZ - Turkey	O_{DZ}	3787781	2737711	1693522

Sources: Ouadah, 2016 ; Banque mondiale, 2010 ; Karam, 2016; Balkiz, 2016

4. DISCUSSION

The results of this study demonstrate that integrated forest management in the Mediterranean is not only ecologically vital but also socially profitable, with at least one management option yielding a positive Net Present Value (NPV) across all pilot sites. However, the economic outcomes must be interpreted through the lens of 'social profitability,' which necessitates a transition from theoretical gains to tangible local benefits. As suggested by the ESMERALDA framework (Brown et al., 2018), moving toward a more holistic decision-making process requires acknowledging that while certain services like recreation and water protection are readily monetized, others like biodiversity function as foundational 'support values' that underpin the resilience of the entire system (Burkhard et al., 2018 ; Burkhard et Maes (Eds.), 2017)). The exclusion of a direct monetary value for biodiversity should not be viewed as a lack of importance, but rather as a methodological safeguard against double-counting and an acknowledgment of its 'insurance value' against the escalating risks of climate change (Baumgärtner, 2007; TEEB, 2010).

The application of economic evaluation methods for pilot sites raises some limitations related to the lack of information, in particular the difficulty in correctly anticipating the dynamics of physical and biological phenomena (water regulation); nevertheless, models have been used to estimate protection against sedimentation and carbon sequestration as a function of vegetation cover (Jebari et al., 2025 ; Bouguerra et al., 2024).

Furthermore, the long-term viability of these socially profitable options depends on addressing the implementation gap. This requires moving beyond static Cost-Benefit Analysis to include dynamic climate scenarios and robust stakeholder engagement. Only through the creation of economic incentives, such as tailored Payments for Ecosystem Services (PES) and the active involvement of local communities in the management cycle, can the social gains identified in this chapter be realized on the ground (Plan Bleu, 2014; Daly-Hassen et al., 2010). It emerges that sustainable management should be encouraged by the establishment of conditional incentives for those who respect usage restrictions. Given that some stakeholders (users) may suffer a short-term loss following the implementation of sustainable forest management interventions, economic instruments should be defined to improve management (Daly-Hassen et al., 2010), once the optimal interventions have been identified and their economic dimension defined. Traditional incentive systems, as well as market-based economic tools such as compensation mechanisms, payments for environmental services (PES) (Wunder, 2005) should be considered. Effective PES design in the Mediterranean should consider “bundling” services (e.g., water protection and carbon storage) to increase financial viability for land managers. Furthermore, following Ostrom’s design principles for the commons, these schemes must incorporate clear ownership rights and conditional rules to ensure that payments translate into actual conservation outcomes in a changing climate.

For the Barbara watershed, Tunisia, the benefits analysis of acacia planting indicates that it is possible to establish a PES mechanism where downstream water users pay upstream farmers (Croitoru & Daly-Hassen, 2010). Recreation is increasingly becoming a remunerative service through the payment of entry fees (Lebanon, Turkey). In addition, this activity generates significant income for the local population when accompanied by catering services, marketing of local products, etc. In Jabal Moussa (Lebanon), the development of ecotourism ensures jobs for direct and indirect beneficiaries (guest houses and local restaurants). This service acts as a compensation measure for populations negatively affected by the creation of protected forests. When entry fees are not applied, the benefits of recreation are threatened by overload and damage (case of Chréa National Park in Algeria). Indeed, the social benefit is very high compared to the cost of the visit. This favorable ratio between social well-being and excursion-related expenses could support the hypothesis of an increase in affluence to the site. On the other hand, this ratio could serve as an argument for the introduction of a payment for this service by visitors. There is also significant potential for financing green jobs to accompany visitors. In addition, taking into account the value of carbon sequestration makes it possible to show the extent to which Mediterranean forests benefit the global community, and to study the possibility of applying climate change mechanisms.

In general, the approaches applied are appropriate in the Mediterranean context. However, data on the various goods and services should be developed by multidisciplinary research teams, with specialists in quantifying forest production, sedimentation, water purification services, carbon sequestration. Biodiversity and land conservation were not assessed in depth due to a lack of data. In addition, the beneficial impact of grazing in reducing forest fires and cutting fire prevention and firefighting costs was not assessed. However, the resulting economic valuation of goods and services and the cost-benefit analysis can be considered important tools for choosing management options for the sites in question and identifying economic instruments to be used. The cost-benefit analysis must be conducted beforehand to assess the available management options and not to justify investments after that.

The difficulty and lack of accurate data often lead to assumptions that need to be underlined in the economic valuations. The risks and uncertainties also need to be considered as much as possible by presenting value ranges if necessary. Overestimating benefits should be avoided as results and even economic valuation as a decision-making tool can lose their credibility.

Considering this analysis, certain recommendations can be formulated:

- Improve knowledge and build capacities: There is a need to support economic valuation studies, teach economic valuation tools at the university level and organize training for forest resource development administrations and project managers (capacity building). The cost of forest degradation (i.e. overgrazing resulting in reduced fodder production, absence of regeneration and soil erosion, impact of cork harvesting on cork oak regeneration, forest fires, etc.) needs to be estimated along with the benefits of the ecosystems.
- Integrate economic valuation as a tool in the forest management process: This tool is used to compare the different management options proposed (as opposed to no action) in terms of costs and benefits, and to identify the most socially advantageous option. This is done by: (i) estimating the net private and social benefits without action (ii) assessing the private costs for each operation (iii) estimating the net private and social benefits with action (iv) comparing the net benefits of the operation as opposed to those with no action. This assessment helps target the operations with the greatest social impact and determine the economic consequences (i.e. gains and losses) for all stakeholders.
- Develop economic instruments to improve forest management / reduce degradation: It involves identifying and developing economic instruments and funding mechanisms leading to a compromise between local users and the off-site beneficiaries of ecosystem services via national assessments. To bridge the implementation gap, forest administrations should prioritize the development of contractual measures—such as co-management agreements and 'bundled' PES schemes—that ensure the financial sustainability of the services while protecting local livelihoods.

5. CONCLUSION

The economic valuations carried out show the utility of these types of results as a decision-making mechanism for choosing management options and raising the awareness of decision-makers when it comes to using economic valuation as a tool for orienting forest ecosystem conservation and development. This analysis showed, for example, that the development of recreational activities and production of NWFP is beneficial for improving the contribution of forest ecosystems to the well-being of local and national populations and reducing damage to ecosystems. The increase of ecosystem services values provided to improve the well-being of the local and national population and the international community justifies investments in conservation, exploitation and development of forest ecosystems.

Assessments of goods and services also serve as a basis for negotiating and reaching compromises between stakeholders. On a regional scale, the link between the economic valuations of goods and services and the participatory governance initiatives allows the identification of win-win situations (such as in the case of Turkey, the State, the municipality, private enterprise, and visitors are beneficiaries), by implementing compromises to resolve conflicts, or even co-management agreements (development of income-generating activities for local populations in order to reduce grazing in Lebanon, plantations of pastoral species to reduce overgrazing in Tunisia). Weaknesses are mainly tied to the methods used and the availability of data. The lack of biophysical (production of firewood and non-timber forest products) and economic (price) data is a major weakness that makes it impossible to accurately assess the value of goods and services. Valuation of some services (recreation, biodiversity protection) requires surveys and modelling over short periods, which also causes problems. Finally, the specific features of sites (different ecosystems, relatively broad use of resources by the local population) makes it difficult to compare the results. Overall, exploring the economic valuation of diverse cases are assets for integrating economic valuation into forest resource management.

ACKNOWLEDGEMENT

This work was funded by the French Global Environment Facility (FFEM) as part of the project "Optimizing the production of goods and services by Mediterranean forests in a context of global change," coordinated by Plan Bleu and the FAO Silva Mediterranea Committee. The case study on Tunisia was also funded by the Bank-Netherlands Water Partnership Program (BNWPP) and the Technical Assistance for the Protection of the Mediterranean Environment (METAP) program and carried out by the World Bank. It was conducted with the contribution of the organizations responsible for forest management at the various sites.

REFERENCES

Acharya, R. P., Maraseni, T. N., & Cockfield, G. (2020). An ecosystem services valuation research framework for policy integration in developing countries: A case study from Nepal. *Sustainability (Basel), 12*(19), 8250. DOI: 10.3390/su12198250

Balkiz, O. (2016). *Assessment of the socio-economic values of goods and services provided by Mediterranean forest ecosystems: Düzlerçami Forest, Turkey*. Plan Bleu.

Baumgärtner, J. (2007). The insurance value of biodiversity in the provision of ecosystem services. *Natural Resource Modeling, 20*(1), 87–127. DOI: 10.1111/j.1939-7445.2007.tb00202.x

Bishop, J. T. (Ed.). (1999). *Valuing forests: A Review of Methods and Applications in Developing Countries*. IIED.

Bouguerra, S., Stiti, B., Khalfaoui, M., Jebari, S., Khaldi, A., & Berndtsson, R. (2024). Modeling Ecosystem Regulation Services and Performing Cost–Benefit Analysis for Climate Change Mitigation through Nature-Based Solutions Using InVEST Models. *Sustainability (Basel), 16*(16), 7201. DOI: 10.3390/su16167201

Brahic, E., & Terreaux, J. P. (2009). *Evaluation économique de la biodiversité: Méthodes et exemples pour les forêts tempérées*. Editions Quae. DOI: 10.3917/quae.brahi.2009.01

Brown, C., Burns, A., & Arnell, A. (2018). A Conceptual Framework for Integrated Ecosystem Assessment. *One Ecosystem, 3*, e25482. DOI: 10.3897/oneeco.3.e25482

Burkhard, B., & Maes, J. (Eds.). (2017). *Mapping Ecosystem Services*. Pensoft Publishers. (The primary ESMERALDA textbook), Sofia, 374 pp DOI: 10.3897/ab.e12837

Burkhard, B., Santos-Martin, F., Nedkov, S., & Maes, J. (2018). *An operational framework for integrated Mapping and Assessmentof Ecosystems and their Services (MAES). One Ecosystem*. Pensoft Publishers.

Centre d'Analyse Stratégique (CAS) (2009). Approche économique de la biodiversité et des services liés aux écosystèmes. Contribution à la décision publique,

Croitoru L. (2007). Valuing the Non-Timber Forest Products in the Mediterranean Region. *Ecological Economics*, 63(4):768-775http://DOI: 10.1016/j.ecolecon.2007.01.014

Croitoru, L., & Daly-Hassen, H. (2010). Using Payments for Environmental Services to improve Conservation in a Tunisian Watershed. *Mountain Forum Bulletin*, (January), 89–91.

Banque mondiale. (2010). *La génération des bénéfices environnementaux pour améliorer la gestion des bassins versants en Tunisie*. Auteurs: Croitoru, L., Daly-Hassen, H., Cherni, A., Sterk, G., Bird, N., Zanchi, G., Frieden, D., Oka, A., & Tran, L. Rapport no. 50192-TN.

Daly-Hassen, H., Annabi, M., & King-Okumu, C. (2019). Social and private profitability of tree-based adaptation options to climate change in a dryland area of Tunisia. *New Medit*, n°2/2019. pp. 89-104.

Daly-Hassen, H., Croitoru, L., Tounsi, K., Aloui, A., & Jebari, S. (2012). *Evaluation économique des biens et services des forêts tunisiennes*. DGF/FAO.

Daly-Hassen, H., Pettenella, D., & Ahmed, T. J. (2010). Economic instruments for the sustainable management of Mediterranean watersheds. *Forest Systems*, *19*(2), 141–155. DOI: 10.5424/fs/2010192-01310

DGF. 2010. *Résultats du deuxième inventaire forestier et pastoral*. Ministère de la Défense Nationale, Ministère de l'Agriculture, des Ressources Hydrauliques et de la Pêche, et Ministère de l'Enseignement Supérieur et de la Recherche Scientifique.

FAO. (2013). *State of Mediterranean forests (2013)*. FFEM, Silva Mediterranea, Plan Bleu. https://www.fao.org/docrep/017/i3226e/i3226e.pdf [15 February 2016]

Gauquelin T., Michon G., Joffre R., Duponnois R., Genin D., 2018. Mediterranean forests, land use and climate change: a social-ecological perspective. *Regional Environmental Change*, 2018, 18 (3), pp.623-636. ff10.1007/s10113-016-0994-3ff. ffhal-01594954f Hanley N. & Barbier E.B. (2009). *Pricing nature: Cost-benefit analysis and environmental policy*, Edward Elgar Publishing, USA.

Heckenhahn, J., & Drupp, M. A. (2024). Relative Price Changes of Ecosystem Services: Evidence from Germany, *Environmental & Resource Economics,* Springer. *European Association of Environmental and Resource Economists*, *87*(3), 833–880.

Jebari, S., Bouguerra, S., Chebil, A., Lamia, B., & Touzi, S. (2025). Quantification de la dégradation du paysage agricole et évaluation économique d'un aménagement approprié: Cas du bassin versant Douimis (Tunisie). *New Medit*, *5*(2025). Advance online publication. DOI: 10.30682/nm2505g

Jumiyati, S. (2025). Economic Valuation of Forest Ecosystem Services: Methods, Applications, and Policy Implications. *Journal of Forestry*, *1*(1).

Karam, B. (2016). *Estimation de la valeur économique et sociale des services rendus par les écosystèmes forestiers méditerranéens, Biosphère de Jabal Moussa, Liban.* Plan Bleu.

Karimi, H., Staudhammer, C. L., Therrell, M. D., Kleindl, W. J., Mungai, L. M., Amanambu, A. C., & Jones, C. N. (2025). Mapping Tradeoffs and Synergies in Ecosystem Services as a Function of Forest Management. *Land. MDPI*, *14*(8), 1–26.

Khalfaoui, M., Daly-Hassen, H., Stiti, B., & Jebari, S. (2020). Toward Decision-Making Support: Valuation and Mapping of New Management Scenarios for Tunisian Cork Oak Forests. *Forests*, *11*(2), 197. DOI: 10.3390/f11020197

Mace, G. M., Norris, K., & Fitter, A. H. (2012). Biodiversity and ecosystem services: A multilayered relationship. *Trends in Ecology & Evolution*, *27*(1), 19–26. DOI: 10.1016/j.tree.2011.08.006 PMID: 21943703

Mastrorilli, M., Rana, G., Verdiani, G., Tedeschi, G., Fumai, A., & Russo, G. (2018). Economic Evaluation of Hydrological Ecosystem Services in Mediterranean River Basins Applied to a Case Study in Southern Italy. *Water (Basel)*, *10*(3), 241. DOI: 10.3390/w10030241

Mavsar, R., Herreros, F., Varela, E., Gouriveau, F., & Duclerq, M. (2014). *Méthodes et outils d'évaluation socio-économiques des biens et services rendus par les écosystèmes boisés méditerranéens.* Projet FFEM/Plan Bleu, Composante 2. CTFC et EFIMED, 113 p.

Merlo, M., & Croitoru, L. (Eds.). (2005). *Valuing Mediterranean forests: Towards total economic value.* CABI Publishing. DOI: 10.1079/9780851999975.0000

Ortaçeşme, V., Özkan, B., & Karagüzel, O. (2002). An Estimation of the Recreational Use Value of Kursunlu Waterfall Nature Park by the Individual Travel Cost Method. *Turkish Journal of Agriculture and Forestry*, *26*, 57–62.

Ouadah, N. (2016). *Estimation de la valeur économique et sociale des services rendus par les écosystèmes forestiers méditerranéens, Parc National de Chréa, Algérie*. Plan Bleu.

Palahi, M., Mavsar, R., & Birot, Y. (2008). Mediterranean forests under focus. *International Forestry Review*, *10*(4), 676–688. DOI: 10.1505/ifor.10.4.676

Paris Corona, P., Quatrini, V., Schirru, M., Dettori, S., & Puletti, N. (2018). Towards the economic valuation of ecosystem production from cork oak forests in Sardinia (Italy). iForest - Biogeosciences and Forestry, 11(5), 660-667. https://doi.org/doi: 10.3832/ifor2558-011

Plan Bleu (2014). *Méthodes et outils d'évaluation socio-économique des biens et services rendus par les écosystèmes boisés méditerranéens*, Rapport technique préparé par le CTFC et EFIMED, Projet FFEM.

Robertson, G., Linser, S., & Köhl, M. (2024). Forest disturbance and damage: Perspectives for forest monitoring and reporting. *Forests Monitor*, *1*(1), 39–65. DOI: 10.62320/fm.v1.i1.10

Russo, F., Maselli, G., & Nestico, A. (2023). Forest ecosystem services: Economic evaluation of carbon sequestration on a large scale. *Valori e Valutazioni*, *33*, 17–30. DOI: 10.48264/VVSIEV-20233303

Sattout, E. J., Talhouk, S. N., & Caligaria, P. D. S. (2007). Economic value of cedar relics in Lebanon: An application of contingent valuation method for conservation. *Ecological Economics*, *61*(2-3), 315–322. DOI: 10.1016/j.ecolecon.2006.03.001

TEEB. (2010). *The Economics of Ecosystems and Biodiversity: Ecological and Economic Foundations* (Kumar, P., Ed.). Earthscan.

Wunder, S. (2005). *Payments for environmental services: Some nuts and bolts*. CIFOR.

Chapter 6
Carbon Sequestration in Moroccan Forest Ecosystems Through a Synthesis of Biomass and Soil Carbon Stocks for Climate Change Adaptation and Mitigation

Mohamed El Mderssa
http://orcid.org/0000-0003-1643-7218
Polydisciplinary Faculty of Beni Mellal, Morocco

Rajaa Timzioura
http://orcid.org/0009-0008-6233-1119
Faculty Polydisciplinary of Beni Mellal, Morroco

Fouad Malki
http://orcid.org/0000-0002-0164-0275
National School of Forestry Engineers, Salé, Morocco

Abderrahim Ettaqy
http://orcid.org/0000-0003-4662-5770
Faculty of Sciences and Technics of Béni Mellal, Morocco

Younes Naciri
Faculty of Sciences and Technics of Béni Mellal, Morocco

Taha El Abdelhakim
Faculty Polydisciplinary of Beni Mellal, Morroco, Morocco

DOI: 10.4018/979-8-3373-6239-7.ch006

Laila Nassiri
Faculty of Sciences of Meknes, Morocco

Jamal Ibijbijen
Faculty of Sciences of Meknes, Morocco

Younes Abbas
http://orcid.org/0009-0003-8681-2172
Faculty Polydisciplinary of Beni Mellal, Morroco

ABSTRACT

Climate change increasingly affects forest ecosystems, but forests remain key components of climate mitigation through carbon sequestration. More than half of terrestrial carbon is stored in forest soils and biomass, making forests strategic carbon sinks. This review presents a national-scale synthesis of carbon stocks in Moroccan forest ecosystems, integrating biomass with soil organic carbon. It is based on a structured analysis of published studies. The results highlight strong variability among ecosystems, with soil organic carbon representing the largest carbon pool, particularly in Mediterranean and semi-arid forests. The highest average carbon stocks were observed in zeen oak and cedar forests (197.88 and 187.16 tC ha^{-1}), followed by holm oak (134.06 tC ha^{-1}), while cork oak ecosystems showed lower values (76.08 tC ha^{-1}). These differences mainly reflect species composition and soil properties. Overall, Moroccan forests show a high potential for carbon sequestration, supporting their role in climate change mitigation and sustainable forest management.

INTRODUCTION

Climate change, widely recognized by the international scientific community, is an undeniable reality that demands urgent action to mitigate its far-reaching impacts (IPCC, 2021). Its effects are already being felt globally, and despite significant efforts to reduce greenhouse gas (GHG) emissions, climate change is expected to intensify, leading to profound economic, social, and environmental consequences across regions and sectors (Arias et al., 2021). Human activities, particularly the emission of greenhouse gases, have been conclusively linked to the warming of the atmosphere, oceans, and land. Carbon dioxide (CO_2) levels (Shah et al., 2024), in particular, saw a significant rise in 2019 (IPCC, 2021), and in 2023 (IPCC, 2021; Shah et al., 2024). Currently, their atmospheric concentration reached 419.3 parts per million, 51% higher than pre-industrial levels.

Morocco, with its diverse climate zones and reliance on weather-sensitive sectors such as agriculture and water resources, is particularly vulnerable to the impacts of climate change (Arabi et al., 2024). The country's climate has always been variable due to its complex topography, which includes deserts, mountains, and coastal areas. Over the past century, however, Morocco has experienced significant climate shifts, including rising temperatures and altered precipitation patterns (Born et al., 2008). Projections suggest that average annual temperatures could increase by 1.5 to 3.5°C by mid-century and may exceed 5°C by 2100 (Hadri et al., 2021; Marchane et al., 2017; Hakam et al., 2024), significantly increasing the number of hot days, particularly in summer (Boudhar et al., 2017). Heatwaves could become 26 days longer by the end of the century, while precipitation already concentrated in the northern regions continues to decline, exacerbating drought conditions (Filahi et al., 2017). Under a moderate emissions scenario, annual precipitation could decrease by 10 to 20% by 2100 (Hadri et al., 2021; Marchane et al., 2017), with declines of up to 30% in the Sahara. The frequency of droughts has already doubled since 1990, threatening both water resources and agricultural productivity (Gumus et al., 2024; Pérez-Cutillas & Salhi, 2024). These different climatic pressures are expected to have profound impacts on the distribution, composition and carbon sequestration capacity of Moroccan forest ecosystems (Benabou et al. 2022; Moukrim et al. 2022).

In response to these challenges, forest ecosystems offer a powerful nature-based solution (NbS) for mitigating and adapting to climate change (Shirkey et al., 2024). NbS strategies that focus on forests aim to: (i) protect ecosystems to sequester CO_2 emissions, primarily by avoiding deforestation (Griscom et al., 2017; Busch et al., 2019); (ii) restore degraded ecosystems to enhance carbon sequestration especially by reducing wood harvesting or grazing pressures; and(iii) implement best management practices that reduce GHG emissions from forests, croplands, and grasslands, while promoting carbon storage (Girardin et al., 2021).

Terrestrial ecosystems, particularly forests, store around 662 Gt of carbon, over half of the global carbon stock in soils and vegetation and currently absorb approximately 30% of global CO_2 emissions each year (Cook-Patton et al., 2020; FAO, 2020).

Moroccan forests are no exception, providing immense ecological, social, and environmental value. These forests play a key role in soil conservation, water regulation, climate stabilization, combating desertification, preserving biodiversity, and supporting livelihoods (Mori et al., 2017; Nandal et al., 2023; Smith et al., 2019). As natural carbon sinks, forests absorb atmospheric CO_2 through photosynthesis, storing carbon in biomass and soil (Farineau & Morot-Gaudry, 2018). Their contribution, in the case of Morocco, amounts to approximately 241 million T of carbon stored solely in forest biomass (FAO, 2015c).

This review highlights the critical role of Moroccan forest ecosystems in climate change mitigation through carbon sequestration. It addresses key scientific gaps, as national studies remain fragmented and rarely integrate biomass, roots, litter, deadwood, and soil carbon across contrasting substrates and bioclimatic gradients.

By synthesizing published carbon storage values in aboveground biomass, below-ground biomass, and soils across Morocco's major forest types, this systematic review provides one of the first integrated and updated national-scale comparisons. It aims to reveal ecological drivers and rank carbon reservoirs under contrasting conditions, which will allow for relevant conclusions to guide restoration priorities and sustainable carbon-oriented forest management strategies aligned with national mitigation pathways. These insights allow for a better understanding of the mitigation potential of Morocco's forest resources in light of its commitments under the UNFCCC, particularly the NDC, and support the development of more effective strategies for climate change adaptation and sustainable forest management. Ultimately, this review offers valuable guidance for decision-makers seeking to enhance carbon sequestration while preserving biodiversity and maintaining essential ecosystem functions.

Carbon Sequestration in Forests: A Key Strategy for Mitigating Atmospheric CO_2 Emissions

Carbon sequestration in forest soils and biomass is a crucial component of climate change mitigation strategies, as outlined in Article 3.4 of the Kyoto Protocol (UN, 1998) and in Article 5 of the Paris Agreement (UN, 2015). This ecosystem service, which plays a central role in climate regulation and biodiversity preservation, is increasingly recognized in global frameworks for ecosystem services. The Common International Classification of Ecosystem Services (Haines-Young & Potschin, 2012) defines carbon sequestration as an ecological function of ecosystems, emphasizing its importance in addressing climate change.

Carbon sequestration in forests represents one of the most cost-effective and efficient strategies for reducing CO_2 concentrations in the atmosphere (Pan et al., 2011; Cheng et al., 2024). Maximizing the carbon sequestration capacity of forest ecosystems while promoting energy conservation is essential to achieving carbon neutrality *(Piao et al., 2022)*. The success of forest-based carbon sequestration largely depends on sustainable forest management practices (Boisvenue et al., 2022). However, over time, forests reach a dynamic equilibrium where the amount of carbon sequestered is balanced by the carbon released through tree mortality and decay (Cheng et al., 2024; *Odum, 1969)*. While some studies suggest that mature forests can continue to sequester carbon, their capacity generally declines with age (Shang et al., 2023).

Increasing carbon storage in forests is an effective nature-based solution for reducing atmospheric CO_2 (Cook-Patton et al., 2020; FAO, 2020). As trees grow, they absorb CO_2 through photosynthesis, making them one of the most efficient natural mechanisms for reducing CO_2 levels in the short term. However, as forests age, sequestration may be released slowly through wood products or more rapidly through natural disturbances such as fires (Van Kooten et al., 1995; Leng et al., 2024). The frequency of wildfires has increased in recent years, exacerbated by climate change, and these fires can release large amounts of CO_2 back into the atmosphere, contributing to global warming (Jones et al., 2022; Williams et al., 2019).

There is ongoing debate over how best to manage forest-based carbon sequestration strategies (Favero et al., 2020). Some researchers argue for the protection of existing forests, suggesting that logging should be restricted to preserve carbon stocks (Klapwijk et al., 2018; Li et al., 2018). Others advocate for utilizing the substitution effect of wood products, promoting the use of sustainable wood in place of carbon-intensive materials and fossil fuels to reduce overall CO_2 emissions (Pingoud et al., 2018). The Intergovernmental Panel on Climate Change (IPCC, 2006) has provided guidelines for accounting carbon in wood products.

In addition to forest biomass, soil carbon stocks play an important role in understanding the carbon storage potential of forest ecosystems. Accurate assessment of soil organic carbon (SOC) is essential for developing strategies to mitigate climate change while preserving biodiversity (Dallahi et al., 2023). However, measuring SOC presents challenges due to its uneven distribution across different soil types, depths, and landscape positions. As a result, a variety of methods for assessing SOC dynamics have been developed. Recent innovations, such as visible and near-infrared reflectance spectroscopy (Vis-NIR) and mid-infrared (MIR) spectroscopy, have proven effective for quickly and cost-efficiently estimating SOC content, while traditional methods such as wet or dry oxidation remain in use for more in-depth analyses (Barré et al., 2020).

For forest biomass, there are several approaches to estimate carbon stocks, including biome averages, forest inventories, and remote sensing techniques. Remote sensing data, when combined with allometric equations, can provide valuable insights into carbon stocks (Ma et al., 2024). Although generic data exists at the global level, such as those provided by the IPCC (Chave et al., 2005), offering a broad framework for assessment, it is always preferable to develop country-specific equations to improve the accuracy of carbon stock assessments.

In the context of climate change, Moroccan forests are vital in mitigating its impacts and contribute significantly to the country's efforts to meet its commitments under the Paris Agreement. According to Articles 4.2 and 4.11 of the Agreement, each Party is required to prepare, communicate and maintain successive Nationally Determined Contributions (NDCs) that it intends to achieve. Morocco has committed to reducing its greenhouse gas emissions through NDCs. The national mitigation potential of NDC 3.0, submitted in 2025, amounts to a 53% reduction in greenhouse gas emissions by 2035, compared to the 45.5% target set in NDC 2.0 for 2030. Forestry actions include approximately twenty unconditional and conditional actions and represent 3.76% of the country's planned mitigation efforts (NDC, 2025).

Diversity and Importance of the Moroccan Forest Ecosystems

In the western part of the Mediterranean basin, Moroccan forest ecosystems cover an area of approximately 9 037 714 ha, including 6 212 056 ha of forest formations and the remainder consisting of alfatiere stands, accounting for 12.7% of the national territory (ANEF, 2024; FAO, 2020). The climate is influenced by the Atlantic Ocean, the Mediterranean Sea, the Sahara Desert, and the topography (Born et al., 2008). The Atlas and Rif mountain ranges form an orographic barrier and a transition zone between the western areas influenced by moisture from the Atlantic Ocean and the eastern border affected by dry air, preventing any substantial maritime influence from reaching eastern Morocco (Laaribya & Belghazi, 2016). All types of bioclimates can be found: per-humid, humid, sub-humid, semi-arid, arid, and Saharan (Quézel & Médail, 2003; Benabid, 2000; Emberger, 1939).

This diversity of bioclimates directly affects the distribution of plant species. Thus, Morocco is considered one of the most important and richest countries in terms of biodiversity in the Mediterranean basin, possessing a rich biological diversity characterized by numerous ecosystems, habitats, and different species (Benabid, 1985). It ranks second in terms of biodiversity, after Anatolia in Turkey, with an overall endemism rate of 20% (Dobignard & Chatelain, 2011). Forest ecosystems are primarily composed of 71% natural formations based on broadleaf trees (especially with holm oak at 31.7%, cork oak at 8.2%, and argan tree at 5.4%), and conifers (including especially thuja at 11.7%, juniper at 7.3%, cedar at 2.9%, pine at 2.1%, and fir at 0.1%) (FAO, 2020). These ecosystems extend from the thermomediterranean level to the oromediterranean level (Benabid, 2000).

Figure 1. Geographical distribution of Moroccan forest ecosystems

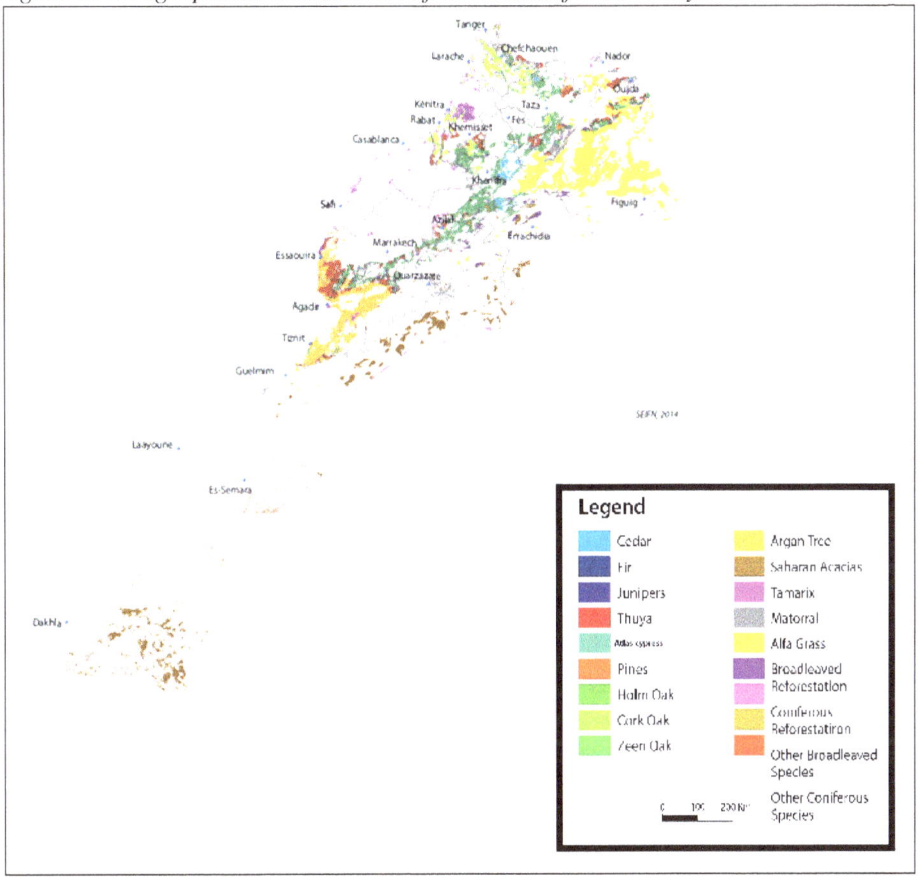

(SEIFN, 2014)

METHODOLOGY

This literature review relies on a structured and comparative synthesis of published studies reporting carbon stocks in Morocco's main forest ecosystems, integrating aboveground biomass (AGB), belowground biomass (BGB) and soil organic carbon stocks (SOCS). The search strategy targeted peer-reviewed articles, national forest inventories and doctoral theses, using specific keywords related to Moroccan forest types (e.g., *Quercus rotundifolia*, *Cedrus atlantica*, *Quercus suber*, *Tetraclinis articulata*, *Argania spinosa*) combined with terms referring to carbon pools and soil depth. Only studies providing numerical carbon stock values, clear SOC sampling depth, and basic environmental descriptors (forest type, substrate, bioclimatic zone and degradation level) were retained. Studies presenting only narrative or unclear quantitative estimates were excluded. The extracted data were harmonized through grouping by major ecosystem types (holm oak, cedar, cork oak, thuja, argan) and interpreted based on key ecological drivers including parent material, climate, stand structure and forest condition. The synthesis emphasized ranges and reported averages rather than recalculating weighted means, due to heterogeneity and inconsistent availability of raw datasets across sources. To contextualize Morocco's carbon sink potential, findings were cross-referenced with comparable studies from Mediterranean, temperate humid and semi-arid forests internationally, enabling a broader interpretation of carbon dynamics and management implications. This approach supports a quantitative-oriented review while ensuring transparency, reproducibility and practical relevance for forest carbon management and policy guidance in Morocco.

CARBON STORAGE IN AGB/BGB OF FOREST ECOSYSTEMS IN MOROCCO: SUMMARY OF PREVIOUS WORK

The carbon storage potential in the aboveground and belowground biomass of forest trees in Morocco varies significantly depending on the type of forest ecosystem, its degree of development, and various other ecological and environmental factors. Numerous studies have quantified carbon stocks across different forest types, considering species composition, tree density, stage of development, and geographical distribution.

In the Middle Atlas, carbon stocks in the aboveground biomass (AGB) of mixed forests, consisting of Holm oak and Atlas cedar, range from 55.5 t ha^{-1} in the Dayat Hachlaf forest to 88.5 t ha^{-1} in the Ajdir forest *(*Boulmane et al., 2015*;* El Mderssa et al., 2019*)*. In pure holm oak stands, such as those in the forests of Reggada, Tafechna, and Ksiba, AGB carbon stocks are reported as 58 t ha^{-1}, 64 t ha^{-1}, and 91 t ha^{-1}, respectively *(*Boulmane et al., 2010*)*. These values are generally lower in the High Atlas, with carbon stocks of 17.1 t ha^{-1}, 24.5 t ha^{-1}, and 40.1 t ha^{-1} observed in the forests of Ait Aamar, Asloul, and Ifghane, respectively *(*Boulmane et al., 2015*)*.

For cedar forests, El Mderssa et al. (2019) report 78.75 t ha^{-1} in the Jbel Aoua South forest*)* and 99.42 t ha^{-1} in the Azrou forest (El Mderssa et al., 2022). These values can vary significantly depending on the forest's condition, with estimates of 206.27 t ha^{-1} and 142 t ha^{-1} for natural and degraded cedar stands in Azrou, respectively *(*Zaher et al., 2020*)*.

In the same area, the annual carbon sequestration of pure cedar in Azrou forest varies between 0.35 and 8.82 tC ha^{-1} yr^{-1}, with an average of 5.48 tC ha^{-1} yr^{-1}, according to predictions based on physiological principles and growth models (Boukhris et al., 2023).

For Cork oak, within the Maamora forest, cork oak (*Quercus suber* L.) stands growing on slightly acidic clay–sandy Arenosols display highly contrasted carbon stocks depending on stand age and structural dynamics. Carbon stored in aboveground biomass (AGB) ranges from 8.91 to 105.37 tC ha^{-1}, while belowground biomass (BGB) varies between 0.25 and 15.26 tC ha^{-1}, following a chronosequence of six stand age classes (<10, 10–20, 20–40, 40–80, 80–120, and >120 years) (Zaher at al., 2025). The lowest carbon stocks are observed in young stands under 10 years old, where organic inputs and biomass stratification remain limited, whereas the highest values are reached in older, denser, and well-structured stands, illustrating a strong long-term carbon accumulation capacity. These synthesis findings demonstrate that, in Maamora's ecological functioning, stand architecture and soil feedbacks, especially litter quality, rooting depth, and organo-mineral stabilization, control the majority of carbon storage. This confirms that conservation of old stands and maintenance of age-class diversity are strategic priorities for strengthening future carbon sink resilience and guiding national-scale climate mitigation and forest management planning.

Cork oak stands in the Maamora forest also exhibit a marked variability in biomass carbon stocks, strongly influenced by stem density. A study conducted on these formations, considering three density classes, shows that total carbon stocks in biomass (AGB + BGB) reach 41.2 tC ha^{-1} in open stands (<100 stems ha^{-1}), 44.2 tC ha^{-1} in intermediate-density stands (100–200 stems ha^{-1}), and 77.7 tC ha^{-1} in dense stands (>200 stems ha^{-1}) (Oubrahim et al., 2015). These findings confirm that increasing stand density promotes greater carbon accumulation per unit area, particularly in well-structured and fully stocked forests, highlighting the importance of stem density as both a key explanatory variable and a practical management lever for optimizing carbon sink strength and guiding restoration and reforestation priorities in the Maamora ecosystem.

Similarly, in the El Harcha forest on the central plateau of Oulmes, carbon stocks are around 48.28 t ha^{-1} *(Bouabid et al., 2020).*

In contrast, the Zitchouine forest, located on a schistose substrate, shows a relatively low carbon stock of approximately 16.60 t ha^{-1} (Ikraoun et al., 2024).

In the Maamora forest, the values of AGB carbon stock of cork oak stands using field data combined with UAV–ALS–TLS data range between 27.97 to 44.76 t ha^{-1} (mean = 33.36 t ha^{-1}), and 26.3 to 39.04 t ha^{-1} (mean = 33 t ha^{-1}), respectively (Fadil et al., 2022).

For the argan ecosystem, a study conducted within the framework of arganiculture, aimed at assessing carbon sequestration as a function of age in argan orchards, estimates a carbon stock ranging between 0.003 and 0.22 t C ha^{-1} for trees aged

2 to 6 years (Oumasst et al., 2024). These values remain markedly lower than the carbon stock measured in argan biomass in natural forests, which reaches 1.62 t ha^{-1} in the Haha forest at the age of 5 years (Belghazi, 2013). This comparison highlights that the carbon storage potential in argan orchards during the establishment phase is constrained by the young age of the trees and their still-shallow root systems, whereas in dense natural stands, growth, enhanced by established root carbon reserves and high-quality litter inputs, enables substantially greater carbon accumulation at equivalent ages.

CARBON STORAGE IN FOREST SOILS IN MOROCCO: INSIGHTS FROM RECENT RESEARCH

Soil organic carbon (SOC) stocks, particularly in the upper 30 cm of soil where more than 80% of the carbon is concentrated (Chowdhury et al., 2023), vary significantly depending on several factors, especially soil properties, climate, organisms in the soil including vegetation, fauna, human activities, land use and management practices, relief, parent materials of soil formation, age or time of soil weathering process and spatial position (El Mderssa et al., 2024).

Regarding land use, forests exhibit the highest soil organic carbon stocks (SOCS), particularly in the 0-20 cm surface layer. In comparison, arable lands show intermediate SOCS levels, while pastures record the lowest values, illustrating the negative impact of tillage and overgrazing on carbon sequestration. These differences underscore the essential role of forests as primary carbon sinks (Danish et al., 2025).

In Morocco's Middle Atlas, SOC values for pure holm oak forests exhibit variability depending on the underlying substrate. Hence, in the Reggada forest, located on dolomitic soil, SOC is approximately 47.2 t ha^{-1}, while in the Tafechna forest, situated on limestone, the value increases to 63 t ha^{-1} (Boulmane et al., 2010). In the Ksiba forest, SOC reaches 58.2 t ha^{-1} (Boulmane et al., 2010). In contrast, the SOC in the Zitchouine forest, located on schist and quartzite sandstone substrates in the central plateau of Oulmes, is lower at 40.07 t ha^{-1} (Ikraoun et al., 2022).

SOC values can also be much higher in certain areas due to the combined effect of several factors. In the Jbel Aoua South forest in the west-central Morocco on a sandy dolomitic substrate, SOC is as high as 174.25 t ha^{-1}, while in the Azrou forest in the middle Atlas, which sits on a basaltic substrate, SOC reaches 264.73 t ha^{-1} (El Mderssa et al., 2019). These variations are linked to differences in forest types, abundance of the shrub layer, soil composition, and management practices.

In mixed forest stands, where holm oak is interspersed with other species like Atlas cedar or Zeen oak, SOC values are also notable. In the Jaaba forest, a stand of holm oak mixed with Zeen oak on basalt soil stores 109.02 t ha^{-1}, while a stand of holm oak mixed with Atlas cedar in the Jbel Aoua South forest stores 119 t ha^{-1} (El Mderssa et al., 2019). These values are comparable to those found in the Hachlaf and Ajdir forests, where SOC values range from 94.5 to 115.6 t ha^{-1} (Boulmane et al., 2015).

However, in the Zitchouine forest, mixed with cork oak, SOC is much lower at around 42.7 t ha^{-1}, and even lower at around 6.71 t ha^{-1} in thuja stands (Ikraoun et al., 2022).

For cedar forests, SOC varies depending on the forest's development stage and condition. In a young Atlas cedar stand in the Jbel Aoua South forest, SOC is 160.86 t ha^{-1}, while in a mature, pure cedar stand in the Azrou forest, it reaches 178.07 t ha^{-1} (El Mderssa et al., 2019). In the same region, *Zaher* et al., reports similar SOC values for natural cedar forests (183.33 t ha^{-1}) and degraded cedar stands (213.46 t ha^{-1}).

In the Maamora forest, where cork oak is the dominant species (Malki et al., 2022a), SOC values in the first 100 cm of soil are estimated at 43.7 t ha^{-1}, 52.2 t ha^{-1}, and 81.2 t ha^{-1} respectively for low, middle, and high stand densities, with more than 87% of total organic carbon concentrated in the upper 40 cm (Oubrahim et al., 2015).

Still under cork oak in the Maamora forest, Zaher et al., reports across the studied stand age classes (<10, 10–20, 20–40, 40–80, 80–120, and >120 years), SOCS up to 40 cm depth increased consistently with stand age, rising from 40 t ha^{-1} in stands younger than 10 years to 106.61 t ha^{-1} in stands older than 120 years. These results confirm that forest maturity enhances soil carbon accumulation by sustaining long-term organic inputs and progressive organo-mineral carbon stabilization.

Similar findings were reported by Boulmane et al. (2017), with SOC in the first 100 cm ranging from 60.4 to 74.3 t ha^{-1} depending on the level of forest degradation (92% in the upper 40 cm). However, Malki et al. (2022b) recorded lower SOC values of 40.12 t ha^{-1} for cork oak stands in the same region.

In conclusion, we can state that the soil in the Maamora forest ecosystem stores almost half of the total forest carbon, with 51% of total carbon stocks present in soil (Oubrahim et al., 2015).

For thuja stands in the Zitchouine forest, SOC is 47.91 t ha^{-1} for pure thuja stands and 26.37 t ha^{-1} when mixed with holm oak and cork oak (Ikraoun et al., 2022).

Estimating SOC stocks, particularly in the upper 30 cm, using large soil datasets, multispectral satellite data, climate data and machine learning approaches, with 30 m resolution maps over the past 32 years, revealed biome-level SOC patterns in Morocco.

Across biomes, the amount of SOC in a forest soil profile (to 1 m depth) ranges from 70 to 92 t C ha^{-1} in temperate and boreal regions, around 60 to 80 t C ha^{-1} in Mediterranean biomes, and can reach 20 to 165 t C ha^{-1} in humid tropical regions (El Mderssa et al., 2024).Another aspect related to the prediction of SOCS, machine learning modeling of SOCS in the Ourika watershed (Moroccan High Atlas) identified Random Forest (RF) and Cubist as the best-performing models among the tested algorithms (Cubist, RF, SVM, GBM). Models were calibrated using 420 field soil samples (0–30 cm) and 88 covariates from climate, topography, remote sensing, and Soil grids. LU/LC (land use/land cover) was the most influential predictor, and prediction uncertainty was lowest in dense forest areas with higher litter and root carbon inputs, while SVM and GBM showed weaker performance (Meliho et al., 2023).Finally, the integration of remote sensing (satellite imagery, drones, LiDAR) with machine learning models could be a very promising approach for estimating forest carbon storage over large areas with high accuracy. Understanding how forests act as carbon sinks is essential for effective forest management and carbon accounting (Fadil et al., 2024; El Mderssa et al., 2024).

Table 1. Carbon stocks in major Moroccan forest ecosystems

Ecosystem	Geographic region	AGB (tC ha^{-1})	SOC (tC ha^{-1})	Sources
Holm oak (*Quercus rotundifolia*)	Middle Atlas	55.5 – 91	47.2 – 63 (up to 174 on dolomitic substrate)	Boulmane 2010, 2015 ; Ikraoun 2022 ; El Mderssa 2019
	High Atlas	17.1- 40.1	24.6 - 60	Boulmane 2015
	Central Plateau	(no AGB values reported dataset)	40.07	Ikraoun 2022
Atlas cedar (*Cedrus atlantica*)	Middle Atlas	78.75–206.27 (depending on stand condition)	160.86 – 183.33 (up to 213.46 in degraded stands)	Zaher 2020 ; El Mderssa 2019,
Mixed stands (*Holm oak–cedar, Holm oak–Zeen oak*)	Middle Atlas	55 – 88	109.02 – 119	Boulmane 2010 ; El Mderssa 2019
Cork oak (*Quercus suber*)	Maamora	41.2 – 77.7 (According to density classes) 8.91 to 105.37 (According to age classes) 26.3 – 44.76 (ALS/UAV/TLS estimates)	43.7-81.2 (According to density classes) 40-106.61 (According to age classes)	Oubrahim 2015 ; Boulmane 2017 ; Boujraf 2023 ; Fadil 2022 ; Malki 2022b
	Central Plateau	16.6	43.02	Ikraoun 2022 ; Ikraoun 2024

continued on following page

Table 1. Continued

Ecosystem	Geographic region	AGB (tC ha^{-1})	SOC (tC ha^{-1})	Sources
Thuja (*Tetraclinis articulata*)	Central Plateau	(no AGB values reported dataset)	47.91 (pure stands), 26.37 (mixed stands)	Ikraoun 2022
Argan (*Argania spinosa*)	Arganiculture et Haha forest	0.003 - 0.22 (arganiculture 2 to 6 years) -1.62 (Natural forest)	(no SOC values reported dataset)	Belghazi 2013 ; Oumasst et al. 2024

AGB: Aboveground Biomass; SOC: Soil Organic Carbon

DISCUSSION: CARBON STOCK VARIABILITY, INTERNATIONAL CONTEXT, AND METHODOLOGICAL CONSIDERATIONS

Drivers of Carbon Stock Variability

The observed variability of carbon stocks in Moroccan forest ecosystems is explained by a combination of interacting ecological drivers highlighted throughout this synthesis. Bioclimatic gradients strongly influence net primary productivity and carbon residence time, with humid and cold montane variants favoring greater carbon accumulation and slower organic matter turnover, whereas semi-arid and warm low-elevation bioclimates limit biomass growth and soil carbon stabilization.

Soil substrate and parent material control carbon retention by shaping texture, pH, clay floor depth, and mineral composition; dolomitic, basaltic and limestone-derived soils in the Middle Atlas, for example, promote higher soil organic carbon (SOC) through shrub layer support and organo-mineral complex formation, while compact clays, marls and saline or hydromorphic soils constrain carbon storage.

Stand structure and dendrometric potential regulate biomass carbon density, as dense, multi-layered, mature and well-stocked stands store substantially more aboveground carbon (AGB) than coppices, young plantations or degraded open formations, consistent with observations that the highest carbon values align with forests showing greater basal area, tree density and individualized relief-driven microclimates.

Finally, forest condition and degradation status act as decisive modifiers; overgrazing, repeated biomass removal, and land-use conversion reduce both AGB and SOC stocks by disrupting litter inputs and soil structure, ultimately limiting future sink capacity.

Together, these factors demonstrate that carbon sequestration in Morocco follows the expected dynamics of Mediterranean and montane forests, while emphasizing that bioclimate, substrate and stand architecture must be jointly considered to guide national carbon-oriented forest management strategies.

The comparative synthesis presented in this review shows that the Moroccan forest ecosystems with the highest carbon sequestration potential are the Atlas cedar forests and the Zeen oak stands, which record the highest mean carbon stocks per unit area, reaching 187.16 tC ha^{-1} and 197.88 tC ha^{-1}, respectively, reflecting a strong capacity for long-term carbon storage in both biomass and soils. Mature pine forests, particularly Pinus pinaster, also demonstrate high sequestration potential, with reported values ranging from 166 to 249 tC ha^{-1}, although these correspond to fast-growing exotic plantations and remain highly dependent on stand age and growth dynamics. In contrast, Holm oak mature forests exhibit intermediate but substantial stocks ($\approx$ 134.06 tC ha^{-1}), while Cork oak ecosystems show comparatively more moderate total stocks ($\approx$ 76.08 tC ha^{-1}), mainly due to edaphic constraints, stand structure and degradation level. Accordingly, mountainous humid and cold bioclimates, stand maturity, and multi-layered forest stratification emerge as the most favorable ecosystem configurations for maximizing Morocco's forest carbon sink potential.

Rather than analyzing AGB and SOC separately by ecosystem, this review emphasizes that forest degradation has a direct negative impact on total carbon stocks by simultaneously reducing biomass inputs and disrupting soil organic carbon stabilization processes. In Morocco, degradation pressures; mainly overgrazing, recurrent biomass removal (fuelwood, charcoal, branches), land conversion, and anthropogenic disturbances; lead to lower litter deposition, weaker root development, loss of stand stratification, and deterioration of soil structure and physicochemical properties, ultimately decreasing both AGB and SOC pools. The synthesis of evidence across Moroccan forests shows that dense, mature, and well-structured stands hold the highest carbon sink capacity, while open, overexploited, or degraded formations experience reductions that can exceed 50% of their initial storage potential, depending on substrate, bioclimate, and disturbance intensity. These findings highlight that degradation level is not only a loss factor, but a key explanatory variable controlling current and future carbon sink strength, and must therefore guide national forest restoration priorities, mitigation planning, and carbon-oriented management strategies under Morocco's NDC and PES pathways.

International and Regional Comparative Analysis

Studies conducted across various international contexts show that soil organic carbon stocks vary widely depending on soil type, climate, vegetation, and edaphic properties. In temperate humid regions, for example, Eglin (2005) and Lecointe et al. (2005) estimated carbon stocks in the organic horizon (<30 cm) ranging from 136 to 153 tC ha^{-1} in British forests; values that remain generally higher than those observed in our case. Similarly, a study carried out in Canada by Jessica (2009) under five types of plant formations reported SOCs values between 61 and 133 tC ha^{-1} in the superficial organic layer, with higher levels under hardwood stands, highlighting the importance of litter quality and plant material in soil carbon accumulation. In the northern shore of the Mediterranean region, soils generally exhibit more modest carbon contents, as reported by Rodriguez-Murillo (2001) and Diaz-Pines et al. (2011), with stocks ranging from 40 to 70 tC ha^{-1} in the upper 50 cm of soil under *Quercus pyrenaica*. At a broader scale, studies conducted in Brazil, Jordan, India, and Kenya (Batjes, 1996; Sombroek et al., 1993; Van Engelen et al., 2005) confirm that, for a soil depth of 1 m, carbon stocks may vary from 12 tC ha^{-1} in arid-region xerosols and arenosols to more than 100 tC ha^{-1} in humid-region podzols. Tropical soils may even reach 150–300 tC ha^{-1}, whereas soils from arid zones remain relatively poor ($\approx$70 tC ha^{-1}). This global variability places the Moroccan SOC values into perspective, showing that they align with the dynamics of Mediterranean and semi-arid soils subject to similar ecological constraints, while highlighting the major influence of climate, soil proprieties, and intrinsic characteristics and density of vegetation type, on carbon sequestration.

Placing the Moroccan findings in a broader international context helps clarify how the country's forest carbon stocks compare with those of other regions and highlights the ecological and silvicultural drivers of variability. Differences in forest type, stand age, growth dynamics, and management practices collectively shape the carbon sequestration potential observed across ecosystems worldwide.

A global comparison of aboveground biomass carbon stock values obtained in Morocco with those of the northern Mediterranean coast reveals that the latter are significantly higher, primarily due to the context and developmental conditions, particularly in terms of climate, stand nature and structure, age, management regimes, and dendrometric characteristics. This observation is consistent with Le Clec'h et al. (2013), who reported that the highest carbon stocks are found in the most densely forested areas. In France, for example, chestnut coppices exhibit relatively high annual carbon sequestration rates, reaching approximately 4.04 tC ha^{-1} yr^{-1} (*Ranger et al., 1988*), whereas Moroccan holm oak coppices rarely exceed 0.81 tC ha^{-1} yr^{-1} in the studied iliçaies (Boulmane et al., 2015). High-forest stands generally store more carbon, particularly silver fir stands (87 tC ha^{-1}) and beech forests (84 tC ha^{-1}), while young Douglas-fir plantations do not exceed 45 tC ha^{-1} (*IF, 2018*). In the Iberian Peninsula, carbon stocks can be significantly higher, especially in mature plantations of fast-growing exotic species. For instance, *Pinus pinaster* Ait. stands may accumulate between 166 and 249 TC ha^{-1} in the same region (Ruiz-Peinado et al., 2013). These findings underscore the crucial influence of growth dynamics, stand maturity, and silvicultural strategy on carbon storage capacity, while providing a meaningful comparative framework for interpreting Moroccan values.

Limitations and Methodological Uncertainties

This review is a literature-based compilation rather than a standardized data-driven meta-analysis, meaning the reported carbon stock values could not be recalculated from a unified raw dataset. As a result, the synthesis inherits the heterogeneity of original studies, including differences in sampling depth, plot size, replication, and the history of disturbance and silvicultural practices, which limits strict numerical comparability. Biomass carbon values rely on variable allometric models, wood density and fraction carbon assumptions, while SOC measurements are affected by methodological differences, especially in carbonate-rich soils where organic and inorganic carbon are not always separately quantified. Others key carbon pools such as litter, deadwood, deep soil layers (>40 cm), and coarse or fine root C stocks remain underrepresented in Moroccan literature, similar to other Mediterranean and semi-arid regions. Additionally, forestry management interventions and disturbance impacts (fire, grazing, logging) are not systematically reported using harmonized degradation indices. Therefore, carbon values should be interpreted as indicative ranges useful for traceability and management guidance rather than fully harmonized national estimates.

Table 2. Key sources of uncertainty in the synthesis

Source of Uncertainty	Description	Impact on carbon estimates	Notes for interpretation
Heterogeneous sampling depth (SOC)	Studies use different soil depths: 30, 40, 50, 100 cm to 1 m	Limits strict comparability, may inflate or reduce total SOC depending on depth considered	Report ranges/means from original studies without recalculation
Allometric model variability (AGB/ BGB)	Different equations, wood density and fraction carbon assumptions, diameter/height measurement protocols	Affects biomass-to-carbon conversion and carbon per hectare estimates	Prefer ecosystem grouping to ensure traceability
Influence of soil carbonates	Some soils contain inorganic carbon not consistently corrected across studies	Possible over- or under-estimation of SOC with oxidation-based methods	No unified carbonate correction protocol available
Stand age and structure differences	Mix of coppice, young plantations, and mature multi-layered stands	Impact biomass quantification: Higher stocks in mature, dense stands; lower in young or open/degraded stands	Basal area, density, stratification not uniformly reported

continued on following page

Table 2. Continued

Source of Uncertainty	Description	Impact on carbon estimates	Notes for interpretation
Disturbance regime inconsistency	Grazing, logging, fire, and drought impacts not quantified with a standard index	Reduces litter and root inputs, disrupts soil structure, weakens future sink capacity	Disturbance severity and history differ by study
Spatial and ecosystem data gaps	Underrepresentation of deep SOC, litter, deadwood, and some species (e.g., fir, deep roots)	Underestimates total ecosystem carbon contribution at national scale	Forward-looking projections (future trends) deserve to be included in the synthesis

MANAGEMENT STRATEGIES TO MAXIMIZE CARBON SEQUESTRATION IN FOREST ECOSYSTEMS

Carbon sequestration in forests represents one of the most cost-effective and efficient strategies for reducing CO_2 concentrations in the atmosphere (Pan et al. 2011*;* Cheng et al. 2024). Maximizing the carbon sequestration capacity of forest ecosystems while promoting energy conservation is essential to achieving carbon neutrality (*Piao et al. 2022*). The success of forest-based carbon sequestration largely depends on sustainable forest management practices (Boisvenue et al. 2022). It is essential for maintaining the multiple functions of forest ecosystems while providing valuable goods and services. Wood production, belowground carbon stocks, and nutrient cycling were generally significantly impacted by forest management compared to the unmanaged stand (Carmona-Yáñez et al. 2023). However, over time, forests reach a dynamic equilibrium where the amount of carbon sequestered is balanced by the carbon released through tree mortality and decay (O*dum Eugene 1969;* Cheng et al. 2024). In the context of climate change, one of the primary goals of forestry is to enhance carbon sequestration, both in forest biomass and soils, through targeted management strategies. By optimizing forest management practices, it is possible to increase carbon stocks and help mitigate the adverse effects of climate change. The primary forestry strategies for maximizing carbon sequestration include:

Reducing Deforestation and Forest Degradation

Preventing the loss of forest cover is essential to safeguard existing carbon stocks and sustain the long-term capacity of forests to function as carbon sinks (Griscom et al., 2017). Although harmonized studies in Morocco jointly quantifying carbon losses from aboveground biomass (AGB) and soil organic carbon (SOC) in relation to deforestation intensity and degradation level remain scarce, global evidence shows that forests can generally lose between 30 and 70% of their carbon reservoirs in biomass and soils, depending on deforestation pressure, degradation drivers, disturbance regimes, and management history. The highest carbon losses are typically reported in open, degraded, or overexploited stands, where litter inputs decline and soil structure deteriorates, reducing carbon residence time and limiting future sequestration potential. Therefore, addressing forest degradation alongside deforestation is critical for prioritizing restoration actions and ensuring the long-term resilience of national carbon sinks. Supporting secondary forest regrowth, halting deforestation, and extending the lifespan of harvested wood products can collectively generate cumulative negative emissions while promoting sustainable land-use strategies and maintaining future sink capacity (Houghton & Nassikas, 2018).

Increasing Forest Area Through Afforestation, Reforestation, and Ecosystem Restoration

Expanding forest areas through afforestation, reforestation, and ecosystem restoration enhances the carbon sequestration capacity of forest ecosystems (*IPCC, 2019*). By increasing tree cover, these strategies not only generate new carbon stocks but also restore ecological balance, soil functions, and biodiversity in degraded landscapes. In Morocco, national restoration programs prioritize biological replanting, assisted natural regeneration, and silvopastoral improvements, progressing at an average rate of 50,000 ha per year (NDC, 2025), thereby reinforcing long-term carbon sink resilience while supporting key forest types such as Atlas cedar, holm oak, cork oak, argan, thuja, and maritime pine. The effectiveness of afforestation depends on the balance between biogeochemical gains (carbon storage) and biogeophysical effects (albedo, roughness, evapotranspiration). Preventing degradation is equally strategic: enforcing ecological carrying capacities and limiting diffuse disturbances helps recover litter inputs, root development, and soil structural stability, mitigating future carbon losses and sustaining the regrowth potential of vegetation capital.

Improved Forest Management Practices

Forest management has positive and negative effects on carbon stocks. Optimizing forest management can enhance the carbon storage capacity of forests. This includes implementing silvicultural practices that maintain or increase carbon stocks in both forest biomass and soil. For example, selective thinning, controlled harvesting, and natural regeneration practices can increase carbon sequestration rates by improving forest health and fostering greater tree growth. In Morocco, synthesis findings show that optimized silviculture under humid and cool montane forests (e.g., cedar and zeen oak) maintains the highest total carbon density, while selective thinning and assisted natural regeneration help recover AGB and strengthen soil C inputs without compromising long-term sink capacity when basal area and canopy continuity are preserved. Moreover, evidence indicates that unmanaged or degraded open stands experience the largest carbon losses and the weakest future sequestration prospects, confirming that forest condition and management intensity must be jointly calibrated to secure stable SOC and resilient biomass-driven carbon gains.

Silvicultural Practices for Carbon Sequestration

Continuous Cover Forestry (CCF) is increasingly recognized as a sustainable and effective alternative to conventional clear-cutting, particularly for its positive impact on carbon sequestration. Unlike clear-cutting, which removes the entire forest cover and disrupts ecological processes, CCF relies on maintaining a continuous canopy through selective tree harvesting. This approach preserves ecosystem functionality while delivering significant climatic and ecological benefits. Studies show that CCF supports soil carbon storage by ensuring consistent litter inputs, improving soil health, strengthening organic matter retention, and reducing carbon losses caused by physical disturbance of surface soil horizons (Ahtikoski et al., 2022). In parallel, thinning intensity strongly influences carbon density stored in both biomass and soil organo-mineral complexes. By reducing inter-tree competition, thinning improves resource access, stimulates the growth of remaining trees, and increases atmospheric carbon uptake, although carbon gains depend on forest type, stand age, and thinning intensity (Ganatsas et al., 2024; Cachinero-Vivar et al., 2021; Erkan et al., 2023). A complementary practice—conversion of coppices to high forests through repeated thinning—can further increase aboveground biomass carbon accumulation and improve root carbon transfer to soil, reinforcing total ecosystem storage capacity (Ganatsas et al., 2024; Pereira et al., 2024). Moreover, in Morocco, the literature synthesis confirms that continuity principles inspired by CCF and well-calibrated thinning are particularly favorable in humid to sub-humid montane bioclimates dominated by Atlas cedar and zeen oak, where higher basal area and multi-layered stratification provide stable organic inputs and longer carbon residence time. In contrast, open, over-exploited, or repeatedly harvested coppices exhibit the strongest AGB and SOC declines, demonstrating that degradation level, stand maturity, and silvicultural intensity must be co-optimized to secure both current and future biomass–soil carbon sink strength and effectively guide national carbon-oriented forest restoration and management strategies.

Afforestation With Fast-Growing Species

Afforestation with fast-growing species is considered a relevant lever of intensive forest management to accelerate carbon uptake, particularly during the early developmental stages of forest stands. Rapid-growth species such as *Populus spp.*, *Eucalyptus spp.*, *Pinus pinaster*, and some managed conifers can sequester substantial carbon amounts per unit area in a short rotation time, making them strategic candidates for boosting AGB accumulation and short-term mitigation targets (Ameray et al., 2021). However, this synthesis also shows that their long-term carbon storage contribution remains conditioned by stand age, rooting depth, litter quality, and management intensity, and may not systematically surpass the soil carbon stabilization potential of native hardwood-dominated ecosystems. In Morocco, although afforestation with fast-growing exotics improves productivity, open or degraded native stands (e.g., low-density cork oak openings storing $\approx$17.58 tC ha^{-1}) illustrate that carbon sink strength drops sharply when litter inputs and stratification decline, whereas dense native montane hardwood and conifer forests sustain stronger organo-mineral carbon transfer and higher total stocks. Therefore, the added value of intensive afforestation in Morocco lies not only in growth speed but in its co-optimization with structural complexity and soil carbon feedbacks, ensuring that productivity gains effectively contribute to national mitigation pathways (NDC, REDD+, and PES) without undermining future SOC residence time and ecosystem resilience.

Maintaining Forest Ecosystems and Biodiversity

Beyond silvicultural and technical interventions, the maintenance of ecosystem integrity and biodiversity emerges as a key long-term driver of forest carbon retention. International evidence shows that forests hosting richer and functionally diverse biotic communities exhibit greater carbon residence time, enhanced productivity of organic inputs, improved soil structure feedbacks, and higher resilience of carbon pools to disturbances such as drought, pests, fires and land-use pressures (Enríquez-de-Salamanca, 2024; Weiskopf et al., 2024). Protecting diverse tree assemblages and understory layers sustains continuous litter and root carbon inputs, reinforces organo-mineral carbon stabilization pathways, and improves resistance and recovery capacity, thereby preserving both current and future carbon sink strength. Although Mediterranean and semi-arid forests—including Morocco's—generally show more moderate stocks than humid temperate or tropical forests, comparative syntheses highlight that biodiversity loss can still markedly reduce national terrestrial carbon storage potential by limiting organic inputs and destabilizing biomass and soil carbon reservoirs, especially in degraded or open stands (Weiskopf et al., 2024). Accordingly, holistic carbon-oriented forest strategies should prioritize mixed-species reforestation, protection of mature multi-layered stands, conservation of native conifer and broadleaf formations, and governance frameworks that jointly address carbon, biodiversity and socio-ecological resilience, supporting national mitigation pathways under Paris agreement, NDC and nature-based carbon enhancement planning.

CONCLUSION

This synthesis highlights the strategic importance of Moroccan forest ecosystems as key components of national and regional climate change mitigation efforts through carbon sequestration in both biomass and soils. By compiling and analyzing a wide range of published studies, this work provides an integrated overview of carbon stocks across the main forest ecosystems of Morocco, including holm oak, Atlas cedar, cork oak, thuja, argan, and mixed stands.

The results clearly demonstrate a strong variability in carbon storage capacity among ecosystems, driven primarily by forest type, stand structure, degree of degradation, substrate, and bioclimatic conditions. In most ecosystems, soil organic carbon represents a major, often dominant, component of total forest carbon stocks, particularly in Mediterranean and semi-arid contexts where aboveground biomass may be limited. Atlas cedar and mixed forest stands show the highest total carbon stocks, reflecting favorable ecological conditions and greater structural complexity, whereas arid-zone ecosystems such as argan woodlands exhibit lower but still ecologically significant carbon storage.

By placing Moroccan forest carbon stocks within a broader Mediterranean and international context, this review confirms that the observed values are consistent with those reported in comparable ecological zones, while also underlining the specific constraints related to climate aridity, land-use pressure, and forest degradation. The synthesis further highlights the critical role of sustainable forest management, restoration of degraded stands, and protection of native forest ecosystems in enhancing long-term carbon sequestration capacity.

From a methodological perspective, this work underscores the complementarity of classical field-based approaches (soil sampling, allometric equations) with emerging tools such as remote sensing and machine learning, which offer promising avenues for improving spatial coverage and accuracy of carbon stock assessments at national scales.

Overall, this synthesis provides a robust scientific basis to support decision-making related to forest conservation, restoration, and climate policy in Morocco, particularly in the context of the country's commitments under the Paris Agreement and its Nationally Determined Contributions. Strengthening forest-based climate solutions while preserving biodiversity and ecosystem services emerges as a key priority for ensuring the resilience and sustainability of Moroccan forest landscapes under ongoing climate change.

REFERENCES

Ahtikoski, A., Rämö, J., Juutinen, A., Shanin, V., & Mäkipää, R. (2022). Continuous cover forestry and cost of carbon abatement on mineral soils and peatlands. *Frontiers in Environmental Science*, *10*, 837878. Advance online publication. DOI: 10.3389/fenvs.2022.837878

Ameray, A., Bergeron, Y., Valeria, O., Montoro Girona, M., & Cavard, X. (2021). Forest carbon management: A review of Sylvicultural Practices and Management Strategies Across Boreal, Temperate and Tropical Forests. *Current Forestry Reports*, *7*(4), 245–266. DOI: 10.1007/s40725-021-00151-w

ANEF. (2024). Forêts en chiffres. http://www.eauxetforets.gov.ma≡

Arabi, M., Mechkirrou, L., El Malki, M., Alaoui, K., Chaieb, A., Maaroufi, F., Karmich, S. (2024). Overview of ecological dynamics in Morocco–Biodiversity, water scarcity, climate change, anthropogenic pressures, and energy resources–Navigating towards ecosolutions and sustainable development. *E3S Web of Conf.*, 527, 01001 (2024). DOI: 10.1051/e3sconf/202452701001

Arias, P. A., Bellouin, N., & Coppola, E. (2021). Technical Summary. In *Climate Change 2021: The Physical Science Basis Working Group I Contribution to the Sixth Assessment Report of the Intergovernmental Panel on Climate Change* (pp. 35–144). Cambridge University Press., DOI: 10.1017/9781009157896.002

Barré, P., Cécillon, L., Chenu, C., Martin, M., Vidal-Beaudet, L., & Eglin, T. (2020). La séquestration de carbone dans les sols agricoles, forestiers et urbains: État des lieux des méthodes d'évaluation et de quantification. *Etude et Gestion des Sols*, *27*, 305–320.

Batjes, N. H. (1996). Total carbon and nitrogen in the soils of the world. *European Journal of Soil Science*, *47*(2), 151–163. DOI: 10.1111/j.1365-2389.1996.tb01386.x

Belghazi, T. (2013). Typologie de l'arganeraie et Estimation de La Biomasse Aérienne Des Taillis d'arganier (Plateau Des Haha-Essaouira, Maroc). Ph.D. Thesis, UCL-Université Catholique de Louvain, Ottignies-Louvain-la-Neuve, Belgium

Benabid, A. (1985). Les écosystèmes forestiers du Maroc. *Forêt Méditerranéenne*, 53–64.

Benabid, A. (2000). *Flore et écosystèmes du Maroc: Évaluation et préservation de la biodiversité*. Kalila Wa Dimna.

Benabou, A., Moukrim, S., Lahssini, S., El Aboudi, A., Menzou, K., Elmalki, M., & Rhazi, L. (2022). Impact of climate change on potential distribution of Quercus suber in the conditions of North Africa. *Biosystems Diversity*, *30*(3), 289–294. DOI: 10.15421/012231

Boisvenue, C., Paradis, G., Eddy, I. M. S., McIntire, E. J. B., & Chubaty, A. M. (2022). Managing forest carbon and landscape capacities. *Environmental Research Letters*, *17*(11), 114013. DOI: 10.1088/1748-9326/ac9919

Born, K., Christoph, M., & Fink, A. H. (2008). Moroccan Climate in the Present and Future: Combined View from Observational Data and Regional Climate Scenarios. In Zereini, F., & Hötzl, H. (Eds.), *Climatic Changes and Water Resources in the Middle East and North Africa. Springer Berlin Heidelberg, Berlin* (pp. 29–45). DOI: 10.1007/978-3-540-85047-2_4

Boudhar, A., Boudhar, S., & Ibourk, A. (2017). Input–output & water relations. *Journal of Economic Structures*, •••, 6.

Boujraf, A., Dallahi, Y., Dahmani, J., Orlando, C. A., & Douira, A. (2023). Effect of plant Communities and Ecological Parameters on Soil Organic Carbon Stocks in the Mamora Forest, Morocco. *Journal of Ecological Engineering*, *24*(5), 128–136. DOI: 10.12911/22998993/161616

Boukhris, I., Lahssini, S., Collalti, A., Moukrim, S., Santini, M., Chiti, T., & Valentini, R. (2023). Calibrating a Process-Based Model to Enhance Robustness in Carbon Sequestration Simulations: The Case of Cedrus atlantica (Endl.) Manetti ex Carrière. *Forests*, *14*(2), 401. DOI: 10.3390/f14020401

Boulmane, M., del Carmen Santa-Regina, M., Halim, M., Khia, A., Oubrahim, H., Abbassi, H., & Santa-Regina, I. (2015). Organic Carbon Storage in Evergreen Oak Forest Ecosystems of the Middle and High Moroccan Atlas Areas. *Open Journal of Forestry*, *5*(3), 260–273. DOI: 10.4236/ojf.2015.53023

Boulmane, M., Makhloufi, M., Bouillet, J. P., Saint-Andre, L., Satrani, B., Halim, M., & El Antry, S. (2010). Estimated Organic Carbon Stock in the Quercus ilex Moroccan Middle Atlas. *Acta Botanica Gallica*, *157*, 451–467. DOI: 10.1080/12538078.2010.10516222

Boulmane, M., Oubrahim, H., Halim, M., & Bakker, M. R. (2017). The potential of Eucalyptus plantations to restore degraded soils in semi-arid Morocco (NWAfrica). *Annals of Forest Science*. 57-74pp. DOI: 10.1007/s13595-017-0652-z

Busch, J., Engelmann, J., Cook-Patton, S. C., Griscom, B. W., Kroeger, T., & Possingham, H.. (2019). Low-cost CO_2 removal through tropical reforestation. *Nature Climate Change*, *9*, 463–466. DOI: 10.1038/s41558-019-0485-x

Cachinero-Vivar, A. M., Palacios-Rodríguez, G., Lara-Gómez, M. A., & Navarro-Cerrillo, R. M. (2021). Thinning Effect of C Sequestration along an Elevation Gradient of Mediterranean Pinus spp. Plantations. *Forests*, *2021*(12), 1583. DOI: 10.3390/f12111583

Carmona-Yáñez, M. D., Lucas-Borja, M. E., Zema, D. A., Jing, X., Kooch, Y., Garrido Gallego, P., Plaza-Alvarez, P. A., Zhou, G., & Delgado-Baquerizo, M. (2023). Influence of management and stand composition on ecosystem multifunctionality of Mediterranean tree forests. *Trees (Berlin)*, *37*(6), 1801–1816. DOI: 10.1007/s00468-023-02462-w

Chave, J., Andalo, C., Brown, S., Cairns, M. A., Chambers, J. Q., Eamus, D., Fölster, H., Fromard, F., Higuchi, N., Kira, T., Lescure, J.-P., Nelson, B. W., Ogawa, H., Puig, H., Riéra, B., & Yamakura, T. (2005). Tree allometry and improved estimation of carbon stocks. *Oecologia*, *145*(1), 87–99. Advance online publication. DOI: 10.1007/s00442-005-0100-x

Cheng, F., Tian, J., He, J., He, H., Bao, G., Yang, Y., Liu, G., & Zhang, Z. (2024). China's future forest carbon sequestration potential under different management scenarios. *Trees, Forests and People*, *17*, 100621. DOI: 10.1016/j.tfp.2024.100621

Chowdhury, A., Naz, A., & Maiti, S. K. (2023). Variations in soil blue carbon sequestration between natural mangrove metapopulations and a mixed mangrove plantation: A case study from the world's largest contiguous mangrove forest. *Life (Basel, Switzerland)*, *13*(2), 271. DOI: 10.3390/life13020271

Cook-Patton, S. C., Leavitt, S. M., Gibbs, D., Harris, N. L., Lister, K., Anderson-Teixeira, K. J., Briggs, R. D., Chazdon, R. L., Crowther, T. W., Ellis, P. W., Griscom, H. P., Herrmann, V., Holl, K. D., Houghton, R. A., Larrosa, C., Lomax, G., Lucas, R., Madsen, P., Malhi, Y., & Griscom, B. W. (2020). Mapping carbon accumulation potential of natural forest regrowth. *Nature*, *585*(7826), 545–550. DOI: 10.1038/s41586-020-2686-x

Dallahi, Y., Malaainine, M. E., Hbiak, I., Boujraf, A., Ould Abidine, M. M., Orlando, C. A., Meliho, M., El Mderssa, M., & Minoubi, A. (2023). Contribution to the Modeling of the Organic Matter of Moroccan Forest Soils within the Context of Global Change – Case Study of the Central Plateau. *Ecological Engineering and Environmental Technology*, *24*(8), 261–271. DOI: 10.12912/27197050/172004

Danish, A., Wazir Aitizaz Ahsan, A. H., Begum, F., Ashiq, H., Chitsan, L., & Hussain, F. (2025). *Assessing the impact of land use and land cover changes on soil properties and carbon sequestration in the upper Himalayan Region of Gilgit.* Sustainable Chemistry One World., DOI: 10.1016/j.scowo.2024.100038

Díaz-Pines, E., Rubio, A., Miegroet, H. V., Montes, F., & Benito, M. (2011). Does tree species composition control soil organic carbon pools in Mediterranean mountain forests? *Forest Ecology and Management*, *262*(10), 1895–1904. DOI: 10.1016/j.foreco.2011.02.004

Dobignard & Chatelain. (2011). Index synonymique de la flore d'Afrique du Nord. *Éditions des conservatoires et jardins botaniques. Genève.* 430 p.

El Mderssa, M., Benjelloun, H., Zaher, H., Zennouhi, O., Nassiri, L., & Ibijbijen, J. (2019). Estimating the sequestration potential of organic carbon in forest soils in the Central Middle Atlas: A tool to fight climate change. *Atlas Journal of Biology*, •••, 603–610. DOI: 10.5147/ajb.v0i0.198

El Mderssa, M., Elmalki, M., Whalen, J. K., Ikraoun, H., Aliyat, F. Z., Dallahi, Y., Abbas, Y., Nassiri, L., & Ibijbijen, J. (2024). Forest stand and soil types determine soil organic carbon storage in the Middle Atlas region of Morocco using machine learning models. *All Earth*, *36*(1), 1–10. DOI: 10.1080/27669645.2024.2400432

El Mderssa, M., Malki, F., Ikraoun, H., Nassiri, L., & Ibijbijen, J. (2022). Determination des parametres d'evaluation du stock de carbone dans les ecosystemes forestiers (Cedrus atlantica Manetti, cedre de l'Atlas au Maroc): Methodes specifiques et generiques. *Bois et Forêts des Tropiques*, *351*, 67–77. DOI: 10.19182/bft2022.351.a36330

Emberger, L. (1939). Aperçu général sur la végétation du Maroc: commentaire de la carte phytogéographique du Maroc 1:1.500.000 *Mémoire hors-série de la Société des Sciences Naturelles du Maroc*

Enríquez-de-Salamanca, Á. (2024). Environmental and social impacts of carbon sequestration. *Integrated Environmental Assessment and Management*, *20*(6), 1812–1838. DOI: 10.1002/ieam.4925

Erkan, N., Güner, Ş. T., & Aydın, A. C. (2023). Thinning effects on carbon stocks in *Pinus brutia* plantations. *Carbon Balance and Management*, *18*(1). Advance online publication. DOI: 10.1186/s13021-023-00226-0

Fadil, S., Sebari, I., Bouhaloua, M., Mohamed, B., Ait El Kadi, K., Kellouch, S., Lfalah, H., & Meghraoui, K. (2022). Biomass and Carbon Stock Quantification in Cork Oak Forest of Maamora Using a New Approach Based on the Combination of Aerial Laser Scanning Carried by Unmanned Aerial Vehicle and Terrestrial Laser Scanning Data. *Forests*, *13*(8), 1211. DOI: 10.3390/f13081211

Fadil, S., Sebari, I., Moulay, M. A., & El kadi, K. A. (2024). Modeling and spatialization of biomass and carbon stock using unmanned Aerial Vehicle Lidar (Lidar-UAV) metrics and forest inventory in cork oak forest of Maamora. *Regional Science Policy & Practice*, *16*(11), 100127. Advance online publication. DOI: 10.1016/j.rspp.2024.100127

FAO. (2015c). *Évaluation des ressources forestière mondiales 2015. Répertoire de données de FRA 2015*. FAO.

FAO. (2020). Global Forest Resources Assessment 2020.

Farineau, J., & Morot-Gaudry, J. F. (2018). La photosynthèse: Processus physiques, moléculaires et physiologiques. *Editions Quae. Hors Collection,* 460.

Favero, A., Daigneault, A., & Sohngen, B. (2020). Forests: Carbon sequestration, biomass energy, or both? *Science Advances*, *6*(13), eaay6792. Advance online publication. DOI: 10.1126/sciadv.aay6792

Filahi, S., Tramblay, Y., Mouhir, L., & Diaconescu, E. P. (2017). Projected changes in temperature and precipitation indices in Morocco from high-resolution regional climate models. *International Journal of Climatology*, *37*(14), 4846–4863. DOI: 10.1002/joc.5127

Ganatsas, P., Tsakaldimi, M., Karydopoulos, T., Papaemannuil, A., & Papadopoulos, S. (2024). Long-term effect of different forest thinning intensity on carbon sequestration rates and potential uses in climate change mitigation actions. *Mitigation and Adaptation Strategies for Global Change*, *29*(1), 3. DOI: 10.1007/s11027-023-10102-4

Girardin, C. A. J., Jenkins, S., Seddon, N., Allen, M., Lewis, S. L., Wheeler, C. E., Griscom, B. W., & Malhi, Y. (2021). Nature-based solutions can cool the planet. *Nature*, *593*(7858), 191–196. DOI: 10.1038/d41586-021-01241-2

Griscom, B. W., Adams, J., Ellis, P. W., Houghton, R. A., Lomax, G., Miteva, D. A., Schlesinger, W. H., Shoch, D., Siikamäki, J. V., Smith, P., Woodbury, P., Zganjar, C., Blackman, A., Campari, J., Conant, R. T., Delgado, C., Elias, P., Gopalakrishna, T., Hamsik, M. R., & Fargione, J. (2017). Natural climate solutions. *Proceedings of the National Academy of Sciences of the United States of America*, *114*(44), 11645–11650. DOI: 10.1073/pnas.1710465114

Gumus, V., El Moçayd, N., Seker, M., & Seaid, M. (2024). Future drought projections in Morocco and potential impact on agriculture. *Journal of Environmental Management, 367*, 122019. DOI: 10.1016/j.jenvman.2024.122019

Hadri, A., Saidi, M. E. M., Saouabe, T., & El Fels, A. E. A. (2021). Temporal trends in extreme temperature and precipitation events in an arid area: Case of chichaoua mejjate region (morocco). *Journal of Water and Climate Change, 12*(3), 895–915. DOI: 10.2166/wcc.2020.234

Haines-Young, R., Potschin, M., & Kienast, F. (2012). Indicators of ecosystem service potential at European scales: Mapping marginal changes and trade offs. *Ecological Indicators, 21*, 39–53. DOI: 10.1016/j.ecolind.2011.09.004

Hakam, O., Ongoma, V., Beniaich, A., Meskour, B., El Kadi, M. A., & Brouziyne, Y.. (2024). Climate change impact on argan distribution. *Modeling Earth Systems and Environment, 10*(4), 5529–5652. DOI: 10.1007/s40808-024-02077-1

Houghton, R. A., & Nassikas, A. A. (2018). Negative emissions from stopping deforestation. *Global Change Biology, 24*(1), 350–359. DOI: 10.1111/gcb.13876

IF. (2018). *L'Institut de France Le changement climatique*. Académie des Sciences.

Ikraoun, H., El Mderssa, M., Aliyat, F.Z., Nassiri, L., Ibijbijen I. (2024). Assessing Carbon Storage and Biomass in Moroccan Cork Oak Forests. *Ecological Engineering & Environmental Technology.* 25(6), 43–53 pp. https://doi.org/DOI: 10.12912/27197050/186550

Ikraoun, H., El Mderssa, M., El Mouridi, Z., Nassiri, L., & Ibijbijen, J. (2022). Carbon stock assessment and anlysis of physiochemeical characteristics under seven woody forest formations in Oulmes Central Plateau, Morocco. *Soil & Environment, 41*(1), 55–62. DOI: 10.25252/SE/2021/222684

IPCC. (2006). *Guidelines for National Greenhouse Gas Inventories* (Vol. 4). Agriculture, Forestry and Other Land Use.

IPCC. (2021). *Climate Change 2021 – The Physical Science Basis*. Cambridge University Press.

Jones, M. W., Abatzoglou, J. T., Veraverbeke, S., Andela, N., Lasslop, G., Forkel, M., Smith, A. J. P., Burton, C., Betts, R. A., van der Werf, G. R., Sitch, S., Canadell, J. G., Santín, C., Kolden, C. A., Doerr, S. H., & Le Quéré, C. (2022). Global and regional trends and drivers of fire under climate change. *Reviews of Geophysics*, 60(3), e2020RG000726. DOI: 10.1029/2020RG000726

Klapwijk, M. J., Boberg, J., Bergh, J., Bishop, K., Björkman, C., & Ellison, D.. (2018). Capturing complexity: Forests & mitigation. *Global Environmental Change*, *52*, 238–247. DOI: 10.1016/j.gloenvcha.2018.07.012

Laaribya, S., & Belghazi, B. (2016). Radial growth of Atlas cedar. *Nature & Technology*, *14*, 19–32.

Le Clec'h, S., Oszwald, J., Jégou, N., Dufour, S., Cornillon, P. A., Miranda, I. de S., Gonzaga, L., Grimaldi, M., Gond, V., & Arnauld de Sartre, X. (2013). Cartographier le carbone stocké dans la végétation: Perspectives pour la spatialisation d'un service écosystémique. *Bois et Forêts des Tropiques*, *316*(316), 35–47. DOI: 10.19182/bft2013.316.a20529

Leng, Y., Li, W., Ciais, P., Sun, M., Zhu, L., Yue, C., Chang, J., Yao, Y., Zhang, Y., Zhou, J., Li, Z., Wang, X., Xi, Y., & Peng, S. (2024). Forest aging limits future carbon sink in China. *One Earth*, *7*(5), 822–834. DOI: 10.1016/j.oneear.2024.04.011

Li, H., Wang, S., Bai, X., Luo, W., Tang, H., Cao, Y., Wu, L., Chen, F., Li, Q., Zeng, C., & Wang, M. (2018). Global carbonate carbon sink. *The Science of the Total Environment*, *643*, 157–170. DOI: 10.1016/j.scitotenv.2018.06.196

Ma, T., Zhang, C., Ji, L., Zuo, Z., Beckline, M., Hu, H., Li, X., & Xiao, X. (2024). Development of forest aboveground biomass estimation, its problems and future solutions: A review. *Ecological Indicators*, *159*, 111653. Advance online publication. DOI: 10.1016/j.ecolind.2024.111653

Malki, F., Al Karkouri, J., Sabir, M., & El Mderssa, M. (2022a). Spatio-temporal evolution of Maamora forest. *E3S Web of Conferences 337 0100, 10.*

Malki, F., Al Karkouri, J., Sabir, M., Ibijbijen, J., El Mderssa, M., Ikraoun, H., & Dallahi, Y. (2022b). Assessment of the storage potential of organic carbon in Maamora forest soils: A strategic guidance tool for reforestation. *Soil & Environment*, *41*(1). Advance online publication. DOI: 10.25252/SE/2022/232782

Marchane, A., Tramblay, Y., Hanich, L., Ruelland, D., & Jarlan, L. (2017). Climate change impacts on surface water resources in the Rheraya catchment (High Atlas, Morocco). *Hydrological Sciences Journal*, *62*(6), 979–995. DOI: 10.1080/02626667.2017.1283042

Meliho, M., Boulmane, M., Khattabi, A., Dansou, C. E., Orlando, C. A., Mhammdi, N., & Noumonvi, K. D. (2023). Spatial Prediction of Soil Organic Carbon Stock in the Moroccan High Atlas Using Machine Learning. *Remote Sensing (Basel)*, *15*(10), 2494. DOI: 10.3390/rs15102494

Mori, A. S., Lertzman, K. P., & Gustafsson, L. (2017). Biodiversity and ecosystem services in forest ecosystems: A research agenda for applied forest ecology. *Journal of Applied Ecology*, *54*(1), 12–27. DOI: 10.1111/1365-2664.12669

Moukrim, S., Lahssini, S., Rhazi, M., Menzou, K., El Madihi, M., Rifai, N., & Rhazi, L. (2022). Climate change impact on potential distribution of an endemic species Abies marocana *Trabut. Ekologia (Bratislava)*, *41*(4), 329–339. https://doi .org/. DOI: 10.2478/eko-2022-0034

Nandal, A., Yadav, S. S., Rao, A. S., Meena, R. S., & Lal, R. (2023). Advance methodological approaches for carbon stock estimation in forest ecosystems. *Environmental Monitoring and Assessment*, *195*(2), 315. DOI: 10.1007/s10661-022-10898-9

NDC. (2025). Contribution Déterminée au niveau National CDN 3.0 du Maroc 2026-2025. Royaume du Maroc, 107. https://unfccc.int/sites/default/files/2025 10/MOROCCO%20NDC%203.0%20_30.9.25.pdf

Odum Eugene, P. (1969). The Strategy of Ecosystem Development. *Science*, *164*(3877), 262–270. DOI: 10.1126/science.164.3877.262

Oubrahim, H., Boulmane, M., Bakker, M. R., Augusto, L., & Halim, M. (2015). Carbon storage in degraded cork oak (Quercus suber) forests on flat lowlands in Morocco. [early view]. *IForest (Viterbo)*. Advance online publication. DOI: 10.3832/ ifor1364-008

Oumasst, A., Tiouidji, F. E., Tabi, S., Zahidi, A., El Mousadik, A., El Finti, A., Aitlhaj, A., & Hallam, J. (2024). Development of Allometric Equations to Determine the Biomass of Plant Components and the Total Storage of Carbon Dioxide in Young Mediterranean Argan Trees. *Sustainability (Basel)*, *16*(11), 4592. DOI: 10.3390/su16114592

Pan, Y., Birdsey, R. A., Fang, J., Houghton, R., Kauppi, P. E., Kurz, W. A., Phillips, O. L., Shvidenko, A., Lewis, S. L., Canadell, J. G., Ciais, P., Jackson, R. B., Pacala, S. W., McGuire, A. D., Piao, S., Rautiainen, A., Sitch, S., & Hayes, D. (2011). A large and persistent carbon sink in the world's forests. *Science*, *333*(6045), 988–993. DOI: 10.1126/science.1201609

Pereira, A. P. A., Cherubin, M. R., de Araujo, A. S. F., Santana, M. C., de Medeiros, E. V., & da Costa, D. P. (2024). Soil quality in mono and mixed eucalyptus. *Sustainability*, *16*(6), 2534. DOI: 10.3390/su16062534

Pérez-Cutillas, P., & Salhi, A. (2024). Long-term hydroclimatic projections and climate change scenarios at regional scale in Morocco. *Journal of Environmental Management, 371, 123254.* Piao S, Yue C, Ding J, Guo Z. (2022). Perspectives on the role of terrestrial ecosystems in the 'carbon neutrality' strategy. Science China Earth Sciences65:1178–1186. https://doi.org/DOI: 10.1016/j.jenvman.2024.123254

Pingoud, K., Ekholm, T., Sievänen, R., Huuskonen, S., & Hynynen, J. (2018). Trade-offs between carbon and harvest. *Journal of Environmental Management*, *210*, 96–103. DOI: 10.1016/j.jenvman.2017.12.076

Quézel, P., & Médail, F. (2003). *Ecologie et biogéographie des forets du bassin méditerranéen.* Elsevier., http://www.lavoisier.fr/livre/environnement/ecologie-et-biogeographie-des-forets-du-bassin-mediterraneen/quezel/descriptif-9782842994518

Rivas Martinez, S. (1982). Etages bioclimatiques, secteurs chorologiques et série de végétation de l'Espagne méditerranéenne. *Ecologia Mediterranea*, *8*(1), 275–288. DOI: 10.3406/ecmed.1982.1954

Rodríguez, M., & Carlos, J. (2001). Organic carbon content under different types of land use and soil in peninsular Spain. *Biology and Fertility of Soils*, *33*(1), 53–61. DOI: 10.1007/s003740000289

Ruiz-Peinado, R.. (2013). Thinning impacts on carbon. *Forest Ecology and Management.*

Sabir, M., Sagno, R., Tchintchin, Q., Zaher, H., & Benjelloun, H. (2020). SOC stocks in Morocco. In *Carbone des sols en Afrique* (pp. 71–90). IRD éditions.

SEIFN. (2014). Service des Etudes et de l'Inventaire Forestier National, Maroc forestier in http://www.eauxetforets.gov.ma/ForetsMarocaines/ForetsChiffres/Pages/Forets-En-Chiffres.aspx

Shah, I. H., Manzoor, M. A., Ji'nhui, W., Li, X., Hameed, M. K., Rehaman, A., Li, P., Zhang, Y., Niu, Q., & Chang, L. (2024). Comprehensive review: Effects of climate change and greenhouse gases emission relevance to environmental stress on horticultural crops and management. *Journal of Environmental Management*, *351*, 119978. DOI: 10.1016/j.jenvman.2023.119978

Shang, R., Chen, J. M., Xu, M., Lin, X., Li, P., Yu, G., He, N., Xu, L., Gong, P., Liu, L., Liu, H., & Jiao, W. (2023). China's current forest age structure will lead to weakened carbon sinks in the near future. *Innovation (Cambridge (Mass.))*, *4*(6), 100515. Advance online publication. DOI: 10.1016/j.xinn.2023.100515

Shirkey, G., Anctil, A., John, R., Kolluru, V., Mungai, L., & Kashongwe, H.. (2024). Opportunities for NbS. *Environmental Research Letters*.

Smith, P., Adams, J., Beerling, D. J., Beringer, T., Calvin, K. V., & Fuss, S.. (2019). Land management options. *Annual Review of Environment and Resources*, *7*, 40.

Sombroek, W. G., Nachtergaele, F. O., & Hebel, A. (1993). Amounts, Dynamics, Sequestering of Carbon in Tropical ans Subtropical Soils. AMBIO. V(22). N°7. 417-426pp

United Nations. (1998). Kyoto Protocol to the United Nations Framework Convention on Climate Change.

United Nations. (2015). Paris Agreement, 27

Van Engelen, V. W. P., Batjes, N. H., Dijkshoorn, J. A., & Huting, J. R. M. (2005). *Harmonized Global Soil Resources Database (Final Report)*. (ISRIC Report 2005/06). ISRIC. https://edepot.wur.nl/46477

Van Kooten, G. C., Binkley, C. S., & Delcourt, G. (1995). Carbon taxes & forest rotations. *American Journal of Agricultural Economics*, *77*(2), 365–374. DOI: 10.2307/1243546

Weiskopf, S. R., Isbell, F., Arce-Plata, M. I., Di Marco, M., Harfoot, M., Johnson, J., Lerman, S. B., Miller, B. W., Morelli, T. L., Mori, A. S., Weng, E., & Ferrier, S. (2024). Biodiversity loss reduces global terrestrial carbon storage. *Nature Communications*, *15*(1), 4354. DOI: 10.1038/s41467-024-47872-7

Williams, A. P., Abatzoglou, J. T., Gershunov, A., Guzman-Morales, J., Bishop, D. A., & Balch, J. K.. (2019). Anthropogenic climate change & wildfires in California. *Earth's Future*, *7*(8), 892–910. DOI: 10.1029/2019EF001210

Zaher, H., Rhoufacha, M., Dahmani, H. (2025). Carbon stock dynamics along a cork oak stand chronosequence: implications for climate-smart forest management. *Journal of the Saudi Society of Agricultural Sciences*. 24-45 pp. DOI: 10.1007/s44447-025-00054-3

Zaher, H., Sabir, M., Benjelloun, H., & Hounzandji, P. I. (2020). Effect of forest land use change on carbohydrates, physical soil quality and carbon stocks in Moroccan cedar area. *Journal of Environmental Management*, *254*, 109544. Advance online publication. DOI: 10.1016/j.jenvman.2019.109544

Zhang, C.. (2019). Effects of thinning on carbon. *Journal of Forestry Research*.

Chapter 7 Assessing Carbon Sequestration Capacity of Eucalyptus in Restored Sub-Mediterranean Lands: Case of Eucalyptus Clones in Morocco

Saliha Diarra
http://orcid.org/0009-0006-0557-0493
Hassan II Institute of Agronomy and Veterinary Medicine, Rabat, Morocco

Mohamed Ankri
National School of Forestry Engineers, Salé, Morocco

Said Lahssini
http://orcid.org/0000-0003-1986-8240
National School of Forestry Engineers, Salé, Morocco

Abdellatif Khattabi
National School of Forestry Engineers, Salé, Morocco

Najib Magri
http://orcid.org/0000-0002-1670-389X
Centre Innovation, Recherche et Formation, Morocco

Said Moukrim
http://orcid.org/0000-0002-3172-3776
Faculty of Sciences, Mohammed V University in Rabat, Morocco

ABSTRACT

Eucalyptus plantations are an important component of Morocco's forest landscapes. They provide numerous socioeconomic benefits and alleviate pressure on natural forests. However, their contribution to carbon sequestration remains poorly docu-

DOI: 10.4018/979-8-3373-6239-7.ch007

mented. This chapter assessed the carbon sequestration capacity of five Eucalyptus clones of different ages in the Gharb region in Morocco. Destructive sampling of 20 trees per clone–age combination was conducted to quantify carbon stocks in tree biomass, litter, and soil through laboratory analyses, and to develop clone-specific allometric models for aboveground carbon estimation. Mean biomass carbon fractions ranged from 53.60 ± 1.27% to 54.96 ± 1.80%, exceeding commonly applied default values. Aboveground carbon stocks varied from 3.86 to 32.87 t C ha^{-1}, while soil represented the largest carbon pool (37.67–74.70 t C ha^{-1}). These findings highlight strong clonal variability and emphasize the need for informed clonal selection and management to enhance climate mitigation strategies through carbon sequestration.

INTRODUCTION

Global warming, driven primarily by anthropogenic greenhouse gas emissions from fossil fuel combustion, deforestation, and land-use change, constitutes a major environmental challenge. The global mean surface temperature has already increased by about 1.1 °C relative to the pre-industrial period (1850–1900), with human influence identified as the dominant driver since the mid-20th century. Projections suggest that warming could reach 1.8 –5.7 °C by the end of the century under different emission scenarios (IPCC, 2021). Its impacts, ranging from rising temperatures and altered precipitation patterns to ecosystem degradation and shift in species distribution (Benabou et al. 2022; Moukrim et al. 2022), affect not only the environment but also the socio-economic fabric of communities worldwide (Leal Filho et al., 2021). In response, carbon sequestration in terrestrial ecosystems has emerged as a crucial strategy to mitigate global warming.

Forests, in particular, play a central role in this regard, storing over half of the carbon in terrestrial ecosystems (Pan et al., 2011; Morgan et al., 2021). Indeed, through photosynthesis, trees absorb CO_2 and convert it into biomass, making forest ecosystems powerful carbon sinks. The Paris Agreement formally recognized this role of forest ecosystems in mitigating climate change and emphasized the need to preserve and enhance forest carbon stocks as part of national climate strategies. According to the IPCC (2021), forests play a major role in climate regulation by sequestering approximately 9.2 Gt of CO_2 per year, representing nearly 33% of global greenhouse gas emissions.

The main carbon pools in forests are aboveground biomass, belowground biomass, and soil. Aboveground biomass includes all living vegetation above the soil surface, such as trunks, branches, leaves, and understory plants, while belowground biomass comprises roots and associated living tissues. According to Pan et al. (2011), approximately 44% of total forest carbon stocks are in soils, while about 42% are in living biomass (above and belowground), with smaller fractions in deadwood (8%) and litter (5%). However, the carbon sequestration capacity of forests is highly variable depending on multiple factors, including forest type, age, management regime, and disturbances like insects, forest fires, and deforestation. Consequently, one of the main strategies to enhance the contribution of forests to climate change mitigation is to conserve natural, high-carbon ecosystems, while scaling up afforestation and reforestation efforts. In many countries, fast-growing species such as those of the genus Eucalyptus are increasingly employed in short-rotation plantations. These species are often favored for their high net primary production and rapid biomass accumulation rates, as well as their economic importance (Hirigoyen et al., 2021; Dai et al.,2023).

In Morocco, Eucalyptus plantations constitute a substantial component of the reforested landscape, representing approximately 31% of the total reforested area, or around 208 568 ha (ANEF, 2025). The dominant species include *Eucalyptus camaldulensis* Dehnh., *Eucalyptus gomphocephala* DC, and *Eucalyptus grandis* W. Hill ex Maiden, alongside several hybrid clones. These plantations provide multiple ecosystem services, including firewood and honey production, soil protection, and carbon sequestration, as well as important socio-economic benefits. Despite their contribution to Morocco's forest cover, the specific role of Eucalyptus plantations in carbon sequestration remains poorly quantified. Understanding carbon accumulation in young plantations is particularly important, as commercial Eucalyptus regimes often employ short-rotation cycles, and the early years correspond to periods of rapid biomass growth.

Moreover, to date, no publicly available carbon models have been specifically calibrated for the Eucalyptus clones used in Moroccan plantations, and information on their carbon sequestration potential and contribution to climate change mitigation remains very limited. This represents a significant knowledge gap. Addressing it is essential to generate carbon estimates that are accurately contextualized to local edaphic and climatic conditions, thereby providing a robust basis for informed policy-making and management strategies aimed at maximizing the mitigation potential of these fast-growing plantations. Focusing on young clonal stands (3 to 7 years) allows for the capture of early growth dynamics and initial carbon accumulation, which is critical both for industrial planning and for modeling long-term carbon budgets under different management and climate scenarios.

Within this context, the present study aims to assess the carbon sequestration potential of five Eucalyptus clones (Clone 53, Clone 397, Clone 583, Clone 3933, and Clone 579) of varying ages, established in the Gharb region. Specifically, we seek to address the following research question: To what extent do these clonal plantations contribute to carbon sequestration efforts, and how do their performances compare in terms of carbon sequestration? Three main hypotheses guide this work. First, we hypothesize that the carbon fraction in dry biomass varies significantly among tree compartments regardless of clone type, and that these values may deviate from the commonly applied generic values of 0.47 and 0.50. Second, the total aboveground carbon stock may differ significantly between clones due to differences in total biomass accumulation and stands characteristics. furthermore, we hypothesize that the soil compartment may represent a significant role in total carbon sequestration efforts in these ecosystems.

To explore these hypotheses, the study aims: (i) to directly quantify the carbon content of the main tree compartments (stem, branches, foliage, and bark) through destructive sampling and laboratory analysis, (ii) to assess aboveground carbon stocks for each clone using clone-specific allometric models, (iii) to compare total aboveground carbon stocks per hectare among clones under a standardized stand density; and (iv) to evaluate the contribution of litter and soil carbon pools to overall ecosystem carbon sequestration. This study is expected to provide data and tools that support informed clone selection for reforestation programs aimed at maximizing carbon sequestration. More broadly, the results may strengthen national forest carbon accounting systems and inform the design of climate change mitigation and carbon offset initiatives.

MATERIALS AND METHODS

Study Area

The study area is located in the Gharb region, in northwestern Morocco, near the Atlantic coast, within the province of Sidi Slimane (Figure 1). The site lies within the EUCAFOREST development perimeter, which is predominantly planted with Eucalyptus, including several hybrid clones. The area is characterized by relatively flat terrain (<18 m altitude) and has a semi-arid to temperate bioclimate with annual rainfall of 550 mm/year and an average annual temperature of 15.8 °C (Boulmane et al., 2017). Most soils are sandy, resting on a layer of clay of varying depth (0.85 to 3.20 m), classified as arenosols according to FAO/IUSS (2006).

History and Management Practices at EUCAFOREST

EUCAFOREST was created in December 1995 by Cellulose du Maroc (the former national paper pulp company), following a reforestation agreement concluded with the Ministry of Agriculture and Land Development. Under this convention, the company was granted a perimeter of approximately 24,000 ha for reforestation and long-term management. The EUCAFOREST perimeter forms part of Morocco's national program for industrial and commercial Eucalyptus plantations, designed both to ensure a sustainable supply of timber and to rehabilitate marginal or degraded lands. The eucalyptus clones deployed in these plantations were developed between 1987 and 1993 through genetic improvement programs conducted by the National Center for Forest Plant Improvement (CNAPF). These efforts aimed to enhance genetic uniformity, increase productivity, and improve adaptation to local environmental conditions (Marien & Azami, 2011).

The establishment of eucalyptus plantations generally follows the following steps:

- Site survey and marking: Detailed soil mapping, assessment of slope and drainage, and delineation of plantation blocks.
- Vegetation clearance: Mechanical removal of shrubs, herbaceous cover, and residual woody material.
- Soil preparation: Mechanical tillage or subsoiling, typically to depths of 30–60 cm to break compacted layers, improve planting conditions, and facilitate deeper root development.
- Planting: Eucalyptus seedlings produced in nurseries, either rooted cuttings or micro-propagated plantlets, are transplanted during the autumn, at the onset of the rainy season, using the designated spacing generally 3x4 (833 plantlets /ha)
- First-year management consists mainly of weed control and the replacement of failed or non-surviving plants.

In coppiced stands, a first thinning is generally conducted at 3–4 years, depending on stand conditions. The plantations are then harvested between 9 and 12 years, according to market demand and site characteristics. This management cycle is repeated for four successive rotations, after which the trees are uprooted at the end of the fourth rotation.

Figure 1. Study area

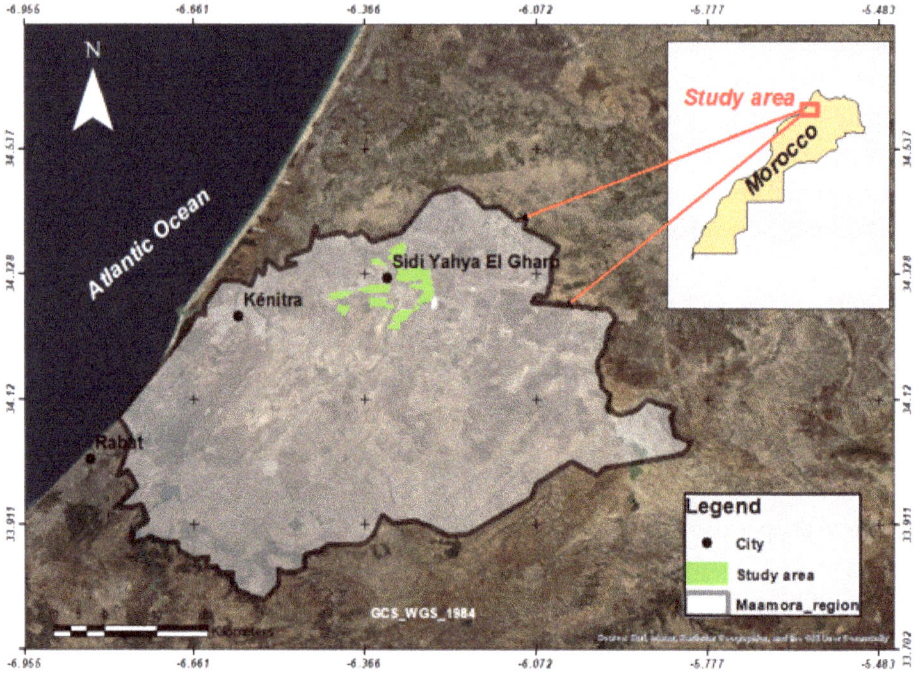

Sampling Strategy

For each clone-age combination, 20 trees were randomly selected, as all stands were even-aged and exhibited strong structural homogeneity. The decision to sample 20 trees was justified by the practical constraints associated with destructive sampling, which is well known to be labor-intensive, costly, and often limited by regulatory constraints (Picard et al., 2012). Within such limitations, a sample size of 20 trees provides a reasonable compromise, as previous studies have shown that 15–25 destructive samples are generally sufficient to obtain stable parameter estimates for classical allometric equations (Picard et al., 2012; Parsapour et al., 2013). The sample trees were selected to reflect the structure of the stands and to ensure adequate representation of all circumference classes (Appendix A). Table 1 summarizes the main characteristics of the studied clonal stands and sample trees.

For soil sampling, a total of 17 pedological profiles were excavated to characterize each soil and collect samples from each horizon for laboratory analyses. Additionally, considering four depth intervals (0–20 cm, 20–40 cm, 40–60 cm, and 60–100 cm), a soil auger was used to collect samples from four plots for each clone–age combination and for each depth level, resulting in a total of 144 soil samples. Regarding litter, samples were collected within 1 m^2 quadrats, with four plots established for each clone–age combination. Importantly, both soil and litter samples were collected under canopy cover as well as in open areas to ensure representative estimates of field conditions.

Table 1. Stands and sample trees characteristics for each clone

Clone N°	*Age (years)*	*Stand density (N/ha)*	*Selected samples trees*	*Mean DBH (cm)*	*Mean height (m)*
Clone 583	3	1768	20	5.68	4.64
Clone 583	5	2091	20	5.83	5.68
Clone 583	7	1565	20	8.76	7.87
Clone 3933	5	630	20	5.88	8.36
Clone 3933	7	2806	20	8.91	10.81
Clone 579	5	1785	20	7.80	10.59
Clone 579	7	1887	20	8.50	11.10
Clone 397	6	748	20	7.85	9.047
Clone 53	6	732	20	10.49	11.545

Data Collected

The biomass of the various components of the tree (stem, bark, branches, and foliage) was determined following a destructive approach (Picard et al., 2012), which involves felling the sample trees. For each sample tree, the circumference at 1.30 m and the total height were measured before felling. Once the tree was felled, the different components (stem, bark, branches, and foliage) were separated and weighed on site using an electronic platform scale with a maximum capacity of approximately 300 kg and an accuracy of approximately 100 g. Then one tree was selected by size class, from which samples were taken from each compartment as follows:

For the stem: 2 to 4 cm thick slices were taken from each meter of the stem;

For the bark: a sample of a few hundred grams was taken;

For dead branches: a sample of a few hundred grams was taken;

For living branches: a sample of a few hundred grams was taken;

For foliage and twigs: a sample of a few hundred grams was taken.

These samples were weighed on site and stored in cool conditions in tightly sealed plastic bags before further analysis in the laboratory. Figure 2 illustrates the main field operations

Figure 2. Main operations in the field

Laboratory Work

Moisture Content Evaluation

Once at the laboratory, the samples were oven-dried at a temperature of 65°C until they reached a constant weight. The temperature of 65°C is the standard value used in several similar studies and allows the results to be compared with those of other studies. After drying, the dry mass of the samples was weighed to calculate their moisture content. This was expressed as a percentage and determined as the difference between the fresh mass and the dry mass, and reported to the fresh mass.

Calculation of Dry Biomass of Tree Compartments

Based on the moisture content of a sample, we are able to obtain the dry mass of the compartment from which the sample originates. This is obtained by multiplying the fresh biomass of this compartment by a conversion coefficient corresponding to one minus the moisture content of the corresponding sample.

Measurement of Carbon Content

The carbon content of the sample trees was assessed using the ash method (Allen et al., 1986). This involves taking subsamples from the dried samples, which are then ground into fine pieces for calcination. then, a few grams from each compartment are placed in an uncovered porcelain tare and placed in a muffle furnace set at 600°C until calcination is complete (4 hours). After calcination, samples are removed and cooled for subsequent weighing. After cooling, the tare with the ashes is weighed, and the percentage of carbon is determined according to the following equations:

$$A = \frac{P3 - P1}{P2 - P1} \times 100$$

$$C = (100 - A) \times 0.58$$

Where A is the ashes content (%); P1 is the weight of the porcelain tare (g); P2 is the weight of the crushed samples with the porcelain tare (g); P3 is the weight of the ashes with the porcelain tare(g); C is carbon content (%); and *0.58 is the average carbon content in the organic matter*;

The carbon stock in each compartment is then calculated by multiplying the dry biomass of that compartment by its carbon content, and the total carbon mass is obtained by summing up each compartment.

Soil and Litter Analysis

For soil samples, bulk density, defined as the ratio between the oven-dried soil mass (after drying at 105 °C for 24 hours) and its fresh volume, was determined (Hao et al., 2008). Soil organic carbon content was then measured following Walkley and Black (1934), which is based on wet oxidation with chromic acid. The organic carbon stock within a given horizon (SOC_i) was calculated using the following formula:

SOC_i = Da × E × C (Ellert et al., 2007)

where SOC_i is the organic carbon stock in horizon i (t ha^{-1}); C is the organic carbon concentration in that horizon (%); Da is the bulk density (g cm^{-3}); and E is the effective thickness of the horizon (cm).

The total soil organic carbon stock down to a depth of 1 m was obtained by summing the SOC stocks of all horizons. Litter samples underwent the same analytical procedures as those applied to the woody biomass samples.

Models Fitting and Validation

The models were adjusted individually for each tree compartment (stem, bark, branches, and foliage), as well as for the entire tree. The various equations were adjusted using the ordinary least squares method. This approach was chosen because of its robustness, ease of application, and widespread recognition in the scientific literature. The functions tested were selected from among those most commonly used in the literature (Picard et al., 2012; Ruiz-Peinado et al., 2011; Parsapour et al., 2013). This ensures both the biological relevance of the models and their comparability with the results of previous studies. Depending on the nature of the relationships to be modeled, two functions of the R language were used: the lm() function for linear models and the nls() function for nonlinear models.

After adjustment, only models with coefficients significant at the 5% threshold were considered. In addition, the models with the lowest Akaike information criterion (AIC) and residual standard error (RSE), and the highest values of coefficient of determination (R2) were selected as the best fitted models. In addition, the main properties of the residuals (normality, homoscedasticity, independence) were assessed using the Shapiro-Wilk normality test, the Breusch-Pagan test, and the Durbin-Watson test, respectively. Finally, due to the limited number of observations (20), we opted for a leave-one-out (LOO) cross-validation of the selected models. The means and standard deviations of the Root Mean Square Error (RMSE) were evaluated as final validation criteria.

RESULTS

Evaluation of Fitted Models

The performance of the models developed to estimate the above-ground carbon stock of the five Eucalyptus clones is presented in Table 2. The coefficient of determination (R^2) ranged from 0.89 to 0.98, indicating a strong correlation between above-ground carbon stock and the predictor variables, namely diameter at breast height (DBH) and total height (Ht). Clone 53 exhibited the best fit ($R^2 = 0.98$), followed by clones 397, 583 and 3933 ($R^2 = 0.97$). In contrast, clone 579 displayed the lowest R^2 value (0.89), suggesting a more limited predictive capacity.

Furthermore, validation criteria confirm the robustness of the best-fitted models. Root mean square error (RMSE) values varied from 0.72 to 1.44, with the lowest errors observed for clones 583 (0.72) and 3933 (0.84). Clone 579 recorded the highest RMSE (1.44), indicating comparatively lower predictive precision. Finally, models that incorporated both DBH and Ht generally performed better than those relying solely on DBH. The P-values for the statistical assumption tests are given in Table 5 in Appendix B.

Table 2. Selected models for estimating aboveground carbon stocks in each clone and their performances

Clone	Model	Adjustment criteria		Validation criteria	
		R^2	AIC	RMSE	Sd RMSE
Clone 397	$AGC = 0.101*(DBH)^{2.16}$	0,97	49,93	0,85	0,77
Clone 53	$AGC = 0.2643799*DBH + 0.01285*(DBH^2 *Ht)$	0,98	60,51	1,08	0,66
Clone 583	$AGC = 0,02919*(DBH^2*Ht^{0,649})$	0,97	141,11	0,72	0,72
Clone 3933	$AGC = 0,03726*(DBH^2 *Ht^{0,5374})$	0,97	113,32	0,84	0,88
Clone 579	$AGC = 0,10734* DBH^{2,0757}$	0,89	154,96	1,44	1,36

Model performance was evaluated using the coefficient of determination (**R^2**), Akaike information criterion (**AIC**), root mean square error (**RMSE**), and standard deviation of RMSE (**Sd RMSE**). **DBH** = diameter at breast height; **Ht** = tree total height; AGC = Tree aboveground carbon stock

Carbon Content in Dry Biomass

Regarding carbon content in dry biomass, the stem compartment exhibited the highest concentrations among all tree components across the studied clones, with values ranging from 55.42 ± 1.77% in Clone 583 at three years to 57.41 ± 0.04% in Clone 3933 at five years (Figure 3). In contrast, the bark compartment showed the lowest carbon contents, varying from 50.56 ± 0.79% in Clone 3933 at five years to 52.99 ± 2.89% in Clone 397 at six years. Branches displayed intermediate carbon concentrations, ranging from 52.86 ± 0.53% in Clone 583 at three years to 54.99 ± 0.18% in the same clone at five years. Foliage carbon content was comparatively stable across clones and ages, with values ranging from 53.35 ± 1.65% in Clone 583 at seven years to 54.88 ± 0.12% in Clone 3933 at five years.

Figure 3. Carbon content variation among tree compartments of Eucalyptus clones at different ages

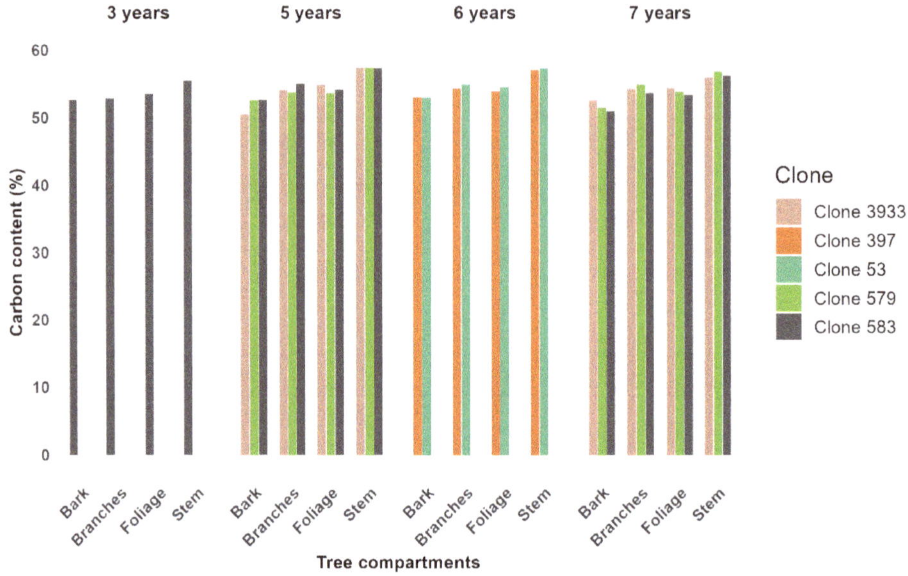

A two-way analysis of variance (ANOVA) was performed to evaluate the effects of clone and biomass compartment on carbon content, as well as their potential interaction. Separate analyses were conducted for six-year-old clones (Clone 53 and Clone 397) and seven-year-old clones (Clone 3933, Clone 583, and Clone 579). For both age groups, the ANOVA results indicated no significant differences in carbon content among clones, whereas a significant effect of biomass compartment was consistently observed. In the six-year-old clones, only the stem compartment differed significantly from the other compartments. In contrast, for seven-year-old clones, three distinct homogeneous groups were identified, corresponding to stems, foliage–branches, and bark (Figure 4).

With respect to the effect of age, only two age classes (5 and 7 years) were available for comparison; therefore, a Welch two-sample t-test was applied. The results showed no statistically significant difference in carbon content at five and seven years old ($p = 0.43$). Mean carbon content was 54.44% at 5 years and 54.06% at 7 years, suggesting that carbon concentration remains relatively stable across these ages.

Figure 4. Homogeneous groups of tree compartments in seven-year-old clones identified by Tukey's HSD test

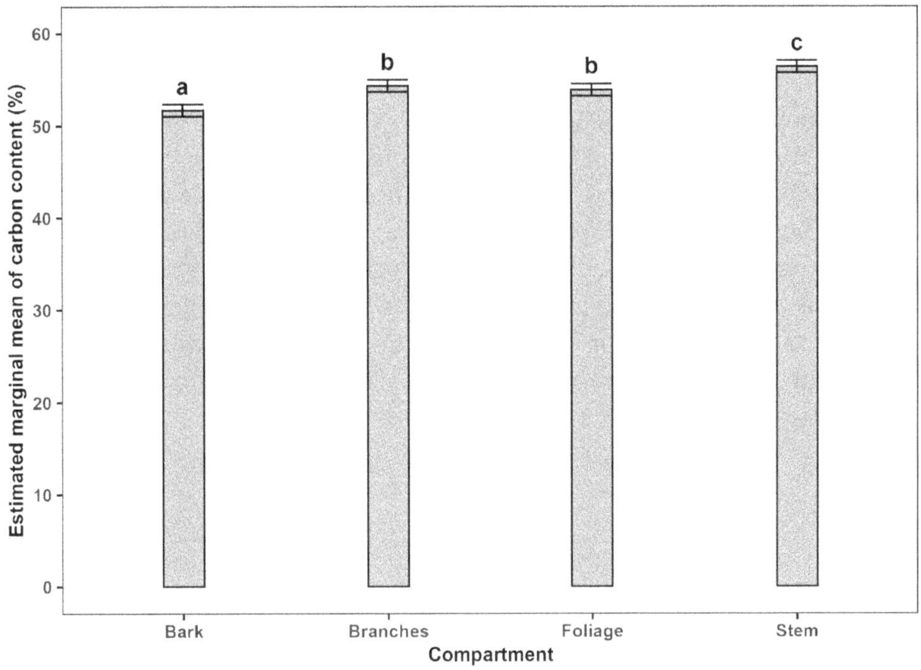

Total Aboveground and Belowground Carbon Stocks

To extrapolate carbon stocks to the stand level, we used the clone-specific allometric carbon models developed in this study. For each clone, the mean DBH and mean height of sample trees within each size class were used as inputs to the allometric model to estimate the average carbon content per tree for that class. Stand-level carbon stock was then calculated by multiplying these class-specific carbon estimates by the corresponding number of trees per hectare and summing the values across all classes to obtain total aboveground carbon per hectare. Belowground carbon stocks were estimated using a root-to-shoot ratio of 0.28, as recommended in the IPCC Guidelines (IPCC, 2019).

A synthesis of above- and belowground carbon stocks is presented in Table 3. Overall, total aboveground carbon stocks were primarily influenced by planting density, stand age, and genotypic performance. Indeed, while the mean carbon fraction in biomass remained relatively stable across, total aboveground carbon stocks showed substantial variability, with a marked positive relationship between stand age and carbon accumulation. For example, aboveground carbon stocks in Clone 583 increased from 5.49 tC·ha^{-1} at age 3 to 13.38 tC·ha^{-1} at age 7. Planting density further accentuated this effect, particularly in the 7-year-old Clone 3933, which exhibited the highest aboveground (32.86 tC·ha^{-1}) and belowground (9.20 tC·ha^{-1}) carbon stocks, likely attributable to its notably high stand density of 2,806 trees/ha compared to the other stands.

In addition, inter-clonal differences were also recorded, suggesting the role of genetics in carbon sequestration potential. For instance, at six year and under comparable stand densities, Clone 53 sequestered nearly twice as much aboveground carbon as Clone 397 (15.01 tC·ha^{-1} vs. 7.49 tC·ha^{-1}), indicating that genotype selection is a key factor in optimizing carbon storage within these eucalyptus plantations. However, as the belowground carbon estimates were derived using a root-to-shoot ratio, their variability is intrinsically linked to that of aboveground carbon stocks. Consequently, it is difficult to attribute these variations directly to the actual root carbon sequestration capacity of the clones.

Table 3. Above- and belowground carbon stocks of eucalyptus clones

Clone	Stand Density	Stand Age (years)	Mean carbon Fraction in biomass (%)	Aboveground Carbon Stock (tC·ha−1)	Belowground carbon Stock (tC·ha−1)
Clone 583	1768	3	53.60±1.27	5.496	1.53888
Clone 583	2091	5	54.74±1.98	8.07	2.2596
Clone 583	1565	7	53.53±2.15	13.383	3.74724
Clone 3933	630	5	54.24±2.82	3.865	1.0822
Clone 3933	2806	7	54.33±1.37	32.867	9.20276
Clone 579	1785	5	54.34±2.10	14.516	4.06448
Clone 579	1887	7	54.31±2.26	18.344	5.13632
Clone 397	748	6	54.56±1.73	7.488	2.09664
Clone 53	732	6	54.96±1.80	15.014	4.20392

Comparison of Carbon Sequestration Performance Among Clones

To enable a meaningful comparison among clones, which originally exhibited different stand densities, a constant stand density was applied. We used the average stand density across all plantations (1557 trees/ha) to minimize the confounding effect of density on aboveground carbon stocks expressed per hectare. Accordingly, the total aboveground carbon stock for each clone was recalculated by multiplying the average aboveground carbon mass of the sample trees by this constant density value of 1557 trees/ha.

This approach minimized the influence of stand density and isolated genotype and stand age as the most drivers of sequestration efficiency, revealing striking disparities in clonal performance (Figure 5). The results highlight Clone 53 as the most productive genotype for carbon sequestration, achieving a maximum carbon stock of approximately 32 tC·ha^{-1} at six years, which is more than double the stock of Clone 397 (14.21 tC·ha^{-1}) at the same age, and even surpasses that of other more mature clones. Temporal trends further distinguish growth strategies among the clones. Indeed, while Clone 579 exhibits early dominance with 16.38 tC·ha^{-1} at five years, its stock plateaued by age seven (17.13 tC·ha^{-1}), allowing Clone 3933 to overtake it through a vigorous late-stage surge that saw its stocks double from ~10 tC·ha^{-1} to over 20 tC·ha^{-1} between five and seven years. Conversely, Clone 583 demonstrates the lowest carbon sequestration potential under these conditions, maintaining the lowest stocks across the timeline (from 7.69 tC·ha^{-1} at three years to only 14.32 tC·ha^{-1} at seven years).

Figure 5. Total aboveground carbon stocks variation between clones based on a standardized stand density

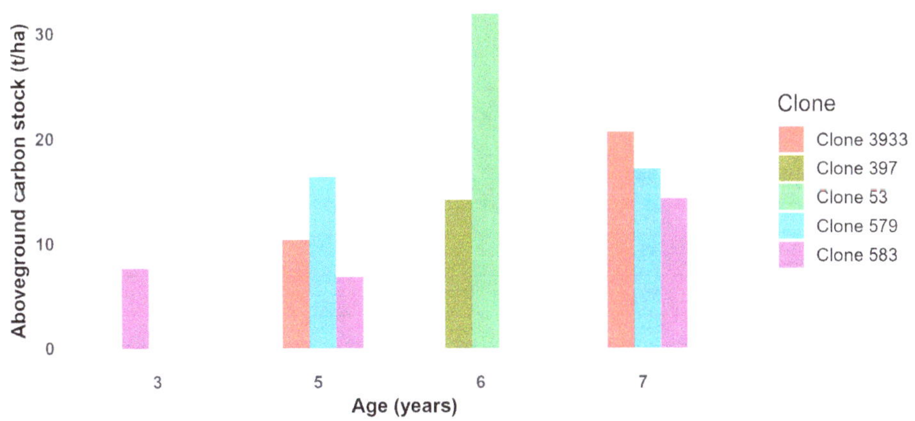

Soil and Litter Carbon Stocks

The analysis of results reveals pronounced heterogeneity in carbon stocks among the different clones, particularly within the soil compartment (Table 4). Clone 3933 stands out significantly, exhibiting the highest soil carbon stock at seven years of age, reaching 74.7 $tC{\cdot}ha^{-1}$, substantially exceeding those of the other clones. In contrast, Clones 53 and 397 show the lowest soil organic carbon stocks at six years, with 37.7 $tC{\cdot}ha^{-1}$ and 38.1 $tC{\cdot}ha^{-1}$, respectively.

When examining the chronosequence, a clear trend of carbon accumulation in both soil and litter is observed with increasing stand age for the studied clones (notably Clones 583, 579, and 3933). The evolution of soil carbon stock is particularly pronounced in Clone 3933, where it increases from 58.2 $tC{\cdot}ha^{-1}$ at five years to 74.7 $tC{\cdot}ha^{-1}$ at seven years, an increase of more than 28% in just two years.

Litter carbon stock follows a similar temporal progression within each clone, although slight interclonal variations are evident. For instance, at six years, Clone 53 exhibits a remarkably low litter carbon stock (0.36 $tC{\cdot}ha^{-1}$) compared with Clone 397 (3.86 $tC{\cdot}ha^{-1}$). This observation strongly suggests substantial differences among clones in litter quality and decomposition rates.

Table 4. Soil and litter carbon stocks for each clone

Clone	Stand Age	Soil bulk density (g/cm^3)	Soil carbon stock ($tC{\cdot}ha^{-1}$)	Litter carbon stock ($tC{\cdot}ha^{-1}$)
Clone 583	3	1.42±0.013	40.7± 2.61	2.57± 1.03
Clone 583	5	1.3±0.09	41.1± 4.27	3.48± 1.71
Clone 583	7	1.44±0.05	46.4± 2.75	3.98± 2.06
Clone 3933	5	1.48±0.21	58.2± 7.07	2.81± 2.32
Clone 3933	7	1.47±0.08	74.7± 4.03	3.77± 1.86
Clone 579	5	1.34±0.07	41± 2.08	1.33± 0.31
Clone 579	7	1.31±0.11	44.6± 2.37	3.67± 1.49
Clone 397	6	1.55±0.06	38.1± 2.57	3.86± 1.21
Clone 53	6	1.59±0.011	37.7± 4.25	0.36± 0.13

Potential Effect of Clone and Age on Soil Organic Carbon Stock

We assessed whether significant differences existed in soil organic carbon (SOC) stocks across clones and age classes. As the assumptions of ANOVA were not met, the non-parametric Kruskal–Wallis test was applied.

The results of the Kruskal–Wallis test revealed no statistically significant differences in soil carbon stock either among age classes ($\chi^2 = 6.16$, df = 3, p = 0.10) or among the studied clones ($\chi^2 = 6.60$, df = 4, p = 0.16). This suggests that, under the conditions of this study, neither age nor clone exerted a significant influence on soil carbon stocks. However, a general trend of increasing soil carbon with stand age was observed, possibly reflecting a gradual effect associated with biomass accumulation and organic matter enrichment over time. The lack of statistical significance could be attributed to the limited sample size per clone and age class, which may have reduced the statistical power of the tests. These exploratory findings highlight the need for further research with larger sample sizes to better characterize the combined effects of genetic material and stand age on soil carbon dynamics.

DISCUSSION

Through a random sampling and breakdown of sample trees by size class, we ensured adequate representativeness of all tree sizes. The biomass and carbon content of the sampled trees were determined using a direct and destructive approach prior to the adjustment of allometric models. Diameter at Breast Height (DBH) alone provides a reliable predictor of total carbon content for all clones, although the inclusion of total height (Ht) contributes to improved predictive accuracy. This is common in the literature, with several other studies (Ruiz-Peinado et al., 2011; Belghazi et al., 2016) highlighting the importance of diameter at breast height and total tree height as very good explanatory variables for biomass production. António et al. (2007) also concluded that total height contributes to greater accuracy in biomass models than the use of diameter alone for *Eucalyptus globulus* Labill. The models for the branch and foliage compartments were the least accurate. This indicates that the carbon content of these two compartments is poorly explained by the DBH and height variables, mainly due to the existence of high variability in the biomass of these two compartments. The same observation has been made in other Eucalyptus species (António et al., 2007; De Oliveira et al., 2021). This variability depends on the species and is influenced by several environmental factors such as temperature, relative humidity, and soil water content (António et al., 2007; Gomat et al., 2024).

Carbon Allocation Among Tree Compartments

Quantifying the carbon content of different tree compartments is essential for accurately estimating aboveground carbon stocks. Our results indicate no significant variation in carbon content among clones, suggesting a relatively stable carbon concentration in biomass. This is consistent with previous studies showing that carbon content tends to be conservative within species and even among species belonging to the same genus (Thomas & Martin, 2012; IPCC, 2019). However, there was a significant variability in carbon content across tree compartments, highlighting a potential genetic variability in carbon allocation strategies. In agreement with others studies (Thomas & Martin, 2012; Zhang et al., 2022; Gomat et al., 2024), the stem compartment exhibits the highest carbon contents among all compartments and across all clones. This predominance likely reflects the high lignin and cellulose content of wood, carbon-rich and relatively stable polymers. Conversely, the bark compartment shows the lowest carbon concentrations, consistent with its composition, which typically includes higher proportions of extractives, water, and ash, thereby reducing its carbon percentage (Brennan et al., 2020).

The allocation of carbon among tree compartments has been widely studied. Some studies have shown that the proportion of carbon allocated to stems generally increases with age, whereas the proportion allocated to leaves and branches tends to decrease (Zhang et al., 2014; Gomat et al., 2024). In the present study, the absence of a significant age effect on carbon content may be attributable to the small sample size and the limited age classes considered (5 and 7 years). Moreover, carbon allocation patterns can vary with genotype and environmental conditions. For instance, in water-limited regions, trees often allocate a larger fraction of carbon to belowground compartments (Basílio et al., 2024).

Above- and Below-Ground Carbon Stocks in Young Clonal Eucalyptus Plantations

Forests, particularly those with fast-growing species, play a central role in climate change mitigation due to their high capacity for rapid carbon sequestration. Our results reveal strong interclonal variation in above-ground carbon stocks. Although these estimates assume equal stand density, observed differences in actual density may influence individual tree growth (Schwerz et al., 2020; Hakamada et al., 2020). Additionally, this variability likely reflects intrinsic clonal traits, including growth rate, resource-use efficiency, and canopy architecture (Binkley et al., 2017).

To contextualize the performance of the Moroccan Eucalyptus clones, it is essential to position their above-ground carbon stocks within a broader global framework. The highest-performing clone in this study (32 $tC{\cdot}ha^{-1}$ at six years) remains considerably below the values typically observed in humid tropical and subtropical regions. For instance, in Brazil, Viera and Rodríguez-Soalleiro (2019) reported 118.45 $tC{\cdot}ha^{-1}$ in 10-year-old Eucalyptus hybrid clones, while Ribeiro et al. (2015) estimated a total stand carbon stock of 73.38 $tC{\cdot}ha^{-1}$ in clonal *Eucalyptus grandis* plantations in southeastern Brazil. Comparable high values have also been documented in Congo, where above-ground carbon stocks in Eucalyptus plantations commonly range from 70 $tC{\cdot}ha^{-1}$ to 110–125 $tC{\cdot}ha^{-1}$ (Saint-André et al., 2010; Kasekete et al., 2022). Across China and India, above-ground carbon stocks in Eucalyptus plantations range from 3.1 tC·ha^{-1} at 1 year old to 70.1 tC·ha^{-1} at 6 years old, and can reach up to 254.50 tC·ha^{-1} depending on site conditions (Chavan and Rasal, 2011; Du et al., 2015). The substantial discrepancy between these values and those observed in the present study can be partly attributed to differences in stand age, planting density, silvicultural practices, and especially climatic conditions (Schwerz et al., 2020; Navarro-Cerrillo et al.,2022; Binkley et al.,2017).

However, when compared with other plantations established under similar semi-arid or Mediterranean conditions, the Moroccan clones perform within a comparable range. In India, Ram et al. (2011) estimated 15.5 tC·ha^{-1} of total carbon at 5 years and 4 months in *Eucalyptus tereticornis*. Similarly, Jana et al. (2009) reported relatively low above-ground carbon stocks in six-year-old plantations: 7.97 tC·ha^{-1} for *Tectona grandis*, 6.26 tC·ha^{-1} for *Albizzia lebbek*, 5.22 tC·ha^{-1} for *Shorea robusta*, and 7.28 tC·ha^{-1} for *Artocarpus integrifolia*. Comparable values were observed in the Mediterranean Basin, where above-ground carbon stocks in Eucalyptus plantations typically range from 13.09 tC·ha^{-1} to 15.8 tC·ha^{-1} for *E. globulus*, to 19–56 tC·ha^{-1} for *Eucalyptus nitens* (González-García et al., 2015; Alegria, 2023; Ferreira et al., 2025). These values fall within the range of our results, confirming that growth in Mediterranean and semi-arid environments is intrinsically lower than in tropical or subtropical zones.

The performance of the young Eucalyptus clones, particularly clone 53, which reached 32 tC·ha^{-1} at six years, illustrates a strong capacity for rapid early-stage carbon sequestration when compared with major Moroccan forest ecosystems. For Atlas cedar (*Cedrus atlantica*), above-ground ecosystem carbon stocks in the Azrou forest is estimated at 78.75 tC·ha^{-1} (El Mderssa et al., 2019), values achieved after several decades. The annual sequestration rates in *Cedrus atlantica* vary between 0.35 and 8.82 tC·ha^{-1}·year^{-1} and average around 5.48 tC·ha^{-1}·year^{-1} in optimal conditions (Boukhris et al., 2023). In contrast, clone 53, sequestering ~5.33 tC·ha^{-1}·year^{-1} over its first six years, therefore performs within the same order of magnitude as optimal cedar productivity. In Oak ecosystems, cork oak (*Quercus suber*) in Maâmora stores between 30.74 and 77.70 tC·ha^{-1} (Oubrahim et al., 2015; Sanaa et al., 2022), while holm oak (*Quercus ilex*) ranges from 51 to 91 tC·ha^{-1} in the Middle Atlas and 17–40 tC·ha^{-1} in the High Atlas (Boulmane et al., 2014, 2015), with slightly higher values reported in Tunisia (82.52 to 151.96 tC·ha^{-1}; Zribi et al., 2016). While clone 53's value (32 tC·ha^{-1}) lies within the lower range of these mature oak systems, its sequestration rate is markedly higher, as oak forests typically accumulate only 0.66–1.78 tC·ha ^{1}·year ^{1}, depending on age and declining over time (Zaher et al., 2025).

While carbon stock data for Moroccan pine species remain limited, studies from comparable Mediterranean environments report substantial above-ground carbon stocks, ranging from 90.96–137.46 tC·ha^{-1} for *Pinus pinaster* (Ruiz-Peinado et al.,

2013); 35.95–68.45 $tC{\cdot}ha^{-1}$ for *Pinus halepensis*; and 47.39–74.65 $tC{\cdot}ha^{-1}$ for *Pinus nigra* (Navarro-Cerrillo et al., 2022), depending on stand age and density. Taken together, these comparisons emphasize that while mature conifer and oak forests accumulate large carbon stocks over longer timeframes, young Eucalyptus clones offer significantly higher short-term sequestration rates, underscoring their strategic relevance for rapid carbon offset initiatives. Therefore, our results for young clonal plantations are promising, suggesting that the potential of the most productive clones could be significantly higher at older ages under optimal management conditions. Nevertheless, additional research is required to determine whether the observed differences in carbon stocks among clones reflect inherent genetic variability or are primarily driven by microsite effects. This assessment would benefit from studies conducted on the same clone under uniform site and management conditions.

Regarding below-ground Carbon stocks (BGC), applying a root-to-shoot ratio of 0.28 for young clonal plantations reveals that below-ground represents a substantial carbon pool. For instance, including BGC would increase the total biomass carbon stock (AGC + BGC) of the Clone 3933 from 32.86 $tC{\cdot}ha^{-1}$ to 42.07 $tC{\cdot}ha^{-1}$. This underscores that relying solely on above-ground estimates underestimates the total mitigation potential of these plantations by approximately 20% to 25%, particularly in water-limited environments where trees may allocate more biomass to roots to access deep soil moisture (Basílio et al., 2024). However, the use of a generic root-to-shoot ratio introduces uncertainty, as it fails to capture genotypic variations and site-specific soil constraints. To improve the precision of future carbon accounting in these plantations, it is recommended to move beyond fixed ratios. In this regard, while destructive sampling (excavation) remains the most accurate approach, it is labor-intensive and costly; the use of Ground-Penetrating Radar (GPR) would be a promising alternative for future research. Some recent studies highlight GPR as a non-destructive, cost-effective tool capable of mapping coarse root architecture and estimating biomass with increasing accuracy in plantation settings (Rocha et al., 2024; Zhang et al., 2024). Therefore, integrating GPR surveys with destructive validation would allow for the development of site-specific allometric models for roots, significantly reducing errors in total carbon stock estimates.

Soil and Litter Carbon Dynamics

The results of this study reaffirm the central role of forest soils as major reservoirs of organic carbon. Across all clones, soil stored the largest proportion of carbon per hectare relative to above-ground biomass, demonstrating that in clonal Eucalyptus systems, total ecosystem carbon is primarily governed by soil organic carbon (SOC). The observed differences in soil and litter carbon stocks among clones highlight the influence of genetic traits, stand density, and age on belowground carbon dynamics.

Clone 3933 with the highest stand density also exhibited the highest SOC value at seven years (74.7 $tC{\cdot}ha^{-1}$), suggesting the capacity of stand density to enhance soil carbon accumulation. Furthermore, beyond bulk density, which was the only soil physicochemical variable quantified in this study along with SOC, numerous other physicochemical properties are known to regulate SOC dynamics and may explain the observed SOC variability. For instance, soil nitrogen and phosphorus availability strongly influence microbial decomposition processes and litter mineralization rates (Manzoni et al., 2010).

Likewise, soil structure and aggregation play a key role in SOC stabilization, with well-aggregated soils enhancing carbon stabilization (Basile-Doelsch et al., 2020). Moreover, soil mineralogical properties such as the cation exchange capacity (CEC), closely linked to clay mineralogy and soil colloids, further contribute to SOC persistence by promoting organo-mineral associations that chemically protect carbon molecules (Liu et al., 2025). Differences in litter and soil carbon stocks among clones likely reflect variations in litter quality, particularly in terms of lignin content, nutrient concentrations, and C/N ratios, which are well-known regulators of litter decomposition rates and SOC dynamics.

A general trend of increasing soil and litter carbon stocks with stand age, particularly evident in Clones 583, 579, and 3933, further supports age-driven carbon accumulation processes. Such trends are widely recognized in forest ecosystems, as maturing stands produce larger biomass inputs, root exudates, and litterfall, thereby contributing to progressively greater SOC pools. The ~28% increase in SOC observed in Clone 3933 between five and seven years underscores the capacity of Eucalyptus plantations to enhance soil carbon over time. Although Kruskal-Wallis statistical tests did not reveal significant differences in soil carbon stock between ages or clones, an upward trend with age was clearly observed. The lack of statistical significance may be attributable to the limited sample size and temporal resolution. Thus, Long-term monitoring combined with expanded soil characterization, including nutrient content, microbial activity, clay fraction, CEC, and soil structural attributes, would enable more robust assessments of how genetic material and stand age interact to shape SOC dynamics in these plantations.

Study Implications and Limitations

This study underscores the role of clonal Eucalyptus plantations in climate change mitigation through carbon sequestration. Substantial interclonal variability in carbon allocation and in above-ground and soil carbon stocks indicates strong potential for optimizing plantation performance.

Policy Implications for National Carbon Accounting and NDC Targets

The findings of this study offer strategic guidance for forest managers and policymakers responsible for operationalizing Morocco's Nationally Determined Contributions (NDCs) under the Paris Agreement. As the country advances its low-carbon development agenda, fast-growing Eucalyptus plantations appear as a promising lever for accelerating carbon removal. The marked interclonal variability observed (32 $tC \cdot ha^{-1}$ for Clone 53 and 14.2 $tC \cdot ha^{-1}$ for Clone 397, both at six years and with similar stand density) highlights the importance of integrating genetic selection into afforestation and reforestation programs. Prioritizing genotypes with high resource-use efficiency and rapid biomass accumulation could significantly enhance short- to medium-term mitigation outcomes, especially given that the estimated sequestration rate of the top-performing clone (~5.33 $tC \cdot ha^{-1} \cdot year^{-1}$) exceeds those reported for slower-growing native species such as oaks.

From a carbon-accounting standpoint, the results reveal important shortcomings in current national greenhouse gas inventories. The widespread use of default IPCC biomass conversion factors (0.47–0.50) may underestimate carbon stocks, as the carbon fractions measured in this study (0.54–0.57) exceed these default values. Updating these parameters in Morocco's national inventories would therefore improve the precision of carbon balance estimates. Moreover, transitioning toward site- and clone-specific carbon mass equations as recommended by the IPCC Guidelines (IPCC, 2019) would reduce reporting uncertainty and better reflect the true contribution of eucalyptus plantations. Such refinement would also strengthen the valuation of carbon assets within both international and domestic carbon markets, thereby enhancing Morocco's ability to mobilize results-based climate finance through mechanisms such as the UNFCCC and the Green Climate Fund.

Finally, the results support the integration of a differentiated, "two-speed" forestry strategy within national climate policy. Indeed, while native ecosystems such as Atlas cedar and oak forests provide irreplaceable biodiversity and long-term carbon storage (El Mderssa et al., 2019), their slow growth limits their potential contribution to near-term climate change mitigation targets. In parallel, high-performing Eucalyptus clones can serve as rapid-response carbon sinks, particularly on marginal or degraded lands where accelerated biomass recovery is desirable. This complementary approach would enable Morocco to simultaneously meet short-term mitigation needs and uphold long-term ecological commitments, ensuring that gains from Eucalyptus plantations reinforce the conservation and restoration of native forest ecosystems.

Effect of Management Practices on Carbon Storage and Management Perspectives

Silvicultural regimes, particularly planting density, site preparation intensity, and rotation length, strongly shape carbon sequestration dynamics in both above- and below-ground pools. Indeed, high initial planting densities generally promote rapid canopy closure and early stand-level biomass accumulation, enhancing short-term carbon storage. However, excessive density intensifies competition for light, water, and nutrients, often requiring early thinning in longer rotations (Hakamada et al., 2020; Gabira et al., 2023). Clonal Eucalyptus trials by Hakamada et al. (2020) further demonstrate density-dependent effects on transpiration and water-use efficiency, with genotype-specific responses linked to drought tolerance. Although reduced spacing can increase above- and below-ground carbon stocks (Schwerz et al., 2020), sustained growth requires regular monitoring and timely thinning to prevent stagnation.

Site preparation practices also influence carbon outcomes. For instance, deep tillage or subsoiling can enhance early stand development by alleviating soil compaction and improving water infiltration, but such disturbances may trigger short-term SOC losses through oxidation (Navarro-Rosales et al., 2023). In addition, repeated short rotations may disrupt soil nutrient balance if nutrient exports are not offset by fertilization or the retention of organic residues (Dai et al., 2023).

This underscores the need for carbon-oriented management of Eucalyptus plantations, beginning with comprehensive soil and site diagnostics (texture, pH, organic matter, nutrient status, and land-use history) that determine baseline SOC and early growth potential. Planting density should be adjusted to site water availability and clonal drought tolerance, with soil-guided fertilization and continuous nutrient monitoring to support adaptive management and reduce divergence in clonal performance. In coppiced systems, stool vigor and sprouting capacity should be evaluated at the clonal level, while rotation length should be tailored to site productivity: although short rotations maximize annual uptake, extending rotations beyond traditional harvest ages (9–12 years) can enhance aboveground biomass and SOC stocks while reducing harvest-related emissions.

Finally, given the dominant contribution of soils to ecosystem carbon storage, conserving litter and harvest residues is essential for long-term SOC formation. Allocating harvested biomass to long-lived wood products rather than short-lived uses or bioenergy can further extend carbon residence time and maximize the climate mitigation potential of Eucalyptus plantations.

Methodological Limitations and Uncertainties

Despite the robustness of the sampling framework, several limitations should be acknowledged. First, the destructive sampling requirement constrained the number of harvestable trees to 20 per clone, which may limit the statistical power and the representativeness of the diameter and height ranges used for fitting the clone-specific allometric equations. Indeed, small sample sizes tend to increase parameter uncertainty, widen confidence intervals, and reduce the stability of nonlinear biomass models, particularly in clones exhibiting large within-stand variability or strong density-mediated growth patterns (Picard et al., 2015). As a result, model fitting uncertainty should be considered when interpreting both individual-tree predictions and their extrapolation to the stand level. In addition, although clone-specific allometric models substantially improve the accuracy of per-tree carbon estimates, the marked differences in stand density among clones may introduce uncertainty when comparing aboveground carbon stocks between clones.

Moreover, belowground carbon stocks were estimated using a root-to-shoot ratio, a common approach in forest carbon accounting. However, this method introduces additional uncertainty because root allocation can vary substantially with tree age, soil properties, genetic differences among clones, and stand density. Consequently, the application of a generalized ratio may oversimplify clone-specific belowground carbon stocks and should be interpreted with caution. Finally, the limited number of soil and litter samples per clone may reduce the representativeness of these measurements and underestimate the variability of soil organic carbon and litter inputs. Taken together, these limitations highlight the need for careful interpretation of statistical outcomes, including non-significant differences or apparent contrasts among clones.

CONCLUSION

The primary objective of this study was to assess the contribution of five Eucalyptus clones to climate change mitigation through their carbon sequestration capacity. This was achieved by directly measuring carbon content in different tree compartments, developing clone-specific allometric models to estimate aboveground carbon stocks, and quantifying carbon stored in litter and soil. Belowground carbon was estimated using a root-to-shoot ratio approach.

The findings show that the mean carbon content of dry biomass across clones (53.53 ± 2.15% to 54.96 ± 1.80%) exceeded the commonly applied generic values of 47–50%. Although differences among clones were not statistically significant, carbon fractions varied significantly among tree compartments, being consistently lower in bark and higher in stems. Aboveground carbon stocks differed markedly among clones and ages, ranging from 3.87 tC ha^{-1} in Clone 3933 at five years to 32.87 tC ha^{-1} at seven years, whereas litter carbon stocks varied from 0.36 to 3.90 tC ha^{-1}. Soil constituted the largest carbon pool (37.67–74.7 tC ha^{-1}), underscoring its dominant role in ecosystem carbon storage. Overall, these values fall within the ranges reported for Eucalyptus plantations globally.

Despite the robustness of the developed models, their applicability remains constrained to ecological conditions similar to those of the study area, and the relatively small sample size may limit statistical power. Nevertheless, the study provides an important basis for identifying clones with high carbon sequestration potential and informing plantation programs aimed at climate change mitigation. Future work should explore interannual and interclonal variability in greater depth, incorporate more precise estimates of belowground carbon, and establish long-term monitoring systems of carbon fluxes across biomass, soil, and litter. Such efforts are essential to better quantify the real contribution of Eucalyptus plantations to national carbon neutrality strategies.

REFERENCES

Alegria, C. (2023). Aboveground Biomass Mapping and Fire Potential Severity Assessment: A case study for eucalypts and shrubland areas in the central inland region of Portugal. *Forests*, *14*(9), 1795. DOI: 10.3390/f14091795

Allen, S. E., Grimshaw, H. M., & Rowland, A. P. (1986). Chemical Analysis. In Moore, P. D., & Chapman, S. B. (Eds.), *Methods of Plant Ecology* (pp. 285–344). Blackwell.

ANEF. (2025). Répartition des superficies reboisées. Agence Nationale des Eaux et Forêts. Retrieved November 12, 2025, from http://www.eauxetforets.gov.ma/DeveloppementForestier/Reboisement/Pages/Reboisement.aspx

António, N., Tomé, M., Tomé, J., Soares, P., & Fontes, L. (2007). Effect of tree, stand, and site variables on the allometry of *Eucalyptus globulus* tree biomass. *Canadian Journal of Forest Research*, *37*(5), 895–906. DOI: 10.1139/X06-276

Basile-Doelsch, I., Balesdent, J., & Pellerin, S. (2020). Reviews and syntheses: The mechanisms underlying carbon storage in soil. *Biogeosciences*, *17*(21), 5223–5242. DOI: 10.5194/bg-17-5223-2020

Basílio, J. J. N., Campoe, O. C., Queiroz, T. B., De Souza, C. R., Carneiro, R. L., Alvares, C. A., & Figura, M. A. (2024). Fine Root Density Dynamics and Carbon Stock of Eucalyptus spp.: Interplay of Age, Genotype, and Edaphoclimatic Conditions. *Plants*, *13*(11), 1503. DOI: 10.3390/plants13111503 PMID: 38891311

Belghazi, T., Ponette, Q., Jonard, M., & Belghazi, B. (2016). Biomasse ligneuse et foliaire des taillis d'arganier dans le plateau des Haha, Maroc. *Forêt Méditerranéenne*, *XXXVII*(2), 157–172.

Benabou, A., Moukrim, S., Lahssini, S., El Aboudi, A., Menzou, K., Elmalki, M., & Rhazi, L. (2022). Impact of climate change on potential distribution of Quercus suber in the conditions of North Africa. *Biosystems Diversity*, *30*(3), 289–294. DOI: 10.15421/012231

Binkley, D., Campoe, O. C., Alvares, C., Carneiro, R. L., Cegatta, Í., & Stape, J. L. (2017). The interactions of climate, spacing and genetics on clonal Eucalyptus plantations across Brazil and Uruguay. *Forest Ecology and Management*, *405*, 271–283. DOI: 10.1016/j.foreco.2017.09.050

Boukhris, I., Lahssini, S., Collalti, A., Moukrim, S., Santini, M., Chiti, T., & Valentini, R. (2023). Calibrating a Process-Based Model to Enhance Robustness in Carbon Sequestration Simulations: The Case of Cedrus atlantica (Endl.) Manetti ex Carrière. *Forests*, *14*(2), 401. DOI: 10.3390/f14020401

Boulmane, M., Oubrahim, H., Halim, M., Bakker, M. R., & Augusto, L. (2017). The potential of Eucalyptus plantations to restore degraded soils in semi-arid Morocco (NW Africa). *Annals of Forest Science*, *74*(3), 57. DOI: 10.1007/s13595-017-0652-z

Boulmane, M., Santa-Regina, I., Khia, A., & Oubrahim, H. (2014). Éstimation du stock de carbone organique dans l'écosystème des iliçaies du Moyen Atlas marocain. *Revue Nature et Technologie, 6*(2), 06-16. https://asjp.cerist.dz/en/article/43400

Boulmane, M., Santa-Regina, M., Halim, M., Khia, A., Oubrahim, H., Abbassi, H., & Santa-Regina, I. (2015). Organic Carbon Storage in Evergreen Oak Forest Ecosystems of the Middle and High Moroccan Atlas Areas. *Open Journal of Forestry*, *5*(3), 260–273. DOI: 10.4236/ojf.2015.53023

Brennan, M., Fritsch, C., Cosgun, S., Dumarcay, S., Colin, F., & Gérardin, P. (2020). Quantitative and qualitative composition of bark polyphenols changes longitudinally with bark maturity in Abies alba Mill. *Annals of Forest Science*, *77*(1), 9. DOI: 10.1007/s13595-019-0916-x

Chavan, B. L., & Rasal, G. B. (2011). Sequestered Carbon Potential and Status of Eucalyptus Tree. *International Journal of Applied Engineering and Technology*, *1*(1), 41–47.

Dai, Q., Wang, T., Wei, P., & Fu, Y. (2023). Effects of Successive Planting of Eucalyptus Plantations on Tree Growth and Soil Quality. *Sustainability (Basel)*, *15*(8), 6746. DOI: 10.3390/su15086746

De Oliveira, T. W. G., Rubilar, R., Sanquetta, C. R., Corte, A. P. D., & Behling, A. (2021). Simultaneous estimation as an alternative to young *eucalyptus* aboveground biomass modeling in ecophysiological experiments. *Acta Scientiarum. Agronomy*, *43*, e52126. DOI: 10.4025/actasciagron.v43i1.52126

Du, H., Zeng, F., Peng, W., Wang, K., Zhang, H., Liu, L., & Song, T. (2015). Carbon storage in a eucalyptus plantation chronosequence in southern China. *Forests*, *6*(6), 1763–1778. DOI: 10.3390/f6061763

El Mderssa, M. E., Belghazi, B., Benjelloun, H., Zennouhi, O., Nassiri, L., & Ibijbijen, J. (2019). Estimation of Carbon Sequestration; Using Allometric Equations; in Azrou Cedar Forests (Cedrus atlantica Manetti) in the Central Middle Atlas of Morocco under Climate Change. *Open Journal of Forestry*, *09*(03), 214–225. DOI: 10.4236/ojf.2019.93011

Ellert, B. H., Vanden Bygaart, A. J., & Bremer, E. (2007). Measuring Change in Soil Organic Carbon Storage. In Carter, M. R., & Gregorich, E. G. (Eds.), *Soil Sampling and Methods of Analysis* (pp. 49–62). CRC Press.

FAO, & IUSS. (2006). *World reference base for soil resources: a framework for international classification, correlation and communication*. FAO, Rome, Italia

Ferreira, L., De Andrade Sandim, A. S., Lopes, D. A., Sousa, J. J., Lopes, D. M. M., Silva, M. E. C. M., & Pádua, L. (2025). Estimating Biomass in Eucalyptus globulus and Pinus pinaster Forests Using UAV-Based LiDAR in Central and Northern Portugal. *Land (Basel)*, *14*(7), 1460. DOI: 10.3390/land14071460

Gabira, M. M., Girona, M. M., DesRochers, A., Kratz, D., Da Silva, R. B. G., Duarte, M. M., De Aguiar, N. S., & Wendling, I. (2023). The impact of planting density on forest monospecific plantations: An overview. *Forest Ecology and Management*, *534*, 120882. DOI: 10.1016/j.foreco.2023.120882

Gomat, H., Bengo, D. M., Ifo, S. A., Mankessi, F., Ossendza, J., Mabiala, P., Zeller, B., & Saint-Andre, L. (2024). What Is the Impact of Ploughing on Eucalyptus Plantations in Bambou Mingali (Plateaux Batéké, Republic of Congo)? *Agricultural Sciences*, *15*(10), 1132–1146. DOI: 10.4236/as.2024.1510061

González-García, M., Hevia, A., Majada, J., Rubiera, F., & Barrio-Anta, M. (2015). Nutritional, carbon and energy evaluation of Eucalyptus nitens short rotation bioenergy plantations in northwestern Spain. *iForest - Biogeosciences and Forestry*, *9*(2), 303–310. DOI: 10.3832/ifor1505-008

Hakamada, R. E., Hubbard, R. M., Moreira, G. G., Stape, J. L., Campoe, O., & De Barros Ferraz, S. F. (2020). Influence of stand density on growth and water use efficiency in Eucalyptus clones. *Forest Ecology and Management*, *466*, 118125. DOI: 10.1016/j.foreco.2020.118125

Hao, X. B. C., Ball, J. L. B., Culley, M. R., Carter, & Parkin, G. W. (2008). "Chapter 57 Soil Density and Porosity." In M. R. Carter & E. G. Gregorich (2nd Eds., *Soil Sampling and Methods of Analysis* (pp: 743–759). Boca Raton, FL: Taylor & Francis Group

Hirigoyen, A., Acuna, M., Rachid-Casnati, C., Franco, J., & Navarro-Cerrillo, R. (2021). Use of optimization modeling to assess the effect of timber and carbon pricing on harvest scheduling, carbon sequestration, and net present value of eucalyptus plantations. *Forests*, *12*(6), 651. DOI: 10.3390/f12060651

IPCC. (2019). *2019 Refinement to the 2006 IPCC Guidelines for National Greenhouse Gas Inventories. IPCC.* https://www.ipcc.ch/report/2019-refinement-to-the-2006-ipcc-guidelines-for-national-greenhouse-gas-inventories

IPCC. (2021). *Climate Change 2021: The Physical Science Basis. Contribution of Working Group I to the Sixth Assessment Report of the Intergovernmental Panel on Climate Change* (Masson-Delmotte, V., Zhai, P., Pirani, A., Connors, S. L., Péan, C., Berger, S., Caud, N., Chen, Y., Goldfarb, L., Gomis, M. I., Huang, M., Leitzell, K., Lonnoy, E., Matthews, J. B. R., Maycock, T. K., Waterfield, T., Yelekçi, O., Yu, R., & Zhou, B., Eds.). Cambridge University Press., DOI: 10.1017/9781009157896

Jana, B., Biswas, S., Majumder, M., Roy, P., & Mazumdar, A. (2009). Carbon Sequestration Rate and Aboveground Biomass Carbon Potential of Four Young Species. *Journal of Ecology and the Natural Environment*, *1*(2), 15–24.

Kasekete, D. K., Ligot, G., Mweru, J. M., Drouet, T., Rousseau, M., Moango, A., & Bourland, N. (2022). Growth, Productivity, Biomass and Carbon Stock in Eucalyptus saligna and Grevillea robusta Plantations in North Kivu, Democratic Republic of the Congo. *Forests*, *13*(9), 1508. DOI: 10.3390/f13091508

Leal Filho, W., Matandirotya, N. R., Lütz, J. M., Alemu, E. A., Brearley, F. Q., Baidoo, A. A., Kateka, A., Ogendi, G. M., Adane, G. B., Emiru, N., & Mbih, R. A. (2021). Impacts of climate change to African indigenous communities and examples of adaptation responses. *Nature Communications*, *12*(1), 6224. DOI: 10.1038/s41467-021-26540-0 PMID: 34711857

Liu, M., Zheng, S., Pendall, E., Smith, P., Liu, J., Li, J., Fang, C., Li, B., & Nie, M. (2025). Unprotected carbon dominates decadal soil carbon increase. *Nature Communications*, *16*(1), 2008. DOI: 10.1038/s41467-025-57354-z PMID: 40011482

Manzoni, S., Trofymow, J. A., Jackson, R. B., & Porporato, A. (2010). Stoichiometric controls on carbon, nitrogen, and phosphorus dynamics in decomposing litter. *Ecological Monographs*, *80*(1), 89–106. DOI: 10.1890/09-0179.1

Marien, J.-N., & Idrissi Azami, A. (2011). Le développement des plantations clonales d'eucalyptus au Maroc: Une dynamique toujours plus actuelle. *Forêt Méditerranéenne*, *32*(1), 31–38.

Morgan, R. B., Herrmann, V., Kunert, N., Bond-Lamberty, B., Muller-Landau, H. C., & Anderson-Teixeira, K. J. (2021). Global patterns of forest autotrophic carbon fluxes. *Global Change Biology*, *27*(12), 2840–2855. DOI: 10.1111/gcb.15574 PMID: 33651480

Moukrim, S., Lahssini, S., Rhazi, M., Menzou, K., El Madihi, M., Rifai, N., & Rhazi, L. (2022). Climate change impact on potential distribution of an endemic species Abies marocana Trabut. *Ekologia (Bratislava)*, *41*(4), 329–339. DOI: 10.2478/eko-2022-0034

Navarro-Cerrillo, R. M., Ruiz-Gómez, F. J., Camarero, J. J., Castillo, V., Barberá, G. G., Palacios-Rodríguez, G., Navarro, F. B., Blanco, J. A., Imbert, J. B., Cachinero-Vivar, A. M., Molina, A. J., & del Campo, A. D. (2022). Long-Term Carbon Sequestration in Pine Forests under Different Silvicultural and Climatic Regimes in Spain. *Forests*, *13*(3), 450. DOI: 10.3390/f13030450

Navarro-Rosales, F., Fernández-Habas, J., Reyna-Bowen, L., Gómez, J. A., & Fernández-Rebollo, P. (2023). Subsoiling for planting trees in dehesa system: Long-term effects on soil organic carbon. *Agroforestry Systems*, *97*(4), 699–710. DOI: 10.1007/s10457-023-00820-8

Oubrahim, H., Boulmane, M., Bakker, M., Augusto, L., & Halim, M. (2015). Carbon storage in degraded cork oak (Quercus suber) forests on flat lowlands in Morocco. *iForest - Biogeosciences and Forestry,* 9(1), 125-137. DOI: 10.3832/ifor1364-008

Pan, Y., Birdsey, R. A., Fang, J., Houghton, R., Kauppi, P. E., Kurz, W. A., Phillips, O. L., Shvidenko, A., Lewis, S. L., Canadell, J. G., Ciais, P., Jackson, R. B., Pacala, S. W., McGuire, A. D., Piao, S., Rautiainen, A., Sitch, S., & Hayes, D. (2011). A large and persistent carbon sink in the world's forests. *Science*, *333*(6045), 988–993. DOI: 10.1126/science.1201609 PMID: 21764754

Parsapour, M. K., Sohrabi, H., Soltani, A., & Iranmanesh, Y. (2013). Allometric equations for estimating biomass in four poplar species at Charmahal and Bakhtiari province. *Tahqiqat-i Jangal va Sinubar-i Iran*, *21*(3), 517–528. DOI: 10.22092/ijfpr.2014.4730

Picard, N., Boyemba-Bosela, F., & Rossi, V. (2015). Reducing the error in biomass estimates strongly depends on model selection. *Annals of Forest Science*, *72*(6), 6. Advance online publication. DOI: 10.1007/s13595-014-0434-9

Picard, N., Laurent, S. A., & Henry, M. (2012). *Manuel de construction d'équations allométriques pour l'estimation du volume et la biomasse des arbres: de la mesure de terrain à la prédiction*. CIRAD-FAO.

Ram, J. C., Dagar, K., Lal, G., Singh, V., Toky, V. S., Tanwar, S. R., & Dar, M.K., & Chauhan. (2011). Biodrainage to combat waterlogging, increase farm productivity and sequester carbon in canal command areas of northwest India. *Current Science, 100*(11), 1673–1680.

Ribeiro, S. C., Soares, C. P. B., Fehrmann, L., Jacovine, L. A. G., & Von Gadow, K. (2015). Aboveground and belowground biomass and carbon estimates for clonal *eucalyptus* trees in southeast brazil. *Revista Árvore, 39*(2), 353–363. DOI: 10.1590/0100-67622015000200015

Rocha, A. A., Borges, W. R., Von Huelsen, M. G., De Oliveira, E., Sousa, F. R., Maciel, S. T. R., De Almeida Rocha, J., & Jacobson, T. K. B. (2024). Imaging tree root systems using ground penetrating radar (GPR) data in Brazil. *Frontiers in Earth Science (Lausanne), 12*, 1353572. Advance online publication. DOI: 10.3389/feart.2024.1353572

Ruiz-Peinado, R., Bravo-Oviedo, A., López-Senespleda, E., Montero, G., & Río, M. (2013). Do thinnings influence biomass and soil carbon stocks in Mediterranean maritime pinewoods? *European Journal of Forest Research, 132*(2), 253–262. DOI: 10.1007/s10342-012-0672-z

Ruiz-Peinado, R., Del Rio, M., & Montero, G. (2011). New models for estimating the carbon sink capacity of Spanish softwood species. *Forest Systems, 20*(1), 1. Advance online publication. DOI: 10.5424/fs/2011201-11643

Saint -André. L., Vallet, P., Pignard, G., Dupouey, J.L., Colin, A., Loustau, D., Le Bas, C., Meredieu, C., Caraglio, Y., Porté, A., Hamza, N., Cazin, A., Nouvellon, Y., & Dhôte, J.F.(2010). Estimating carbon stocks in forest stands: 1. Methodological developments. In D. Loustau (ed.), *Forests, carbon cycle and climate change* (pp. 79-100). Versailles: Ed. Quae

Sanaa, F., Imane, S., Mohamed, B., Kenza, A. E. K., Souhail, K., Lfalah, H., & Khadija, M. (2022). Biomass and carbon stock quantification in Cork Oak Forest of Maamora Using a new approach based on the combination of aerial laser scanning carried by unmanned aerial vehicle and terrestrial laser scanning data. *Forests, 13*(8), 1211. DOI: 10.3390/f13081211

Schwerz, F., Neto, D. D., Caron, B. O., Tibolla, L. B., Sgarbossa, J., Eloy, E., Elli, E. F., & Carvalho, L. G. (2020). Carbon stocks, partitioning, and wood composition in short-rotation forestry system under reduced planting spacing. *Annals of Forest Science, 77*(3), 67. DOI: 10.1007/s13595-020-00974-w

Thomas, S. C., & Martin, A. R. (2012). Carbon Content of Tree Tissues: A Synthesis. *Forests, 3*(2), 332–352. DOI: 10.3390/f3020332

Viera, M., & Rodríguez-Soalleiro, R. (2019). A Complete Assessment of Carbon Stocks in Above and Belowground Biomass Components of a Hybrid *Eucalyptus* Plantation in Southern Brazil. *Forests*, *10*(7), 536. DOI: 10.3390/f10070536

Walkley, A., & Black, I. A. (1934). An examination of the Degtjareff method for determining soil organic matter, and a proposed modification of the chromic acid titration method. *Soil Science*, *37*(1), 29–38. DOI: 10.1097/00010694-193401000-00003

Zaher, H., Rhoufacha, M., & Dahmani, H. (2025). Carbon stock dynamics along a cork oak stand chronosequence: Implications for climate-smart forest management. *Journal of the Saudi Society of Agricultural Sciences*, *24*(7), 45. Advance online publication. DOI: 10.1007/s44447-025-00054-3

Zhang, H., Song, T., Wang, K., Du, H., Yue, Y., Wang, G., & Zeng, F. (2014). Biomass and Carbon Storage in an Age-Sequence of Cyclobalanopsis Glauca Plantations in Southwest China. *Ecological Engineering*, *73*, 184–191. DOI: 10.1016/j.ecoleng.2014.09.008

Zhang, L., Zhang, Z., Guo, L., Cui, X., Butnor, J. R., Li, S., Cao, X., & Chen, X. (2024). A simple method for estimating the coarse lateral root biomass of shrubs using ground-penetrating radar: Validation by Caragana microphylla Lam. in Inner Mongolia. *The Science of the Total Environment*, *919*, 170897. DOI: 10.1016/j.scitotenv.2024.170897 PMID: 38346659

Zhang, P., Cui, Z., Liu, X., & Xu, D. (2022). Above-Ground biomass and nutrient accumulation in ten *eucalyptus* clones in Leizhou peninsula, southern China. *Forests*, *13*(4), 530. DOI: 10.3390/f13040530

Zribi, L., Chaar, H., Khaldi, A., Hanchi, B., Mouillot, F., & Gharbi, F. (2016). Estimate of biomass and carbon pools in disturbed and undisturbed oak forests in Tunisia. *Forest Systems*, *25*(2), e060. DOI: 10.5424/fs/2016252-08062

APPENDIX I

Stands and sample trees structure for each clone: (a = stand structure, b = sample trees structure)

Figure 6.

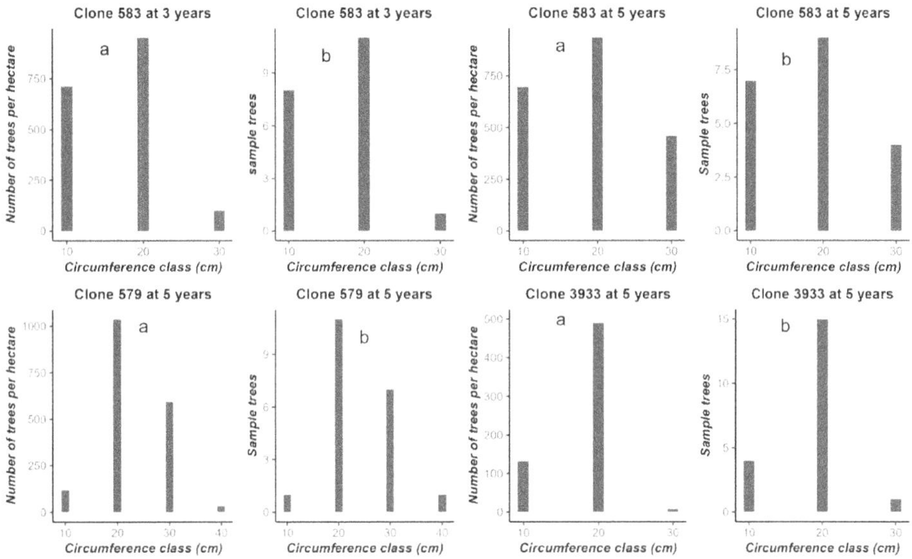

Figure 7.

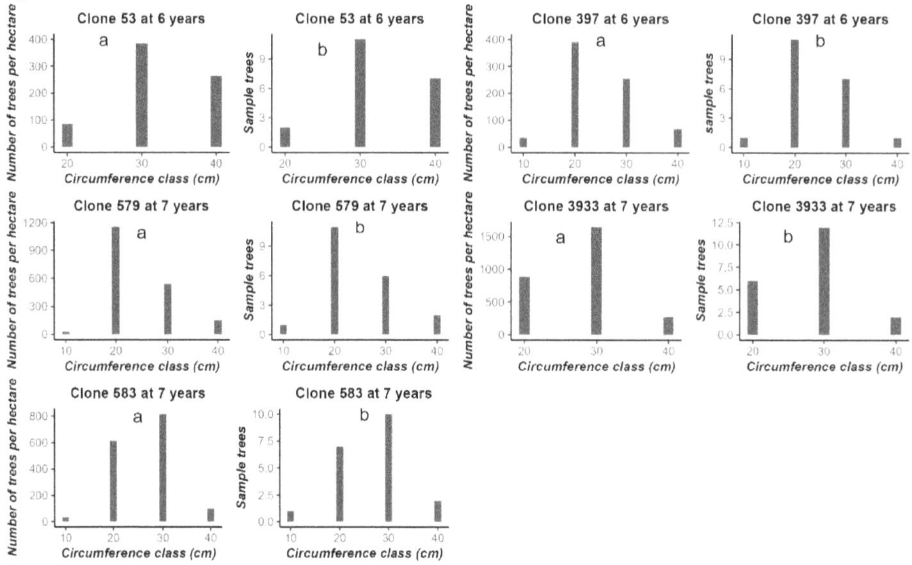

Table 5. P-values of the statistical assumption tests for each clone

Clone	Normality (Shapiro-Wilk test)	Homoscedasticity (Breusch-Pagan test)	Independence (Durbin-Watson test)
Clone583	0.4410799	0.3250327	0.3104322
Clone3933	0.11579937	0.934358156	0.930338108
Clone579	0.07947273	0.423162496	0.385530588
Clone397	0.3687	0.09893	0.716
Clone53	0.2906	0.475	0.84

Section 3
Monitoring and Innovation in Forest Management

Chapter 8
Remote Sensing and Artificial Intelligence for Forest Governance in Morocco:
Advances and Applications Toward Forestry 4.0 and 5.0

Ali Azedou
http://orcid.org/0000-0003-1053-2513
National School of Forestry Engineering, Morocco

ABSTRACT

Remote sensing and artificial intelligence are transforming forest monitoring and governance in Morocco's climate-vulnerable Mediterranean ecosystems. This literature review synthesizes advances in machine learning, satellite imagery, and geospatial analysis for monitoring deforestation, carbon stocks, wildfire risk, and land degradation across Morocco's 9 million hectares of forest. Through systematic analysis of multi-temporal change detection, machine learning algorithms, and Unmanned Aerial Vehicles (UAV) applications, we identify a persistent gap between technical feasibility and operational integration constrained by legal constraints, data fragmentation and capacity gaps hindering operational adoption. The chapter situates Morocco's transition within the Forestry 5.0 paradigm, integrating automated monitoring with participatory governance and adaptive management, and proposes policy recommendations bridging technical capabilities with governance innovation. Findings offer replicable frameworks for climate-resilient forest stewardship across Mediterranean regions globally.

DOI: 10.4018/979-8-3373-6239-7.ch008

INTRODUCTION

The Southern Mediterranean area remains a major biodiversity hotspot, characterized by high species diversity and endemism rates that are among the world's highest. These areas are famous for their dramatic adaptation to periodic droughts and display extraordinary biodiversity with over 25,000 plant species (Scarascia-Mugnozza et al., 2000), of which half are endemic to the area (Hrdina & Romportl, 2017). The value of these areas goes beyond protecting biodiversity and resisting climate change, given their primary roles in ecosystem services such as protecting watersheds, soil protection through prevention of erosion, and considerable carbon sequestration. Carbon concentration in Mediterranean old-growth forests differs significantly, with values ranging from 198 Mg C/ha in successional island forests to as high as 600 Mg C/ha in protected sites characterized by varied vegetation (Badalamenti et al., 2019; Di Matteo et al., 2023).

However, such important ecosystems are faced with threats that are unprecedented and arise from a combination of various factors. Climate change projections forecast increased drought patterns and an enormous increase in the incidence of wildfires (Schwörer et al., 2024); the area of influence of the affected regions will increase by 30% at the turn of the century (Ruffault et al., 2020). On the other hand, the impact of human activities such as excessive grazing, irrational logging, and the construction of urban areas is threatening the continuity of such important forest regions. The Southern Mediterranean forest regions are ranked among the 36 hotspots of the global biosphere, and this is confirmed by the International Union for the Conservation of Nature (IUCN).

Morocco's forests, confined within a landscape that is 93% arid (Sarraf & Jorio, 2010), are increasingly vulnerable to overlapping anthropogenic and climatic stresses, particularly in the sub-Mediterranean and Atlas regions. The national forest estate spans approximately 9 million hectares, comprising 5.8 million ha of woodlands, 2.5 million ha of steppe and pre-forest formations, and 700,000 ha of artificial plantations. However, gross forest cover remains low at 12.8%, and only 8.1% if limited to true forest formations—with stark regional disparities in density and structure (Ellatifi, 2012; Milian, 2007). Forests are nearly entirely state-owned (domaniale) and governed by the century-old Dahir of 1917, with operational oversight by the National Agency of Water and Forests (ANEF). Despite outdated legal frameworks and governance fragmentation, forests contribute directly to 30% of national wood demand, 30% of energy needs, and 17% of livestock feed in mountainous rural zones (Benzyane, 2007).

Despite their critical status, these ecological hotspots are under severe degradation. The rate of deforestation exceeds 17,000 hectares per year, mainly due to undeniably unsustainable fuelwood harvests (10,000,000 cubic meters a year), which are triple the ecological maximum (Benzyane, 2007; CESE, 2022). The degree of over-grazing exceeds ecological thresholds within forests, particularly for pasture that outpaces natural production capacity, particularly for cedar, cork, and argan forests, indicating nearly uniform failure in natural regeneration potential due to excessive grazing (Naggar, 2018). Further degradation from unauthorized logging, urban sprawl, as well as changes in land-use patterns compromises habitats, reducing interconnections between forests within a broader landscape matrix that maintains specific ecological functions, including water cycles, carbon sequestration, as well as certain biodiversity features (Pullar, et al., 2017). Case studies from the Middle Atlas region reflect canopy degradation, senile stands, and pest epidemics in cedar forests due, inter alia, to insufficient secondary forest development, combined with decreased resilience to environmental stressors like inadequate silvicultural treatment after logging (Machouri et al., 2020). In coastal zones like Mamora and Loukkos, cork oak stands suffer from soil compaction, overharvesting, and declining acorn regeneration. Climate-induced stress, especially recurrent droughts, triggers tree dieback, insect infestations (e.g., processionary moth), and amplifies wildfire frequency and severity, threatening keystone species such as Atlas cedar (Cedrus atlantica) and argan tree (Argania spinosa) (FAO, 2021). These converging degradation drivers—often acting cumulatively—erode forest structure, biodiversity, and ecosystem services. As shown across Mediterranean analogs, such multi-stressor dynamics can lead to abrupt ecological transitions, jeopardizing forests' role as carbon sinks, water regulators, and socio-economic lifelines for vulnerable rural communities.

Innovation in forest governance is now a necessity for Moroccan and Mediterranean forests, where conventional approaches often fail in dealing with complex challenges. Remote Sensing (RS) technologies, with the help of high-resolution satellite imagery and LiDAR systems, now enable tracking forest cover changes, biomass estimation, and early detection of degradation with unprecedented spatial and temporal scales. The application of Artificial Intelligence (AI) and Machine Learning (ML) algorithms has shown remarkable capabilities in processing these vast datasets with proposals for the automated detection of forest threats, species distribution mapping, and fire risk assessment at high accuracy rates. The emergence of Forestry 4.0 and 5.0 frameworks, which propose Internet of Things (IoT) sensors, cloud computing, and blockchain technologies, allows for cost-effective real-time forest monitoring and transparent resource management. Such digital innovations have considerably contributed to the arid Mediterranean contexts where resource constraints, coupled with extensive areas to be monitored, have historically posed difficulties for effective forest governance. The case studies published recently in Morocco also highlighted that AI-driven monitoring systems can reduce operational efficiency while improving the detection accuracy of illegal logging and health problems in forests (Ben-Said et al., 2025), demonstrating the scalability and adaptability of such technologies to diverse Mediterranean landscapes.

This chapter attempts to critically discuss how RS, ML, and data-driven solutions may help not only with the restore, monitor, and adaptive management of Moroccan forestry, but also how it may be further incorporated into the innovation agenda to fulfill the innovation visions of both forestry 4.0 and 5.0 paradigms. This chapter intends to fulfill all these by:

- Assessing the current state and emerging trajectories of RS and AI in Moroccan forestry, with emphasis on sub-Mediterranean and mountainous zones;
- Identifying and analyzing cross-cutting challenges that hinder large-scale adoption of these innovations, including data fragmentation, skill gaps, infrastructure limitations, and institutional inertia;
- Situating these approaches within the broader transformation toward Forestry 5.0, where ecological, technological, and socio-economic systems are interlinked through smart monitoring, adaptive planning, and participatory governance involving both institutions and local communities.

Before we explore the specific challenges that exist in Morocco, as well as their related technical solutions, let's establish a conceptual framework for our chapter as a whole. The terms "Forestry 4.0" and "Forestry 5.0" are drawn from a broader model for industrial revolution that has helped determine technological change for nearly two centuries (Holzinger et al., 2024).

These paradigms from the industry have continually been applied to the natural resources sector, generating forestry paradigms based on stages in the digital transformation of forest management in the following way:

- Forestry 1.0: Traditional, field-based methods using manual inventories and judgment.
- Forestry 2.0: GIS Implementation and Early Use of RS in Mapping and Planning.
- Forest Management 3.0: Regular satellite observation and decision aids in digital databases.
- Forestry 4.0 relates to the use of Internet of Things sensor technology, automated systems for forest observation, artificial intelligence, and data-based decision-making in forestry management practices. This model also emphasizes: real-time or near-real-time monitoring functions, forest threat analysis (forest fires, illicit cutting of wood, infestations), predictive analysis related to

risk assessments, cloud-based data platforms that facilitate the rapid sharing of data and reduced manual work involving data collection.

- Forestry 5.0 goes beyond technological automation into embracing human-centered, participatory governance models. This new paradigm places major emphasis on: democratization of access to forest data for civil society and the communities themselves, co-production of knowledge that integrates traditional ecological knowledge with technical systems, transparent, accountable structures of governance, technology serving social-ecological resilience, rather than purely efficiency goals and participatory monitoring and adaptive management frameworks.

These theories map well with the concepts discussed in the larger body of digital transformation literature in the context of natural resource administration. These concepts include the notion of precision forestry (Holopainen et al., 2014), smart governance models (Bolívar & Meijer, 2016), and the concept of socio-technical transition models (Geels, 2019). This chapter goes beyond the technical RS literature review because it describes how RS and AI, not only what they can do, but how they can transform governance, specifically in the context of forest administration in Morocco.

STATE OF MOROCCAN FORESTS – PRESSURES AND GAPS

Moroccan Forest Ecosystems

The diversity of forest environments in Morocco can be considered a reflection not only of the country's biogeography but also of its functional particularism in the Mediterranean environment as a whole. In terms of the diversity of forest environments, it is important to highlight that it is a consequence of the country's specific orography and Mediterranean climate, characterized by opportunities for altitudinal and bioclimatic variations from inframediterranean to oromediterranean biotopes. This contributes to a diverse flora, with over thirty significant timber tree species, forming forest, pre-forest, and pre-steppe vegetation areas covering nearly five million hectares. There is, for instance, the argan forests, restricted to arid to semi-arid environments; cork oak forests, limited to siliceous soils in the Atlantic zone; cedar forests, limited to the Middle and High Atlas ranges; and thuja-holly oak (green oak) forests, occupying significant altitudinal bands. The above table describes the forest environments in Morocco in terms of size, geographical extension, and significance.

Table 1. Main forest ecosystems present in Morocco

Forest Type / Vegetation	Approximate Area (ANEF, 2024)	Key Ecosystem Services	Conservation Status (IUCN 3.1) (IUCN,2025)
Atlas cedar (*Cedrus atlantica*)	**180,150 ha**	● **Regulating:** Carbon sequestration and storage, microclimate regulation. ● **Supporting:** Biodiversity habitat for endemic species (e.g., Barbary macaque). ● **Provisioning:** Timber.	**Endangered (EN)**
Cork oak (*Quercus suber*)	**509,390 ha**	● **Provisioning: Cork**, non-timber forest products (acorns). ● **Regulating:** carbon sequestration.	**Least Concern (LC)**
Holm oak (*Quercus rotundifolia*)	**1,969,220 ha**	● **Provisioning** firewood and charcoal, ● **Supporting**: soil formation, ● **Regulating** climate, ● **Cultural:** recreational, protective functions in high forests.	**Least Concern (LC)**
Zeen oak (*Quercus faginea*)	**18,640 ha**	● **Supporting** high diversity and endemism and soil retention), ● **Regulating** maintaining mesophilous conditions, hydrological buffer in depressions), ● **Provisioning** (wood, fodder).	**Least Concern (LC)**
Arar (*Tetraclinis articulata*)	**726,810 ha**	● **Provisioning** high-value wood, ● **Regulating** resilience to drought, good for regeneration and erosion control.	**Least Concern (LC)**
Junipers (*Juniperus spp.*)	**453,480 ha**	● **Supporting** crucial ecological role in high-mountain ecosystems, ● **Soil retention** and slope stabilization, ● **Provisioning** (wood for heating, cooking, construction, winter livestock fodder), ● **Biodiversity habitat**.	**Least Concern (LC)** and **Vulnerable (VU)**
Argan woodlands (*Argania spinosa*)	**335,450 ha**	● **Provisioning** (Argan oil - high-value cosmetic/culinary product, wood, fodder), ● **Regulating** (combating desertification, soil erosion prevention), ● **Cultural** (UNESCO Intangible Cultural Heritage, women's co-operatives).	**Vulnerable (VU)**

continued on following page

Table 1. Continued

Forest Type / Vegetation	Approximate Area (ANEF, 2024)	Key Ecosystem Services	Conservation Status (IUCN 3.1) (IUCN,2025)
Pines (*Pinus spp.*)	**130,450 ha**	● **Provisioning** (timber, firewood, resin, pine nuts, edible mushrooms), ● **Regulating** (erosion control, hydrological regulation, carbon sequestration), ● **Supporting** (soil protection, rehabilitation).	**Endangered (EN)** and **Least Concern (LC)**)
Moroccan fir (*Abies pinsapo var. marocana*)	**6,212 ha**	● **Biodiversity habitat** (endemic species, unique ecosystems in the Rif Mountains), ● **Regulating** (microclimate regulation, hydrological balance in mountain areas).	**Endangered (EN)**
Acacias (*Acacia spp.*)	**1,011,000 ha**	● **Provisioning** (livestock fodder, firewood, gum from some species), ● **Regulating** (drought resistance, soil binding, crucial for desertification control), ● **Supporting** (microclimate for associated species).	**Least Concern (LC)** and **Critical/Endangered**)
Others species	**111,300 ha**	● Diverse services dependent on specific sub-species, including timber, firewood, local non-wood products, and habitat provision.	Varies by species.

Forests in Morocco play a crucial ecological and socio-economic role. Ecologically, they regulate hydrological cycles, protect soils from erosion, conserve threatened and endemic biodiversity, and offer substantial carbon storage. Besides these ecological services, they support a broad range of ecosystem goods, from timber, fuelwood, cork, fodder, and high-value non-timber forest products such as argan oil, to sustaining cultural practices and nature-based tourism. These forests continue to be fundamental sources of energy and subsistence for rural communities, directly contributing to household food security in mountainous areas and arid zones where alternatives are limited. However, their continued provision is increasingly threatened by unsustainable exploitation, heavy grazing, and the increasingly severe impacts of climate change, making the need for integrated and adaptive management urgent.

Major Anthropogenic Pressures

The Moroccan forests are increasingly facing very harsh human-induced pressures that can negatively affect their resilience possibilities. Deforestation and land allocation continue to be serious issues, as they correlate with agricultural expansion, urbanization, as well as unauthorized wood extraction (Benabid, 1985). Overgrazing also stands out as an important threat: grazing animals, especially goats, which tend to eat tree leaves, often exceed the biological capacity of the land. In pre-forest, pre-steppe, as well as pre-mountain areas, usually with marginal climatic conditions for forests, as is the case with the Rif, these processes have contributed to very serious degradation, with erosion processes among the most serious cases internationally, making soil irremediable due to irreversible erosion processes, making forest regeneration impossible (Naggar & Lahssini, 2017). In addition, historical over-exploitation, led by clearing related to colonization in cedar, oak, and pine forests, as well as years of irresponsible cut-downs for fuelwood, have also impacted negatively on forest structure and biodiversity. Land tenure disputes, still unresolved with no formal delimitation of state forest boundaries, have contributed to an incidence of illegal logging as well as clearing. Recent research suggests that internationally recognized biological formations, like the cork oak woodlands at Maamora or the argan tree woodlands at Souss, have also lost considerable extent and density with biological consequences for the diversity of Morocco's countryside (Fraval et al., 1997). Indeed, these interlocking factors highlight how degradation in Morocco's forests is as socio-economic as biological, dealing with conditions of poverty, population growth, and a lack of alternative livelihoods.

Climatic Pressures

Increasing climatic stresses are presently threatening the Moroccan forests, with drought identified as one of the most widespread sources of biological stress. The climate of Morocco is known to be arid and characterized by limited and unpredictable rainfall, exhibiting inter- and intra-annual variability (Karrou, 2006). Through both history and dendrochronological analyses, it is proved that droughts are recurring and becoming progressively frequent, which was earlier estimated at one event per decade to double or triple that rate during the latter decades (Agoumi & Debbarh, 2025). Due to such arid conditions, reduced soil saturation and groundwater recharge ultimately cause agricultural droughts, and eventually consequent droughts of the hydrological and economic types. In the Tabular Middle Atlas region, the area of irrigation aggrandizement has increased drought stress due to a reduced groundwater and water resource pool (El-Bouhali et al., 2024). Climate change is also responsible for an increase in the occurrence and severity of forest fires, particularly in the Rif and Mediterranean coastal areas, where the prolonged drying season, coupled with the accumulation of combustible materials, increase the risk of fires, with the burned forest area in 2022 amounting nearly eight times the average area over the long term (> 3000 ha/year) (ANEF, 2024; FAO, 2020). Notwithstanding the short-term effects, such events also impede forest regeneration processes, biodiversity, and ecosystem services, reproducing, as in the Mediterranean region, the complex interplay between drought, biotic disturbance, and fires, reinforcing the need for forward-looking, technologically supported forest observation and management practices in the case of Moroccan forests.

Institutional & Legal Frameworks

The forestry management system in Morocco has structurally remained governed by the century-old Dahir of 10 October 1917. This established a centralized, protectionist regime of state ownership and criminalized sanctions that continue to influence legal frameworks. The text formalized the domanial status of all Moroccan forests by codifying that most of the wooded area encompassed the public domain. Furthermore, it established a strict regulatory framework that governed the use, access, and exploitation of the land. It criminalized any unlicensed cutting of trees, pasturing of animals, or utilization of timber. It introduced the regulation of authorized timber, firewood harvesting, hunting, and pasture rights. It established the framework of state forestry institutions by vesting them with police and judicial powers to enforce conservation regulations and prosecutions of offenders. While conceived to meet colonial needs of resource management and extraction of revenue, it instead formalized a regime of exclusion that has remained to date.

In modern Morocco, issues with Dahir application fall prey to an incongruity because its structural underpinning remains rooted in the colonial era while facing increasing needs for local communities' adaptation (Aubert, 2013). Although there have been numerous inclusive projects initiated in the 1990s to modernize the Forest Code, there remains little legislative adaptation to enable more democratic forest resource regulation. Dahir remains dedicated to enforcing state control over forests through their characterization as domanial land, making any local access to forests highly limited as they are solely regulated by forest administration. Notably, this system, which was structurally ingrained under colonial ruling, also premises local populations to continue misusing these forests' resources, as local communities' needs might eventually lead to overuse. Another inappropriate component is related to an application obstacle because the concept of "domaniality" presumes land to be under natural vegetation as governmental territory, causing persistent disputes between local communities' needs for land access versus state-domain concepts over forest land, whose application continues to increase as local communities' populations grow, making land access highly dependent as stated above.

The process of demarcation, which is a significant process for the legal establishment of forest boundaries, exacerbates this problem. The inefficiencies of the administrative system, which lack geodetic points and the skills of geodetic knowledge, contribute little to the definition of the boundaries, this too adding to the confusion regarding the use of the land (Mohammed, 2008). Thus, even the demarcated forests experience invasion and deforestation, further adding to the complications related to governance and, therefore, litigation. The growing population and subsistence living patterns, which exert more pressure on the use of the forest land, show the inflexibility of the Dahir policy to be counterproductive to conservation as well as empowerment.

Over the intervening years, a series of reforms has pursued a policy of modernization that culminated in the establishment of the Agence Nationale des Eaux et Forêts (ANEF), and subsequent modifications of the same law aiming at a consolidation of institutional roles and presence at the sub-national levels—but all of which coexist with a dense network of former decrees. Notwithstanding the efforts at institutional modernization, governance reviews continue to point to institutional fragmentation, overlapped roles and responsibilities, and policy inertia (FAO, 2001). While there have been decentralization and participatory management attempts-from local committees and compensation-for-protection schemes to the pilot transfers of management responsibilities-these have produced mixed results because although they demonstrate a potential for local stewardship, they are often hampered by unclear tenure arrangements, limited fiscal transfers, and weak technical support (Roumane et al., 2010). In addition, an outdated legal framework-which is full of archiving colonial era norms-institutional discontinuities, and capacity gaps in relevant government structures intersect in a way that prevents strategic visions, such as "Forêts du Maroc 2020-2030", from being translated into effective on-the-ground governance (FAO, 2020). Overcoming these limitations will require a streamlined Code forestier, clear tenure and revenue-sharing mechanisms, and an increase in targeted investment in the regional capacity of the ANEF, drawing on best international practice to enable adaptive, participatory, and accountable management.

Monitoring Gaps: Limits of Traditional Inventories, Data Fragmentation, Lack of Integration

The forest monitoring system in Morocco has experienced much change, and the key framework has been the National Forest Inventory, which took place between 1990 and 2005 and created a sound cartographic and digital database of forest resources. This forest monitoring system has been useful in providing detail about the geographical location and overall state of forest ecosystems in Morocco, ranging from the Mediterranean forests of the northern part of the country to the pre-Saharan forests in the southern regions (HCEFLCD, 2018). Although the traditional forest monitoring system has limitations in the monitoring of quickly changing forest ecosystems influenced by ecological, societal, and economic factors, the forestry industry has been embracing innovations to improve the monitoring system. The use of innovative technologies such as UAV surveys, in addition to the use of satellite images, is applied by the Moroccan water and forest service to complement the traditional forest inventory approach (Fadil et al., 2020). These forest management approaches and tools are extremely important for the forest management plan and the specific investments made in the sector because they improve the ability to track the dynamics and degradation processes of the forest accurately (Bijaber & Ahlafi, 2005). Nonetheless, there is room for improvement in the forest monitoring approach through the combined use of high-resolution RS, low-cost drone technology, and modern data management systems to track the dynamic changes that occur in the forest, especially in remote regions. The use of drone technology is promising to the African forestry sectors because of the cost-effective approach that is provided by the technology for the regular forest monitoring and forest inventory updates (Berie & Burud, 2018). The modernization approach in methods indicates that Morocco is committed to the evidence-based approach in forest management but realizes that continuous modernization is necessary because of the changing challenges faced in the forest ecosystem monitoring and evaluation (HCEFLCD, 2018).

RS & AI POTENTIAL

Evolution of RS in Forestry (From Aerial Photos to LiDAR, UAV, Satellites)

The use of RS in the field of forestry has experienced an enormous change in the past century. It started with the use of black-and-white aerial photography in the 1950s, which provided a completely new synoptic approach for foresters to view canopy layers and stand boundaries and landscapes with restricted spectral knowledge (Hildebrandt, 1987; Wulder et al., 2019). Photo-interpretation started with skepticism and was only gradually institutionalized for the purposes of forest inventories and land use classification (Singh et al., 2002). The 1970s and 1980s marked a beginning with the launch of satellite programs like Landsat and SPOT, and a new era started with multispectral regional and global coverage (Magnussen et al., 2018). Although there was an initial attitude problem with satellite remote sensing data, the data started restricting existing knowledge used for land cover change and forest disturbance measurements and culminated in the development of global forest monitoring products (Hansen et al., 2013; White et al., 2017). A timeline outlining the launch of major satellites' sensing technologies is provided in **Figure 1**.

Figure 1. Timeline of the launch of the main satellites for optical, multispectral and radar sensors (Kuenzer et al., 2014)

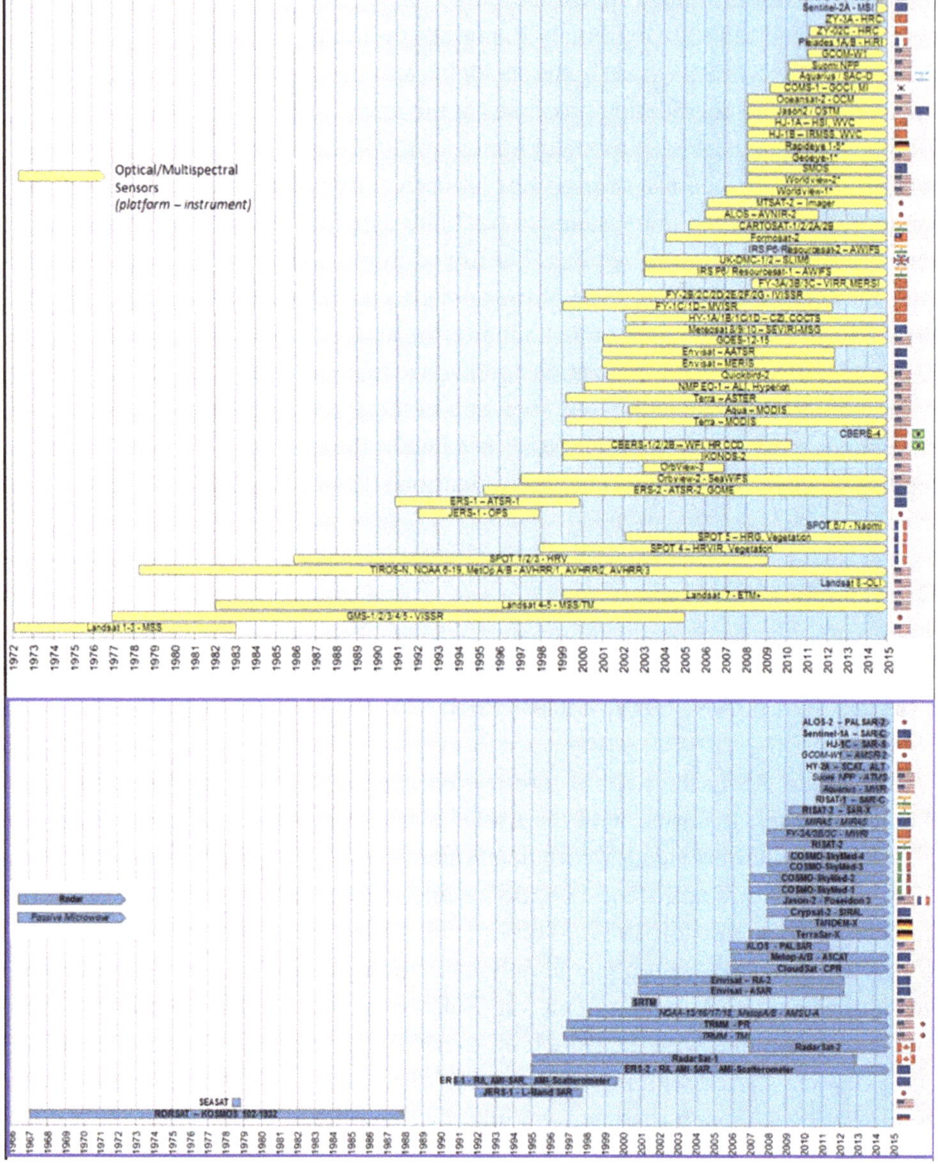

The next decades saw an evolution in terms of diversification with regard to the platforms and sensors used. The radar remote sensing platforms and technologies succeeded in expanding their solutions for monitoring through cloud penetration and delivery of structural data regarding canopy roughness and water content (Ryan et al., 2014). The airborne light detection and ranging (LiDAR) remote sensing platforms dramatically changed forest measurement practices through the creation of 3D cloud point data that allowed for direct determination of tree height and above-ground biomass (Nelson, 2013). The digital aerial photogrammetry (DAP) solution was introduced as an alternative approach with accurate creation of 3D point cloud data enhancing forest inventory attributes (White et al., 2017). The very high-resolution optical satellite and micro-n satellite systems such as Planet and Sentinel have evolved to allow for rapid assessment of fire through monitoring (Dalponte et al., 2020).

Most current innovation has been in the domain of unmanned aerial vehicles (UAVs) in providing access to ultra-fine spatial data, thereby allowing assessments of a target subject at a centimeter resolution for identification, regeneration, and health analysis of specimens (Goodbody et al., 2017). The versatility of these systems can especially benefit areas with fragmented ownership or lack of accessibility towards costly methods of airborne assessments.

Along this path, RS became the primary tool in forestry. With every successive innovation—starting from aerial photography through multispectral satellites, SAR, LiDAR, and then UAVs—the increasing intensity of observation regarding forest characteristics gathered pace. At present, RS is no longer useful for forest footprint assessments but is actually the basis for the use of the climate convention, REDD+ measurement, biodiversity protection, and adaptive forest management. Thus, the increasing significance of RS is recognized to be a backbone to the science of forestry across the world (Coops et al., 2021; Wulder et al., 2019).

The capabilities of the RS in Morocco have continuously been strengthened since the establishment of the Centre Royal de Télédétection Spatiale (CRTS) in 1989. The CRTS began with the use of Landsat images for land cover analysis and drought mapping; however, the center now utilizes satellite-based indicators for drought analysis and training activities. The introduction of the Sentinel series of satellites has made it common practice for the use of SAR images and high-repeat frequency multispectral images for agricultural analysis and risk analysis. The cloudy periods of the regions have now been made accessible. The launch of the two high-resolution optical satellites of the Mohammed VI-A constellation on 8 November 2017 and the Mohammed VI-B on 21 November 2018 from Airbus Defence and Space/Thales Alenia Space with the sun-synchronous orbit deployment of the Vega rocket of the French launch vehicle developer and operator Arianespace. The two satellite systems for the Moroccan government support a revisit strategy with multiple tasks ranging from cadastral mapping and environmental protection to surveillance of the borders and disaster mapping. This doubles the institutional use of CRTS with the imaging capabilities of the Mohammed VI satellite system and supports the growing shift in the use of the country from project-by-project initiatives towards more operational multi-sensor observation for forestry, water, and land.

Key RS Applications in forestry

Geographic Information Systems (GIS) and RS has emerged as a vital tool for comprehensively managing forests, including providing spatially explicit, temporally consistent, and scalable information for decision-making for operational forest management as well as strategic policy formulation. Traditionally used for mapping land cover, developments in satellite and airborne technology have used RS for almost all realms of forest monitoring functions, including, but not limited to, conventional applications such as mapping forest cover, land-use change, and support for country-level forest inventories, as well as more sophisticated applications like biomass, biodiversity, and ecosystem service mapping **(Figure 2)**.

Figure 2. Application of GIS and Remote Sensing (RS) technologies in forestry (Kerry et al., 2022)

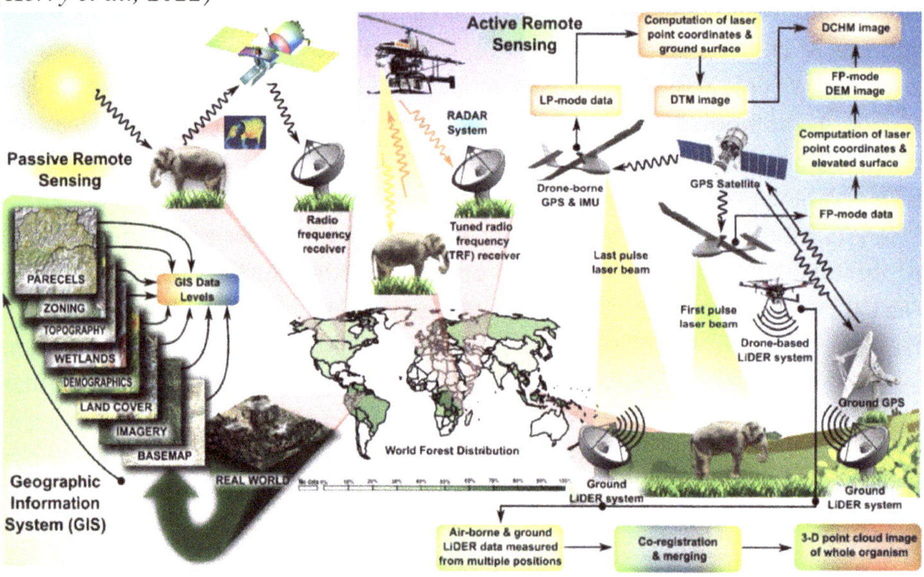

RS has completely altered forest mapping and inventory by providing consistent and spatially explicit data. Mapping of forest cover and LULC is a very popular application (Foody, 2002), and this has been made possible by Landsat imagery, Sentinel-2, and MODIS satellite images that provide capability to map forest boundaries, density, and change over a period of time.

RS improves forest inventories by combining satellite data with plot data so that updates can be done more frequently than before and even in inaccessible areas can be considered. Change detection and disturbance mapping applications (Hirschmugl et al., 2017) have improved in order to update information about clear cutting, selective cutting, storm damage, and other disturbances in forests annually or seasonally. Also, forest resource estimation is increasingly based on multispectral, LiDAR, and radar data that capture canopy height, density, and structure (Alvites et al., 2022). ML algorithms increase the precision of distinguishing forest types and age classes (Zadbagher et al., 2023), while biomass and carbon stock quantification contribute to climate mitigation by scaling field measurements across large areas (Tian et al., 2023). RS also monitors non-timber forest products like argan and medicinal plants, which provide a significant socio-economic value in Moroccan forests.

Risk and disturbance monitoring represents a dynamic RS application area. Fire risk mapping through spectral indices and thermal sensors allows the prediction of fire-prone areas; post-fire severity assessments (Kurbanov et al., 2022) determine the extent of burning and recovery needs—of particular relevance to Morocco's Rif and Mediterranean zones. RS allows for early detection of pests and diseases through changes in canopy reflectance (Aziz et al., 2025) and the monitoring of drought stress, especially in sensitive ecosystems like Atlas cedar forests (Dallahi et al., 2023). Ecological applications support biodiversity conservation through species distribution and habitat modeling, providing suitability maps for conservation planning and wildlife corridors. Landscape metrics describe the quantification of fragmentation, patch size, and connectivity, and watershed and erosion monitoring for risk identification in mountainous forests is important for watershed regulation.

Similarly, RS improves the governance of forests by identifying unlawful logging and incursions, and by facilitating compliance with international treaties. Urban forestry management is improved by RS by relating green infrastructure with urban development. ANEF disseminates maps of fire risks to land classified according to risk levels, which are determined by forest type, topography, and climate (ANEF, 2025). Finally, RS can be applied for the valuation of ecosystem services by relating biophysical measures to socio-economic values, and in this way it further cements its foundation in adaptive and transparent forest management.

METHODOLOGICAL ADVANCES IN THE USE OF GEOSPATIAL DATA FOR MOROCCAN FORESTRY MANAGEMENT & CASE STUDIES

In this case discussion, geospatial methods that have been adopted in the Moroccan context through case studies will be considered. There has been an evolution in

RS technology, ML algorithms, and GIS systems that has changed the way forest resources are managed in Morocco. From deforestation analysis in space and time to wildfire prediction systems that incorporate AI algorithms, Morocco has adopted modern approaches that can facilitate decision-making approaches in forest management. The present topic will discuss some of the technological methods that have been adopted in Morocco's forest management

Monitoring Forest Dynamics

The techniques used for analyzing deforestation, degradation, and forest dieback over time within Morocco have greatly improved, fetching very useful information regarding the dynamics and efficiency of forest management. Variations of deforestation levels within the High Atlas Mountains, the Northern Forest, and the unique Argan environments have shown to be significant (Chebli et al., 2018; Hammi et al., 2010; Laaribya et al., 2021).

Such models have influenced the management of forested regions and natural resources, showing the complexity of relationships between traditional and conservation efforts. Studies conducted in the Tangier-Tetouan-Al Hoceima Region have been able to establish the main drivers of deforestation, including forest fires (35.2%), agricultural lands (30.6%), and wood extraction (13.2%) (Boubekraoui, Attar et al. 2024).

Recent RS technologies and geospatial analysis have significantly enhanced LULC classification, species discrimination, and change detection. ANEF and CRTS employ satellite imagery from Landsat 5-TM and SPOT-XS platforms for national forest inventory mapping. Forests cover approximately 13.5% of Morocco, concentrated in the Atlas and Rif Mountains (Thaiki et al., 2023). In the Middle Atlas, 11% comprises coniferous species, primarily Atlas cedar, while 40% encompasses mixed and broadleaf forests (Mohajane et al., 2017). These baseline data inform management strategies supporting Morocco's goal to restore 600,000 hectares of degraded land by 2030 under the "Stratégie forêts du Maroc."

Technical developments combine different geospatial data sources for analysis. Supervised and unsupervised classifiers use spectral signatures for land cover mapping with differing levels of precision. ML models such as SVM and Random Forest reached a classification accuracy of 86% for Argan forest environments (Elmoussaoui et al., 2021), but a GA-weighted ensemble classifier for Talassemtane National Park reached 96.2% (Azedou et al., 2025). Integrating optical, SAR data, and DEM enhances Argan tree mapping precision (Elmoussaoui et al., 2021), but Spectral indices (NDVI, NDWI, NDBI) are vital for ML classifiers such as Random Forest, SVM, and KNN (Simou et al., 2024).

Problems still exist, especially in spectral confusion among similar species/habitats. Adding auxiliary data such as DEMs and climate information improves accuracy, but data availability is still variable in rural areas.

These technological improvements have significant outcomes in the management of natural resources. Knowledge of forest composition can aid in making decisions regarding sustainable land use (Mouafik et al., 2024). A combination of RS technology with community engagement methods and more exploration in uninvestigated areas might lead to improvements in forest management, ensuring methods are valid and accepted by society.

Assessing Argan Ecosystem Integrity

RS has proven to be an essential means of controlling the argan forests of Morocco (Argania spinosa) as well as other dry-wooded areas that face growing pressures from the environment, including land degradation, desertification, and climate change (El Moussaoui et al., 2025). The area of argan woods, mainly situated in the Souss-Massa area, has been identified as an important ecological resource that, apart from Morocco, has been protected by UNESCO as a Biosphere Reserve. The application of RS technology, such as satellite imagery, LIDAR, and aerial photos, has paved the way for scientists to obtain information on the concerned areas of interest (El Moussaoui et al., 2021; Moukrim et al., 2018).

Recent studies have used high-resolution satellite imagery, including Sentinel-2 and Landsat missions, to quantify the distribution and quality of argan woods. For example, the study shows the use of NDVI time series and ML classifiers like Support Vector Machines as effective in identifying the distinct phenology of argan trees. A combination of spectral and elevation information has shown classification accuracy above 92%, providing an efficient means of quantifying the distribution of the argan species (Sebbar et al., 2022).

Moreover, RS has enabled the exploration of environmental and climatic influences that impact argan range sites, illustrating alarming patterns for climate change and tree conditions. In specific research utilizing MaxEnt, along with actual and projected climate information, it was foretold that argan range sites could lose 47%-81% of suitable habitat in Southern Morocco by 2080 due to up to 4°C warming and reduced seasonal rainfall (Hakam et al., 2024). The application of ML algorithms in multispectral imagery has also indicated a 97% accuracy in yield prediction and disease detection, further highlighting increased disease resistance in non-irrigated trees than in irrigated trees, thereby identifying that AI-driven technology can accurately and effectively track argan sites for management (Saddik et al., 2024).

Apart from these larger monitoring efforts, recent ground-truth analyses have also improved the understanding of argan tree responses to environmental stresses.

Controlled comparisons of ecotype differences originating from different habitats showed argan populations, despite their differences, share similar physiological traits to drought, maintaining photosynthesis even when water is scarce, with a very sensitive stomata response to vapor pressure deficit (Chakhchar et al., 2017). Genetic analysis also affirmed a close relation among these populations, which holds promise for improvement through gradual selection for drought tolerance. Related field studies involving ecophysiological, biochemical, and morphological criteria among populations have affirmed an increase in intra-specific diversity with progressing aridity, also pointing to sensitive areas as shown, and the need for region-specific conservation efforts (Afi et al., 2024). Soil infiltration tests have also shown a marked deterioration among inter-tree spaces as opposed to argan-covered ground, underscoring the importance of tree cover (Kirchhoff et al., 2019). The outcomes, when couched within the context of RS technology, can help improve model calibration and species range mapping, and even improvement in restoration planning efforts through the use of spectral information linked to ecophysiological, biochemical, and morphological criteria.

Soil Erosion And Land Degradation Processes

The RS technologies for the monitoring of soil erosion and land degradation in Morocco have become increasingly important given its susceptibility to desertification and the progressive deterioration of the forest ecosystem. Geographically, Morocco is situated at an interface between Mediterranean and Arid climatic zones, hence very prone to degradation processes, where deforestation is one of the most important triggers leading to accelerated soil erosion. RS platforms, from medium-resolution satellites to high-resolution commercial imagery, provide the ability for quantifying the spatial and temporal dynamics of land cover change and associated degradation patterns across diverse Moroccan ecosystems such as El Jazouli et al. (2019), Modeste et al. (2016), and Simonneaux et al. (2015). More specifically, the capability of these technologies lies in their detection of forest canopy loss through the use of multispectral indices acting as proxies for vegetation health and biomass. Additionally, the addition of SAR sensors has expanded its capabilities of monitoring through cloud penetration and structural information about vegetation that allows year-round assessments, even in frequently cloud-covered mountainous regions of Morocco such as the Rif region, to be taken into consideration.

The relationship between deforestation and secondary soil erosion in the Moroccan scenario has been effectively established via multi-temporal analysis of RS data in conjunction with erosion modeling techniques. Researchers utilizing the Revised Universal Soil Loss Equation model in combination with remotely sensed data have identified that deforested regions of Morocco have an exponential level of soil erosion, especially in mountainous regions with extreme rainfall patterns (Hamouch et al., 2025; Okacha et al., 2023). Deforestation alters the role of vegetation as a natural barrier, and the resultant bare soil is directly struck by rainfall, thereby increasing runoff and sediment transport rates. RS-based land cover classification maps have been highly useful in the demarcation of the erosion-prone regions, especially in the Oum Er-Rbia and Moulouya River catchments (Barakat et al., 2023; Hlal et al., 2025), where the extensive destruction of forested areas due to agricultural land use and the practice of intense grazing practices have led to the formation of intense gullies and reservoir sedimentation. Topographic elements and the demarcation of the erosion-prone regions can be determined using multispectral satellites and digital elevation models (DEMs) achieved through missions such as the Shuttle Radar Topography Mission (SRTM) and ALOS PALSAR, showing high accuracy (Mourjane et al., 2024).

Climate Change Response

Carbon measurement and quantification of forest biomass in Moroccan forests have become ever so essential for climate change mitigation efforts and the country's greenhouse gas accounts under the Paris Agreement. However, the measurement of forest carbon content is obtained primarily through three approaches: allometric modeling, the integration of RS, and forest inventory systems. Mathematical models based on the use of tree measurements such as breast height diameter and tree height have shown that multiple regression models on the logarithmic scale can effectively calculate the value of the forest carbon content in whole trees, stems, and branches in the cedar stands of Morocco (El Mderssa et al., 2019). RS methods have been improved for intercomparison of vegetation indices for Landsat 8 satellite data, with the best values of the coefficient of determination (R^2) of 0.77 for estimating rangeland biomass in the Middle Atlas region (Boukrouh et al., 2024). In the recent past, new technologies like LiDAR UAV, as well as the use of the laser scanner for the local study of carbon biomass, have been made possible with 80% accuracy compared to the actual measurement in the field (Fadil et al., 2024; Sanaa et al., 2022).

However, despite these advances in methodology, several challenges exist in determining the accuracy of carbon assessments in Morocco. Lack of calibration data in the region creates uncertainty in both allometric equations and estimates using technology. Ecosystem variability in Morocco, including Mediterranean wooded ecosystems, mountainous coniferous ecosystems, and degraded shrub ecosystems, requires stratified sampling techniques because forest biomes with generic calibration data are prone to absolute biases. Research attention remains focused on northern forests and specific ecosystems, such as Argan and cedar forests, and thus knowledge gaps persist in drier regions where vegetation may be patchy but ecologically important. Near-term research priorities should establish permanent sample plot networks with periodic remeasurements, allometric equations for major tree species for different environmental conditions, and the integration of emergent technologies, such as unmanned aerial vehicles and upcoming spaceborne missions like the ESA's BIOMASS. Increased institutional capacity to analyze geospatial data and integration of traditional ecological knowledge with new technological systems is an urgent requirement for integrated national carbon accounting frameworks supportive of sustainable forest management goals and climate change mitigation objectives.

Governance Applications

These technologies have now become very vital enablers toward evidence-based decision-making and participatory resource management in Moroccan forest governance. GIS, RS, and spatial data infrastructures basically allow for more transparent and efficient governance structures that link the traditional with the modern aspects of management.

Geospatial technology has immense applications in the conservation and valuation of habitats within protected areas. In Souss-Massa National Park, a supervised classification technique using Sentinel 2A images was able to accurately classify 11 natural habitats and 6 disturbed habitats with a level of perfection that was unmatched (Elbahi et al., 2023). Comparative analysis showed that more than 88% of fenced habitats were natural, while more than 51% of unfenced habitats were disturbed, thus making a case for the effectiveness of fencing.

Decision support systems that combine geospatial solutions have been applicable in dealing with difficult governance challenges. The Morocco habous land decision support system, which involves agricultural land, urban land, and cultural land, provided evidence of how the application of open-source solutions assists in optimizing the management of resources (Harroucha & Chaouni, 2023). Geo-decisional support solutions that governed the Agriculture Development Fund within the Green Morocco Plan assisted policymakers in analyzing the geographic analysis of the allocation of funds to land (Adous et al., 2019).

Nevertheless, critical gaps limit the potential of geospatial technology: issues of data accessibility, data quality, coordination, and the needs of capacity building for open-source technologies and participatory approaches. Moving ahead, the agenda should include enhancing spatial data infrastructures, institutionalizing participatory monitoring approaches, or the design of decision support systems that can provide a mix of the potential of technology with traditional knowledge for socially accepted approaches for forest governance.

Uncertainty, Validation, and Quality Assurance

RS and AI-driven inclusion in the operational forestry management process also need the management of the uncertainty associated with such systems and technologies in forestry management. There is also the need for transparency regarding what monitoring systems can and cannot be reliably determined as part of the evidence-based and inclusive form of forestry management occurring through Forestry 5.0. Within this context, RS and AI applications face multiple uncertainty sources **(Figure 3)**:

Figure 3. The main sources of uncertainty in RS and AI applications for forest management

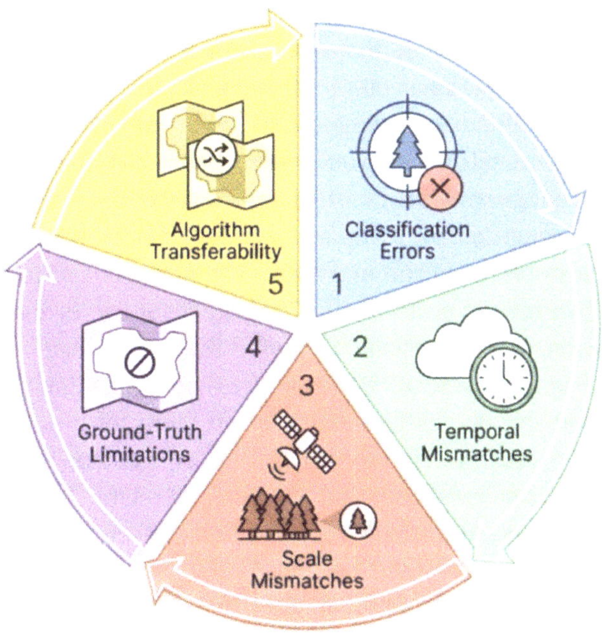

(1) **Classification Errors**: Even high-performing models exhibit commission (false positive) and omission (false negative) errors.
(2) **Temporal Mismatches**: Cloud cover gaps in optical imagery create monitoring blind spots during critical periods (early fire season, post-disturbance assessment windows)
(3) **Scale Mismatches**: Satellite-based monitoring excels at landscape patterns but may miss small-scale disturbances (individual tree mortality, localized illegal

cutting) significant for enforcement or community concerns. Conversely, UAV-derived data provide exceptional local detail but limited spatial coverage.

(4) **Ground-Truth Limitations**: Model validation depends on field reference data that themselves contain measurement errors, spatial imprecision, and temporal lags. Moroccan forests often lack dense permanent sample plot networks for robust validation.

(5) **Algorithm Transferability**: Models trained in one region or season may perform poorly when applied elsewhere due to spectral variability, phenological differences, or land-use patterns.

Economic Feasibility

Although technical feasibility has been established in a variety of applications, economic feasibility in relation to the adoption of RS and AI technology remains an important factor in operationalization in settings such as Morocco where budgets for forests compete for funding within various sectors of development.

The economics of digital forestry would involve direct investment and cost savings. The direct investments would involve obtaining data access for satellites, computer infrastructure for storing data in clouds, software for algorithms and development of algorithms, training of staff, and ground truthing campaigns. The direct cost savings would involve lowering the cost of actual fieldwork for surveying, early warning of potential hazards of a larger scale, increased revenue for carbon projects from accurate monitoring and verification, and targeted investment in restoring land.

There are no available studies regarding the cost of fully functional RS/AI systems implemented in Moroccan forestry practices. Yet, comparison studies around the globe serve as significant references and guidelines in establishing cost-saving and efficient practices in forestry. Manual forest inventories are extremely expensive and time-consuming because of the need to employ large teams of workers to plot samples across wide regions. Satellite imagery and efficient sampling have proven cost-saving efforts of 50-70% while also allowing regular updates (Bernknopf et al., 2019; Foster et al., 2024). It is important for Morocco to develop the systematic assessment of cost savings for traditional as well as RS-assisted inventories in forestry practices.

However, challenges like operational expenses, transition opportunity costs, and economic interests driven by gaps in monitoring have to be remedied. There are potential financing options like the incorporation of digital infrastructure into the state budget, climate finance, and public-private collaboration as potential paths forward. In order to fully integrate the high-impact apps effectively, there have to be stages of implementation and cost-sharing strategies between sectors and cost-benefit reporting of the Moroccan institutions to ensure sustainability and integration into the system. Without maintaining the objective of cost-feasibility in projects, even if technically successful, the projects might not extend beyond proof-of-concept installations for sustainable scaling-up into fully-fledged system implementation.

OVERCOMING STRUCTURAL BARRIERS TO ADOPTION: THE CONTESTED TRANSITION TOWARD FORESTRY 4.0 AND 5.0

Beyond technical skills and institutional structures, deeper barriers to implementing RS/AI technologies include power relationships, economic interests, and organizational cultures—often more difficult to surmount than the limits imposed by budget or technology. To Morocco's Forestry 4.0, an approach that integrates IoT sensors, automated real-time monitoring, and data-driven decisions, and eventual Forestry 5.0, a transition to people-centered, participatory governance enabled through technology, resistance arises because such a reformed paradigm challenges long-established patterns of access to resources and controlling management of those resources. If the policy documents have identified necessary changes, the slow progress in implementation of reform in the forest sector may be incomplete and sometimes ineffective; this is essentially not an administrative problem, but rather fundamental problems associated with a shift towards transparency, accountability, and shared governance that typifies modern digital forestry.

The Forestry 4.0 approach of real-time, automated monitoring fundamentally changes how forest governance works. Historically, different levels of access to water resources, agricultural and urban expansion into forest areas, and local resource dependency have operated within systems where periodic monitoring leaves gaps in time and information. Satellite surveillance that detects resource use patterns objectively and makes data publicly available represents more than just a technical improvement-it shifts who has access to information. Moving from Forestry 3.0's occasional, expert-only assessments to Forestry 4.0's continuous, automated detection closes these gaps. Forestry 5.0's participatory data platforms go further by letting civil society groups, community monitors, and international partners access the same evidence that government agencies see-redistributing information that has traditionally been concentrated in official hands. Therefore, the deployment of monitoring technologies needs to pay careful attention to these governance aspects in order to avoid two problems: underuse (where technical investments result in little actual adoption) or inequitable implementation (where improved surveillance falls disproportionately on small-scale users as larger-scale operations find ways of complying more easily).

Implementation on the ground faces workforce challenges rarely discussed in technical plans. Forest staff in remote areas often work with limited resources and pay structures that haven't changed much in decades. Shifting from manual verification—where field staff use professional judgment about what to report—to digital monitoring that creates systematic records requires not only technical training but fundamental changes in how workers are motivated and rewarded. Forestry 5.0's focus on human-centered technology must therefore address the practical realities of forest administration: competitive salaries that provide fair compensation, performance systems that reward measurable conservation results, and career paths that value technical skills. Without addressing these workforce issues, advanced Forestry 4.0 systems may face resistance from the very staff needed to make them work.

These are also institutional challenges that affect organizational structures and the way information has traditionally been managed. The historic role of ANEF as the main assessor of forest conditions, based on specialized technical knowledge, has created institutional cultures around centralized data control. The approach of Forestry 5.0 democratizes access to data in a way that fundamentally alters this arrangement: public dashboards displaying management results, restoration success rates, and patterns of enforcement expose institutional performance to wider scrutiny. Morocco's experience reflects global patterns whereby forestry agencies often rapidly embrace the efficiency gains of Forestry 4.0, where reduced costs through automated inventory and better resource allocation, slowly adapt to the governance implications of Forestry 5.0, with its community access to data, civil society participation, and independent verification. Such transitions are successful if the political leadership has framed the change as governance modernization, improving institutional credibility on the basis of evidence-based transparency, thus reinforcing legal frameworks that favor open data but also protect professional judgment in complex management situations.

It appears that modernization of legal frameworks is the most critical need relative to attaining goals set by Forestry 4.0 and 5.0. The uniform enforcement approach inherent in the Dahir of 1917-representing a governance philosophy predating electronic monitoring capacity-is wholly inappropriate as a framework for introducing Forestry 4.0's advanced detection. To the extent that automated systems detect subsistence activities-fuelwood collection, medicinal plant gathering, and traditional grazing-with the same technical precision as large-scale commercial extraction, Legal reforms would be required to establish use-right categories, proportional compliance approaches, community management authorities supported by monitoring data, and benefit-sharing arrangements. In this way, technology may support rather than impede conservation goals. The Forestry 5.0 vision of technology enabling fair resource governance-an institutional challenge entailing sustained political commitment beyond mere installation of technical systems-will depend fundamentally on parallel legal modernization.

Lastly, the discrepancy between the improved detection capabilities and enforcement capabilities represents technical problems that need to be addressed with strategic thinking. For example, if the automation monitoring tool is producing thousands of alerts per day, this may be a burden on already strained judicial and administrative enforcement agencies, resulting in response delays that may undermine the effectiveness of the entire system. In this sense, it is clear that a successful technological innovation such as Forestry 4.0/5.0 has to be accompanied by strategic institutional development with improved capacity, simplified compliance systems, and flexibility in management. The strategic allocation of priority targets with tools that monitor minute details to identify situations of great potential impact, priority ecosystems, and response opportunities would become a critical lever for effecting change with improved detection capabilities.

CONCLUSION

Integrating RS and AI represents a fundamental transformation in Moroccan forest governance—from reactive, resource-intensive models toward proactive, adaptive, transparent systems. This chapter demonstrated that data-driven approaches spanning deforestation monitoring, fire prediction, carbon assessment, and participatory governance offer unprecedented capabilities for addressing complex challenges amid accelerating climate change and anthropogenic pressures.

Morocco's forests deliver vital ecosystem services yet face converging threats: high deforestation rates, chronic overgrazing, unsustainable extraction, intensifying drought, and escalating wildfire frequency. While Morocco's geospatial infrastructure provides sovereign earth observation capabilities, critical gaps persist: limited operational integration into management workflows, insufficient biodiversity monitoring capacity, weak linkages between geospatial intelligence and participatory governance, and fragmented spatial data systems.

The transformative potential is substantial. Real-time early warning systems enable rapid response, spatially explicit models optimize restoration investments, and transparent monitoring empowers civil society oversight. However, realizing this potential requires moving beyond pilot demonstrations toward systematic operational integration—demanding legal modernization, institutional adaptation, sustained capacity building, and genuine participatory governance.

Morocco stands at a critical juncture. The convergence of sovereign satellite capabilities, established infrastructure, and ambitious strategies creates unprecedented opportunity. The evidence is compelling, technologies proven, and institutional foundations exist—what remains is political will and collaborative commitment to forge a sustainable forest future.

REFERENCES

Adous, F. E., Ghouat, M. A. E., & Ibannain, F. (2019). A Geo-Decisional Tool for the Management of the Agricultural Development Fund Under the Green Morocco Plan. In *Geospatial Technologies for Effective Land Governance* (pp. 160–173). IGI Global Scientific Publishing., DOI: 10.4018/978-1-5225-5939-9.ch010

Afi, C., Telmoudi, M., Labbassi, S., Chabbi, N., Hallam, J., Msanda, F., & Ait Aabd, N. (2024). Assessing the Impact of Aridity on Argan Trees in Morocco: Implications for Conservation in a Changing Climate. *Resources*, *13*(10), 135. DOI: 10.3390/resources13100135

Agoumi, A., & Debbarh, A. (2025). Ressources en eau et bassins versants du Maroc: 50 ans de développement (1955–2005). 50 Ans de Développement Humains. *Perspectives.*

Alvites, C., Marchetti, M., Lasserre, B., & Santopuoli, G. (2022). LiDAR as a tool for assessing timber assortments: A systematic literature review. *Remote Sensing (Basel)*, *14*(18), 4466. DOI: 10.3390/rs14184466

ANEF. (2024). Forêts En Chiffres. http://www.eauxetforets.gov.ma/ForetsMarocaines/ForetsChiffres/Pages/Forets-En-Chiffres.aspx

ANEF. (2024). ANEF: Recul de 70% des superficies forestières parcourues par le feu. *Le Desk.* https://ledesk.ma/encontinu/anef-recul-de-70-des-superficies-forestieres-parcourues-par-le-feu/

Aubert, P.-M. (2013). Les évolutions de la politique forestière au Maroc: Entre réappropriation du modèle forestier français et idéalisation de la tribu. *Revue Forestière Française*, *LXV*(4), 305–316. DOI: 10.4267/2042/53627

Azedou, A., Amine, A., Kisekka, I., & Lahssini, S. (2025). Genetic algorithm optimization of ensemble learning approach for improved land cover and land use mapping: Application to Talassemtane National Park. *Ecological Indicators*, *177*, 113776. DOI: 10.1016/j.ecolind.2025.113776

Aziz, D., Rafiq, S., Saini, P., Ahad, I., Gonal, B., Rehman, S. A., Rashid, S., Saini, P., Rohela, G. K., Aalum, K., Singh, G., Gnanesh, B. N., & Nabila Iliya, M. (2025). Remote sensing and artificial intelligence: Revolutionizing pest management in agriculture. *Frontiers in Sustainable Food Systems*, *9*, 1551460. Advance online publication. DOI: 10.3389/fsufs.2025.1551460

Badalamenti, E., Battipaglia, G., Gristina, L., Novara, A., Rühl, J., Sala, G., Sapienza, L., Valentini, R., & La Mantia, T. (2019). Carbon stock increases up to old growth forest along a secondary succession in Mediterranean island ecosystems. *PLoS One*, *14*(7), e0220194. DOI: 10.1371/journal.pone.0220194 PMID: 31339941

Barakat, A., Rafai, M., Mosaid, H., Islam, M. S., & Saeed, S. (2023). Mapping of Water-Induced Soil Erosion Using Machine Learning Models: A Case Study of Oum Er Rbia Basin (Morocco). *Earth Systems and Environment*, *7*(1), 151–170. DOI: 10.1007/s41748-022-00317-x

Ben-Said, M., Chemchaoui, A., Etebaai, I., & Taher, M. (2025). Land use and land cover changes in Morocco: Trends, research gaps, and perspectives. *GeoJournal*, *90*(1), 44. DOI: 10.1007/s10708-024-11214-3

Benabid, A. (1985). Les écosystèmes forestiers, préforestiers et presteppiques du Maroc: Diversité, répartition biogéographique et problèmes posés par leur aménagement. *Forêt Méditerranéenne*, *7*(1), 53–64.

Benzyane, M. (2007). La gestion durable des ressources forestières au Maroc: Quelle stratégie? *Forêt Méditerranéenne*, *28*(1), 47–54.

Berie, H. T., & Burud, I. (2018). Application of unmanned aerial vehicles in earth resources monitoring: Focus on evaluating potentials for forest monitoring in Ethiopia. *European Journal of Remote Sensing*, *51*(1), 326–335. DOI: 10.1080/22797254.2018.1432993

Bernknopf, R., Kuwayama, Y., Gibson, R., Blakely, J., Mabee, B., Clifford, T. J., Quayle, B., Epting, J., Hardy, T., & Goodrich, D. (2019). The cost-effectiveness of satellite Earth observations to inform a post-wildfire response. *Resources for the Future*. https://media.rff.org/documents/Valuables_Wildfires.pdf

Bijaber, N., & Ahlafi, Z. (2005). *Études détaillées pour la cartographie des changements dans le cadre de l'Inventaire Forestier National*. Projet FORMA.

Bolívar, M. P. R., & Meijer, A. J. (2016). Smart Governance: Using a Literature Review and Empirical Analysis to Build a Research Model. *Social Science Computer Review*, *34*(6), 673–692. DOI: 10.1177/0894439315611088

Boubekraoui, H., Attar, Z., Maouni, Y., Ghallab, A., Saidi, R., & Maouni, A. (2024). Forest loss drivers and landscape pressures in a Northern Moroccan protected areas' network: Introducing a novel approach for conservation effectiveness assessment. *Conservation*, *4*(3), 452–485. DOI: 10.3390/conservation4030029

Boubekraoui, H., Maouni, Y., Ghallab, A., Draoui, M., & Maouni, A. (2024). Deforestation drivers in northern Morocco: An exploratory spatial data analysis. *Environmental Research Communications*, *6*(7), 071005. DOI: 10.1088/2515-7620/ad5ad6

Boukrouh, S., Bouazzaoui, Y., El Aich, A., Mahyou, H., Chikhaoui, M., Ait Lafkih, M., N'Dorma, O., & Alados, C. (2024). Estimation of standing crop biomass in rangelands of the Middle Atlas mountains using remote sensing data. *African Journal of Range & Forage Science*, *41*(4), 244–259. DOI: 10.2989/10220119.2024.2360991

Chakhchar, A., Haworth, M., El Modafar, C., Lauteri, M., Mattioni, C., Wahbi, S., & Centritto, M. (2017). An assessment of genetic diversity and drought tolerance in argan tree (Argania spinosa) populations: Potential for the development of improved drought tolerance. *Frontiers in Plant Science*, *8*, 276. DOI: 10.3389/fpls.2017.00276 PMID: 28303146

Chebli, Y., Chentouf, M., Ozer, P., Hornick, J.-L., & Cabaraux, J.-F. (2018). Forest and silvopastoral cover changes and its drivers in northern Morocco. *Applied Geography (Sevenoaks, England)*, *101*, 23–35. DOI: 10.1016/j.apgeog.2018.10.006

Coops, N. C., Achim, A., Arp, P., Bater, C. W., Caspersen, J. P., Côté, J.-F., Dech, J. P., Dick, A. R., van Ewijk, K., Fournier, R., Goodbody, T. R. H., Hennigar, C. R., Leboeuf, A., van Lier, O. R., Luther, J. E., MacLean, D. A., McCartney, G., Pelletier, G., Prieur, J.-F., & Woods, M. E. (2021). Advancing the application of remote sensing for forest information needs in Canada: Lessons learned from a national collaboration of university, industrial and government stakeholders. *Forestry Chronicle*, *97*(02), 109–126. DOI: 10.5558/tfc2021-014

Dallahi, Y., Boujraf, A., Meliho, M., & Orlando, C. A. (2023). Assessment of forest dieback on the Moroccan Central Plateau using spectral vegetation indices. *Journal of Forestry Research*, *34*(3), 793–808. DOI: 10.1007/s11676-022-01525-x

Dalponte, M., Marzini, S., Solano-Correa, Y. T., Tonon, G., Vescovo, L., & Gianelle, D. (2020). Mapping forest windthrows using high spatial resolution multispectral satellite images. *International Journal of Applied Earth Observation and Geoinformation*, *93*, 102206. DOI: 10.1016/j.jag.2020.102206

Di Matteo, G., Luzzi, G., Basile, A., Sposato, A., Bertini, G., Neri, U., Pennelli, B., Napoli, R., & Nardi, P. (2023). Carbon concentrations and carbon storage capacity of three old-growth forests in the Sila National Park, Southern Italy. *Journal of Forestry Research*, *34*(1), 233–242. DOI: 10.1007/s11676-022-01549-3

El-Bouhali, A., Amyay, M., & Ech-Chahdi, K. E. O. (2024). Combined impact of drought and land use changes on water resources in the Tabular Middle Atlas, Morocco. *Revista de Estudios Andaluces*, *48*(48), 202–220. DOI: 10.12795/rea.2024.i48.10

El Jazouli, A., Barakat, A., Khellouk, R., Rais, J., & El Baghdadi, M. (2019). Remote sensing and GIS techniques for prediction of land use land cover change effects on soil erosion in the high basin of the Oum Er Rbia River (Morocco). *Remote Sensing Applications: Society and Environment, 13*, 361–374. DOI: 10.1016/j.rsase.2018.12.004

El Mderssa, M., Belghazi, B.,, , Benjelloun, H., Zennouhi, O., Nassiri, L., & Ibijbijen, J. (2019). Estimation of Carbon Sequestration; Using Allometric Equations; in Azrou Cedar Forests (Cedrus atlantica Manetti) in the Central Middle Atlas of Morocco under Climate Change. *Open Journal of Forestry*, *09*(03), 214–225. DOI: 10.4236/ojf.2019.93011

El Moussaoui, E. H., Moumni, A., Khabba, S., Er-Raki, S., Ait Hssaine, B., Chehbouni, A., & Lahrouni, A. (2025). Bibliometric and review analysis of argan trees studies: Global research trends and challenges. *Agroforestry Systems*, *99*(5), 132. DOI: 10.1007/s10457-025-01228-2

Elbahi, A., Dugon, M., Oubrou, W., El Bekkay, M., Hermas, J., & Lawton, C. (2023). A geospatial approach to assess habitat diversity and inform conservation efforts using a case study from North Africa. *African Journal of Ecology*, *61*(2), 422–450. DOI: 10.1111/aje.13130

Ellatifi, M. (2012). L'économie de la forêt et des produits forestiers au Maroc: Bilan et perspectives. *Pessac: Thèse de Doctorat En Sciences Économiques, Université Montesquieu-Bordeaux IV*. https://www.audentia-gestion.fr/economie/PDF/Thse_-_Ellatifi.pdf

Elmoussaoui, E., Moumni, A., & Lahrouni, A. (2021). Cartography of moroccan argan tree using combined optical and sar imagery integrated with digital elevation model. *The International Archives of the Photogrammetry, Remote Sensing and Spatial Information Sciences, XLVI-4-W5-2021*, 211–217. ISPRS TC IV (WG IV-1)
The 6th International Conference on Smart City Applications - 27–29 October 2021, Virtual Safranbolu, Turkey. DOI: 10.5194/isprs-archives-XLVI-4-W5-2021-211-2021

Fadil, S., Sebari, I., Ajerame, M. M., Ajeddour, R., El Maghraoui, I., Ait El kadi, K., Zefri, Y., & Jabrane, M. (2024). An Integrating Framework for Biomass and Carbon Stock Spatialization and Dynamics Assessment Using Unmanned Aerial Vehicle LiDAR (LiDAR UAV) Data, Landsat Imagery, and Forest Survey Data in the Mediterranean Cork Oak Forest of Maamora. *Land (Basel)*, *13*(5), 688. DOI: 10.3390/land13050688

Fadil, S., Sebari, I., Bouhaloua, M., & Kadi, K. A. E. (2020). Opportunités d'utilisation de la technologie drone au niveau des écosystèmes forestiers. *Revue Marocaine des Sciences Agronomiques et Vétérinaires*, *8*(4). https://www.agrimaroc.org/index.php/Actes_IAVH2/article/view/888

FAO. (2001). *Forestry Curriculum Development and Revision*. https://www.fao.org/4/y1842e/y1842e00.htm#TopOfPage

FAO. (2020). *Gestion intégrée des incendies de forêts au Maroc. Stratégie nationale intersectorielle & plan d'action 2020-2030.*

FAO. (2021). *Gestion intégrée des incendies de forêts au Maroc Stratégie nationale intersectorielle & plan d'action 2020-2030—Google Search*. https://www.google.com/search?client=safari&rls=en&q=Gestion+int%C3%A9gr%C3%A9e+des+incendies+de+for%C3%AAts+au+Maroc+Strat%C3%A9gie+nationale+intersectorielle+%26+plan+d%E2%80%99action+2020-2030&ie=UTF-8&oe=UTF-8

Foody, G. M. (2002). Status of land cover classification accuracy assessment. *Remote Sensing of Environment*, *80*(1), 185–201. DOI: 10.1016/S0034-4257(01)00295-4

Foster, A., Rahimzadeh-Bajgiran, P., Daigneault, A., & Weiskittel, A. (2024). Cost-effectiveness of remote sensing technology for spruce budworm monitoring in Maine, USA. *Forests Monitor*, *1*(1), 66–98. DOI: 10.62320/fm.v1.i1.14

Fraval, A., Bourret, J., Guy, P., Arbez, M., & Brenot, C. (1997). *Forêts*. Editions Quae. https://books.google.com/books?hl=fr&lr=&id=rz7lmDAefo0C&oi=fnd&pg=PA6&dq=Fraval+A,+Villemant+C.+Les+actions+anthropiques.+La+Ma%C3%A2mora+et+ses+ennemies.+Les+dossiers+de+l%E2%80%99environnement+de+l%E2%80%99INRA.+For%C3%AAts+1997&ots=Qz6sBwXVsg&sig=H5ZMJjf2jWcT8M0O-hx1LAmjQ5U

Geels, F. W. (2019). Socio-technical transitions to sustainability: A review of criticisms and elaborations of the Multi-Level Perspective. *Current Opinion in Environmental Sustainability*, *39*, 187–201. DOI: 10.1016/j.cosust.2019.06.009

Goodbody, T. R. H., Coops, N. C., Marshall, P. L., Tompalski, P., & Crawford, P. (2017). Unmanned aerial systems for precision forest inventory purposes: A review and case study. *Forestry Chronicle*, *93*(01), 71–81. DOI: 10.5558/tfc2017-012

Hakam, O., Ongoma, V., Beniaich, A., Meskour, B., El Kadi, M. A., Brouziyne, Y., Hssaisoune, M., Tairi, A., Labbaci, A., & Bouchaou, L. (2024). Assessment of the impact of climate change on Argan tree in the Mediterranean GIAHS site, Morocco: Current and future distributions. *Modeling Earth Systems and Environment*, *10*(4), 5529–5552. DOI: 10.1007/s40808-024-02077-1

Hammi, S., Simonneaux, V., Cordier, J. B., Genin, D., Alifriqui, M., Montes, N., & Auclair, L. (2010). Can traditional forest management buffer forest depletion? Dynamics of Moroccan High Atlas Mountain forests using remote sensing and vegetation analysis. *Forest Ecology and Management*, *260*(10), 1861–1872. DOI: 10.1016/j.foreco.2010.08.033

Hamouch, C., Chaaouan, J., & Bouiss, C. (2025). Spatial-temporal assessment of soil erosion using the RUSLE model in the upstream Inaouène watershed, Northern Morocco. *Natural Hazards Review*, *5*(1), 148–156. DOI: 10.1016/j.nhres.2024.08.002

Hansen, M. C., Potapov, P. V., Moore, R., Hancher, M., Turubanova, S. A., Tyukavina, A., Thau, D., Stehman, S. V., Goetz, S. J., Loveland, T. R., Kommareddy, A., Egorov, A., Chini, L., Justice, C. O., & Townshend, J. R. G. (2013). High-Resolution Global Maps of 21st-Century Forest Cover Change. *Science*, *342*(6160), 850–853. DOI: 10.1126/science.1244693 PMID: 24233722

Harroucha, R., & Chaouni, A.-A. (2023). A GIS-based decision support system for the management of the Moroccan habous property. *GeoJournal*, *88*(4), 3905–3922. DOI: 10.1007/s10708-023-10847-0

HCEFLCD. (2018). *Rapport des activités 2018, Haut-Commis-sariat aux Eaux et Forêts et à la Lutte contre la Désertification, Rabat Chellah, Maroc.*

Hildebrandt, G. (1987). TOY or TOOL—Fernerkundung aus dem Weltraum: Spiel- oder Werkzeug für die Forstwirtschaft? *Forstwissenschaftliches Centralblatt*, *106*(1), 141–168. DOI: 10.1007/BF02741143

Hirschmugl, M., Gallaun, H., Dees, M., Datta, P., Deutscher, J., Koutsias, N., & Schardt, M. (2017). Methods for Mapping Forest Disturbance and Degradation from Optical Earth Observation Data: A Review. *Current Forestry Reports*, *3*(1), 32–45. DOI: 10.1007/s40725-017-0047-2

Hlal, M., El Monhim, B., Chenal, J., Munyaka, J.-C. B., Azmi, R., Sbai, A., Cwick, G., & Hichou, B. B. (2025). Application of Deep Learning and Geospatial Analysis in Soil Loss Risk in the Moulouya Watershed, Morocco. *Water (Basel)*, *17*(9), 1351. DOI: 10.3390/w17091351

Holopainen, M., Vastaranta, M., & Hyyppä, J. (2014). Outlook for the next generation's precision forestry in Finland. *Forests*, *5*(7), 1682–1694. DOI: 10.3390/f5071682

Holzinger, A., Schweier, J., Gollob, C., Nothdurft, A., Hasenauer, H., Kirisits, T., Häggström, C., Visser, R., Cavalli, R., Spinelli, R., & Stampfer, K. (2024). From Industry 5.0 to Forestry 5.0: Bridging the gap with Human-Centered Artificial Intelligence. *Current Forestry Reports*, *10*(6), 442–455. DOI: 10.1007/s40725-024-00231-7 PMID: 39464642

Hrdina, A., & Romportl, D. (2017). Evaluating Global Biodiversity Hotspots – Very Rich and Even More Endangered. *Journal of Landscape Ecology*, *10*(1), 108–115. DOI: 10.1515/jlecol-2017-0013

Karrou, M. (2006). *Evaluation des dispositifs d'alerte précoce à la sécheresse à l'échelle nationale (cas du Maroc).* [Rapport d'expertise nationale effectuée dans le cadre du projet SMAS au Maroc (LIFE05TCY/TN/000150).].

Kerry, R. G., Montalbo, F. J. P., Das, R., Patra, S., Mahapatra, G. P., Maurya, G. K., Nayak, V., Jena, A. B., Ukhurebor, K. E., Jena, R. C., Gouda, S., Majhi, S., & Rout, J. R. (2022). An overview of remote monitoring methods in biodiversity conservation. *Environmental Science and Pollution Research International*, *29*(53), 80179–80221. DOI: 10.1007/s11356-022-23242-y PMID: 36197618

Kirchhoff, M., Engelmann, L., Zimmermann, L. L., Seeger, M., Marzolff, I., Aït Hssaine, A., & Ries, J. B. (2019). Geomorphodynamics in Argan Woodlands, South Morocco. *Water (Basel)*, *11*(10), 2193. DOI: 10.3390/w11102193

Kuenzer, C., Ottinger, M., Wegmann, M., Guo, H., Wang, C., Zhang, J., Dech, S., & Wikelski, M. (2014). Earth observation satellite sensors for biodiversity monitoring: Potentials and bottlenecks. *International Journal of Remote Sensing*, *35*(18), 6599–6647. DOI: 10.1080/01431161.2014.964349

Kurbanov, E., Vorobev, O., Lezhnin, S., Sha, J., Wang, J., Li, X., Cole, J., Dergunov, D., & Wang, Y. (2022). Remote sensing of forest burnt area, burn severity, and post-fire recovery: A review. *Remote Sensing (Basel)*, *14*(19), 4714. DOI: 10.3390/rs14194714

Machouri. N., Laouina, A., & Sfa, M. (2020). *Les Suberaies Atlantiques et leurs espaces periforestiers: Problematique de degradation et Possibilite de l'elaboration d'une Strategie Participative De Developpement Durable.* Retrieved August 5, 2025, from https://lerma-flsh.uca.ma/wp-content/uploads/2025/06/Les-Suberaies-Atlantiques-et-leurs-espaces.pdf

Magnussen, S., Nord-Larsen, T., & Riis-Nielsen, T. (2018). Lidar supported estimators of wood volume and aboveground biomass from the Danish national forest inventory (2012–2016). *Remote Sensing of Environment, 211*, 146–153. DOI: 10.1016/j.rse.2018.04.015

Milian, J. (2007). Le dilemme entre développement et protection dans les montagnes du Maroc-le cas des parcs du Moyen Atlas. *Géocarrefour, 82*(4), 177–186. DOI: 10.4000/geocarrefour.3002

Modeste, M., Abdellatif, K., Nadia, M., & Zhang, H. (2016). *Impact of land use and vegetation cover on risks of erosion in the Ourika watershed (Morocco).* https://idl-bnc-idrc.dspacedirect.org/bitstream/10625/55962/1/IDL-55962.pdf

Mohajane, M., Essahlaoui, A., Oudija, F., El Hafyani, M., & Cláudia Teodoro, A. (2017). Mapping Forest Species in the Central Middle Atlas of Morocco (Azrou Forest) through Remote Sensing Techniques. *ISPRS International Journal of Geo-Information, 6*(9), 275. DOI: 10.3390/ijgi6090275

Mohammed, M. (2008). Les tendances en matière de propriété forestière, de mode de faire-valoir des ressources forestières et d'arrangements institutionnels. *Cbneih*, 175.

Moroccan Forest Ecosystems: Risks, Challenges and Opportunities. (n.d.). *CESE.* Retrieved December 30, 2025, from https://www.cese.ma/docs/ecosystemes-forestiers-du-maroc-risques-defis-et-opportunites/

Mouafik, M., Chakhchar, A., Fouad, M., & El Aboudi, A. (2024). Remote Sensing Technologies for Monitoring Argane Forest Stands: A Comprehensive Review. *Geographies, 4*(3), 441–461. DOI: 10.3390/geographies4030024

Moukrim, S., Lahssini, S., Alaoui, H. M., Rifai, N., Arahou, M., & Rhazi, L. (2018). Modélisation de la distribution spatiale des espèces endémiques pour leur conservation: Cas de l'Argania spinosa (L.) Skeels. *Revue d'Écologie, 73*(2), 153–166. DOI: 10.3406/revec.2018.1923

Mourjane, M., Tabyaoui, H., & Hammichi, F. E. (2024). Integration of DInSAR and SBAS radar techniques to delineate the main areas of river recharge by sediments using ALOS-PALSAR-2 data: Oued Ghis case in Morocco. *BIO Web of Conferences, 109*, 01028. DOI: 10.1051/bioconf/202410901028

Naggar, M. (2018). Stratégie sylvopastorale et renouveau des pratiques pastorales en forêt au Maroc. *Revue Forestière Française, 70*(5), 487–501. DOI: 10.4267/2042/70133

Naggar, M., & Lahssini, S. (2017). L'aménagement forestier et les enjeux de restauration des écosystèmes forestiers au Maroc. *Forêt Méditerranéenne, 38*(1), 49–62.

Nelson, R. (2013). How did we get here? An early history of forestry lidar[1]. *Canadian Journal of Remote Sensing, 39*(sup1), S6–S17. DOI: 10.5589/m13-011

Okacha, A., Salhi, A., Arari, K., El Badaoui, K., & Lahrichi, K. (2023). Soil erosion assessment using the RUSLE model for better planning: A case study from Morocco. *Modeling Earth Systems and Environment, 9*(3), 3721–3729. DOI: 10.1007/s40808-023-01731-4

Roumane, A., Bhukuth, A., Bazin, D., Roumane, A., Bhukuth, A., & Bazin, D. J. (2010). *La gestion des ressources forestières au Maroc: Une approche par le capital social.* https://www.academia.edu/download/97757843/6570607.pdf

Ruffault, J., Curt, T., Moron, V., Trigo, R. M., Mouillot, F., Koutsias, N., Pimont, F., Martin-StPaul, N., Barbero, R., Dupuy, J.-L., Russo, A., & Belhadj-Khedher, C. (2020). Increased likelihood of heat-induced large wildfires in the Mediterranean Basin. *Scientific Reports, 10*(1), 13790. DOI: 10.1038/s41598-020-70069-z PMID: 32796945

Ryan, C. M., Berry, N. J., & Joshi, N. (2014). Quantifying the causes of deforestation and degradation and creating transparent REDD+ baselines: A method and case study from central Mozambique. *Applied Geography (Sevenoaks, England), 53*, 45–54. DOI: 10.1016/j.apgeog.2014.05.014

Saddik, A., Hssaisoune, M., Belaqziz, S., Labbaci, A., Tairi, A., Meskour, B., & Bouchaou, L. (2024). Assessing the health and yield of argan trees in Morocco's unique ecosystem: A multispectral and machine learning approach. *International Journal of Remote Sensing, 45*(24), 9192–9220. DOI: 10.1080/01431161.2024.2407560

Sanaa, F., Imane, S., Mohamed, B., Kenza, A. E., Souhail, K., Lfalah, H., & Khadija, M. (2022). Biomass and Carbon Stock Quantification in Cork Oak Forest of Maamora Using a New Approach Based on the Combination of Aerial Laser Scanning Carried by Unmanned Aerial Vehicle and Terrestrial Laser Scanning Data. *Forests, 13*(8), 1211. DOI: 10.3390/f13081211

Sarraf, M., & Jorio, A. (2010). Land Degradation: The Case of Morocco. *The Cost of Environmental Degradation*, 75.

Scarascia-Mugnozza, G., Oswald, H., Piussi, P., & Radoglou, K. (2000). Forests of the Mediterranean region: Gaps in knowledge and research needs. *Forest Ecology and Management, 132*(1), 97–109. DOI: 10.1016/S0378-1127(00)00383-2

Schwörer, C., Morales-Molino, C., Gobet, E., Henne, P. D., Pasta, S., Pedrotta, T., Van Leeuwen, J. F. N., Vannière, B., & Tinner, W. (2024). Simulating past and future fire impacts on Mediterranean ecosystems. *Journal of Ecology*, *112*(5), 954–970. DOI: 10.1111/1365-2745.14293

Sebbar, B., Moumni, A., Lahrouni, A., Chehbouni, A., Belghazi, T., & Maksoudi, B. (2022). Remotely Sensed Phenology Monitoring and Land-cover Classification for the Localization of the Endemic Argan Tree in the Southern-west of Morocco. *Journal of Sustainable Forestry*, *41*(10), 1014–1028. DOI: 10.1080/10549811.2021.1897018

Simonneaux, V., Cheggour, A., Deschamps, C., Mouillot, F., Cerdan, O., & Le Bissonnais, Y. (2015). Land use and climate change effects on soil erosion in a semi-arid mountainous watershed (High Atlas, Morocco). *Journal of Arid Environments*, *122*, 64–75. DOI: 10.1016/j.jaridenv.2015.06.002

Simou, M. R., Houran, N., Benayad, M., Maanan, M., Loulad, S., & Rhinane, H. (2024). Monitoring forest degradation over four decades using remote sensing and machine learning classification algorithms in bouskoura, morocco. *The International Archives of the Photogrammetry, Remote Sensing and Spatial Information Sciences*, *XLVIII-4-W9-2024*, 337–342. GeoAdvances 2024 – 8th International Conference on GeoInformation Advances - 11–12 January 2024, Istanbul, Türkiye. DOI: 10.5194/isprs-archives-XLVIII-4-W9-2024-337-2024

Thaiki, M., Bounoua, L., & Dekkaki, H. C. (2023). Using Satellite Data to Characterize Land Surface Processes in Morocco. *Remote Sensing (Basel)*, *15*(22), 5389. https://ntrs.nasa.gov/citations/20230017899. DOI: 10.3390/rs15225389

The IUCN Red List of Threatened Species. (n.d.). IUCN Red List of Threatened Species. Retrieved October 31, 2025, from https://www.iucnredlist.org/en

Tian, L., Wu, X., Tao, Y., Li, M., Qian, C., Liao, L., & Fu, W. (2023). Review of Remote Sensing-Based Methods for Forest Aboveground Biomass Estimation: Progress, Challenges, and Prospects. *Forests*, *14*(6), 1086. DOI: 10.3390/f14061086

Valavanidis, A., & Vlachogianni, T. (2011). Ecosystems and biodiversity hotspots in the Mediterranean basin threats and conservation efforts. *Sci Adv Environ Toxicol Ecotoxicol*, *10*, 1–24.

White, J. C., Wulder, M. A., Hermosilla, T., Coops, N. C., & Hobart, G. W. (2017). A nationwide annual characterization of 25 years of forest disturbance and recovery for Canada using Landsat time series. *Remote Sensing of Environment*, *194*, 303–321. DOI: 10.1016/j.rse.2017.03.035

Wulder, M. A., Loveland, T. R., Roy, D. P., Crawford, C. J., Masek, J. G., Woodcock, C. E., Allen, R. G., Anderson, M. C., Belward, A. S., Cohen, W. B., Dwyer, J., Erb, A., Gao, F., Griffiths, P., Helder, D., Hermosilla, T., Hipple, J. D., Hostert, P., Hughes, M. J., & Zhu, Z. (2019). Current status of Landsat program, science, and applications. *Remote Sensing of Environment*, *225*, 127–147. DOI: 10.1016/j.rse.2019.02.015

Zadbagher, E., Marangoz, A. M., & Becek, K. (2023). Characterizing and estimating forest structure using active remote sensing: An overview. *Advances in Remote Sensing*, *3*(1), 38–46.

KEY TERMS AND DEFINITIONS

Artificial Intelligence (AI): A computational field focused on developing algorithms and systems capable of learning, reasoning, and adapting from data to perform tasks such as classification, prediction, and pattern recognition in forest monitoring.

Remote Sensing (RS): The acquisition of environmental information about the Earth's surface from satellite or airborne sensors, enabling large-scale observation of forest cover, biomass, and ecosystem changes without direct ground contact.

Machine Learning (ML): A subset of AI that uses statistical and computational models to detect patterns and relationships within large datasets, improving predictive accuracy for forest dynamics, wildfire risk, and vegetation health.

Forestry 5.0: An emerging paradigm integrating digital technologies, artificial intelligence, automation, and participatory governance to achieve sustainable, adaptive, and transparent forest management.

Sustainable Forest Management: A management approach balancing ecological, economic, and social objectives to ensure forest resources meet present needs without compromising their future regenerative capacity.

Chapter 9
Mapping Carob Trees Dieback Caused by Magot-Monkey Using UAV-Based Multispectral Data, Biomass-Chlorophyll Indices, and Machine Learning

Abderrazak Bannari
Space Pix-Map International Inc., Gatineau, Canada

Ahmed Selouani
Geofly, Temara, Morocco

Abderrahmene El Ghmari
http://orcid.org/0000-0002-2773-6048
Geomatic, Geo-Resources and Environment Laboratory, Department of Earth Sciences, Université Sultan Moulay Slimane Béni Mellal, Morocco

ABSTRACT

This study investigates the effectiveness of combining UAV multispectral imagery, biomass-chlorophyll indices, and an unsupervised machine learning classifier (UMLC) to map carob tree dieback caused by the Barbary macaque. The data were rigorously preprocessed, twelve biomass-chlorophyll indices were derived and UMLC was implemented, analyzed, and validated. The results show that among the tested

DOI: 10.4018/979-8-3373-6239-7.ch009

indices, TDVI and CIG exhibited similar and superior performance with excellent dynamic range, and the map generated from the TDVI threshold showed a clear spatial distribution of carob classes. Similarly, UMLC proved its ability to classify individual trees according to their degree of dieback in a complex environment, achieving an excellent overall classification accuracy of 93%. However, since the tested approaches provide almost alike results, the method based solely on spectral indices is recommended for its simplicity, facility to transfer between sensors, and it constitute an accurate alternative compared to the sophisticated and complex methods such UMLC.

1. INTRODUCTION

Since antiquity, the carob tree (Ceratonia siliqua L.), also called the bread of Saint John the Baptist or the Egyptian fig tree has been cultivated in most countries of the Mediterranean basin. This species is environmentally and economically important (Catarino, 1996; Batlle & Tous, 1997; Naggar & Lahssini, 2015). It is cultivated in semi-arid and dry regions and can grow in various soils, including poor sandy soils, rocky hillsides, and deep soils, but it prefers sandy, well-drained loams and calcareous soils (Mahdad, 2013). Carob tree plantations play an important role in carbon sequestration, mitigating the effects of climate change through CO_2 fixation (Palacios-Rodríguez et al., 2022). Its value has been recognized by the ancient Greeks, who brought it from its native Middle East to Greece and Italy, and by the Arabs, who disseminated it along the North African coast and into Spain and Portugal. More recently, it spread to regions such as the USA, Mexico, Chile, and Argentina by Spaniards; to parts of Australia and New Zealand by Mediterranean emigrants; and to South Africa and India by the English. Due to low management requirements, carob trees are suitable for part-time farming and are useful for ornamental purposes, landscaping, and reforestation. They grow as evergreen sclerophyllous shrubs or trees up to 10 m tall, with a wide semi-spherical crown and a thick trunk with rough brown bark and sturdy branches. The leaves are 10-20 cm long, alternating with or without a terminal leaflet 3-7 cm long, with oval to elliptical shapes (Diamantoglou and Mitrakos, 1981). Carob pods are rich in sweet pulp and are used for human nutrition and animal feed, while the wood is suitable for fuel. Currently, the main interest is the production of seeds for gum extraction and other uses, including chemical and paper manufacture, cosmetics, and pharmaceutical applications addressing various human health issues.

Normally, the carob tree is free from serious insect and disease problems and does not receive pesticide sprays. However, many pests and diseases impede carob trees. Gugliuzzo et al. (2019) summarize the main bio-ecology aspects and natural enemies of carob pests in different Mediterranean locations. For instance, Lepidopteran (butterflies and moths) and coleopteran (beetles and weevils) species are the most common insect pests recorded on carob (Gugliuzzo et al., 2019). While, the most harmful insect is the polyphagous larvae of the leopard moth (Zeuzera pyrina L.), which attacks the wood of trunk and branches, causing serious damage, particularly to young trees. In addition, black aphids (Aphis fabae) mainly attack the terminal shoots of young trees (Tous et al., 1996). The downy mildew disease caused by Oidium ceratoniae C. attacks the pods, leaves and branches at different times of the year. Bonsignore (2012) reported that adult Apate monachus beetles cause damage to carob trees, forming tunnels in the trunk and branches. Likewise, other pests are small rodents such as ground squirrels (Pitymys spp.), gophers (Geomyidae) and rats (Rattus spp.) cause severe damage to carob trees and severely damage the root system (Thomson, 1977; Batlle and Tous, 1997). They can remove the bark not only from young shoots but also from the older ones, which leads to the death of the branches or even of the tree. Moreover, among mammalian pests, Magot monkeys (Macaca sylvanus, Cercopithecidae) become a new source of trouble because they can damage seriously carob leaves, pods in storage structures, and equally they gnaw through the outer bark layer. Unfortunately, this activity disrupts the tree's ability to transport nutrients and water between the leaves and roots, thus leading to the death of the portion or the totality of the carob tree. These harmful activities of the Magot monkey are related to the impacts of climate change and global warming during the last half of the 20th century (Giannakopoulos et al., 2005), which increased the environmental and ecosystem changes.

In the Atlas Mountains of Morocco, which are the main sources supplying several regions with water, a trend of increasing rainfall deficits and decreasing runoff was observed (Chaponnière & Smakhtin, 2006). Undoubtedly, this situation has strongly influenced the availability of water, and it is the most important driver of change in several ecosystems. Indeed, in animal and vegetation ecosystems associated with mountainous regions, especially in isolated areas, species have limited options to find favorable conditions for survival due to climate change (Thuiller et al., 2005). For these reasons, the Magot monkey has left his natural environment because his food (i.e. acorns, saw palmetto fruit, insects, wild herbaceous plants, etc.) at high altitudes has become increasingly scarce and, consequently, he takes up residence at low altitudes valleys humid and rich in carob trees. In the mountainous regions of the Béni-Mellal and Azilal provinces (Morocco), several species of mammals are common to several wooded formations, including the Magot monkey that occupies a large forest area of Holm Oak offering a wide food diversity. Regrettably, due to climate warming, an isolated population of Magot monkeys recently found refuge in the Taghzirt-valley (Béni-Mellal Province), and they are aggressively attacking the carob forest causing its dieback, serious damage, and probably its decline in this region. Dieback manifests itself at the tree level by showing visible damage to the canopy, such as reduced foliage or branches mortality or a complete death of the trees. Consequently, the carob forests on the mountainsides around the Taghzirt-valley region are undergoing deforestation and habitat degradation. To measure the economic, ecological and landscape impacts of monkeys on this natural resource, it is necessary to assess the damage caused by this primate. Obviously, remote sensing science and technologies have been applied in forest surveys and the detection of disturbances caused by damage related to biotic and abiotic factors (Wulder et al., 2006; Bahri et al., 2008; Coops et al., 2009; Gomez et al., 2020).

Over the last half-century, satellites remote sensing were more frequently used for the detection of forest outbreaks, infestation, defoliation, dieback, and damage; as well as, time-series monitoring at the local, regional and global scales. Indeed, medium spatial and spectral resolutions such as the Landsat sensor series (MSS, TM, ETM+, and OLI) and the series of SPOT and Sentinel satellites sensors have been commonly used but they have a limited capability for operational forest health management at the tree-crown level compared to aerial detection or field-ground surveys (Rullan-Silva et al., 2013). However, new commercial and advanced generations of satellites and sensors with high spatial, spectral, and radiometric resolutions acquired by IKONOS, Pleiades-HR, World-View series, QuickBird, GeoEye, RapidEye, etc. are suitable and more promising solutions for forest management and detection of pest infestations at very large scale (Dennison et al., 2010; Bannari et al., 2016, 2017 and 2018). Obviously, the detection, mapping and monitoring of forest dieback; as well as different stages of outbreaks and damage must consider data sources with scales-hierarchy oscillating from medium to finer scale for analysing, respectively, from a large area to the crown of the individual tree (Wulder et al., 2006; Coops et al., 2009). Indeed, the tree crown is the main forest component to be observed and analysed for estimating damage conditions by assessing, particularly, foliage discoloration, defoliation, and disturbances of physiological and biological variables. These are related to stress factors and are considered reliable parameters for assessing, monitoring, inventorying, and mapping forest canopy conditions (Innes, 1993).

Furthermore, remote sensing is a dynamic technology that continuously improves sensors, methods, products, data availability and accessibility (Rullan-Silva et al., 2013). Since these new sources of satellite data are still very expensive to cover small areas, the use of unmanned aerial vehicle (UAV) as an alternative has sparked the interest of the scientific community and planners in monitoring forest health and disturbance (Yao et al., 2019). Definitely, UAV-multispectral imagery can complement satellite data to monitor vegetation cover extent and changes (Dash et al., 2018). The first studies carried out using the UAV data focused on precision agriculture and forestry as a tool for collecting field data for sampling, calibrating and validating results extracted from satellite data (Zhang and Kovacs, 2012; Salamí et al., 2014; Candiago et al., 2015; Lehmann et al., 2015; Torresan et al., 2017; Otsu et al., 2018). Nowadays, UAV-multispectral high spatial resolution applications have become rapidly expanding in the field of remote sensing, among others we mention monitoring forest health, defoliation and stress conditions (Manfreda et al., 2018; Otsu et al., 2019). According to Näsi et al. (2015), this technology can be efficient in carrying out small-area inspections and, therefore, analyzing the reflectance characteristics at the tree crown at a high level of detail. Otsu et al. (2019) evaluated the potential of UAV-multispectral data for coniferous forest trees defoliation by moth (*Thaumetopoea pityocampa*) and Pine species discrimination in Spain using simple threshold analyses of several vegetation indices.

Moreover, Cardil et al. (2017) utilized the synergy between the maximum likelihood classification and UAV-RGB data acquired from Pin Forest in Spain to quantify moth defoliation at the individual tree crown level. Based on vegetation indices, different classification methods were used to detect infestation in Oak stands (*Quercus sp.*) by splendor beetle (*Agrilus biguttatus Fab*) in Germany (Lehmann et al., 2015). As well as, to detect infested Picea abies in Finland (Näsi et al., 2015), to assess Alnus glutinosa defoliation in Belgium (Michez et al., 2016), and to predict Pinus radiata discoloration classes in New Zealand (Dash et al., 2017). In Norway, Brovkina et al. (2018) demonstrated the capability of UAV for monitoring Spruce health conditions and the identification of dead tree class; as well as, for separating Spruce and Silver-Fir using vegetation indices. These results highlighted that anomaly in the tree crown are visible from a bird's eye view, hence the interest in UAV technology in forestry. Accordingly, the objective of the present research is the development of a simple and innovative method for mapping the carob dieback and damage caused by the Magot monkey based on the synergy between UAV-multispectral data, digital photogrammetry methods, biomass-chlorophyll indices and unsupervised machine learning classifier (UMLC) as a subset of artificial intelligence.

2. MATERIALS AND METHODS

Figure 1 describes the methodology used in the present research, which is structured in four interrelated steps: 1) UAV mission planning and data acquisition; 2) data preprocessing; 3) data processing; and 4) results analysis and validation. The UAV flight was performed over a carob forest located in a valley of a watershed in the Middle Atlas Mountains. The preprocessing steps were rigorously applied for geometric corrections, topographic rectification, and radiometric calibration. Then, the spectral and radiometric performance of the DJI Phantom-4 camera (onboard UAV) was analysed to identify anomalous trees individually. Regarding the processing steps, two methods were considered. First, the performance and potential of the most popular and widely used biomass (NDVI, SAVI, TDVI, EVI, and VARI) and chlorophyll (MRESR, CIG, CIRed-Edge, GNDVI, SRPI, PSRI, and CARI) indices were tested to discriminate and map dieback and damage in carob forests. Afterwards, a histogram threshold classification was applied to the index that offered the best performance. Second, the ability of UMLC based on K-Means clustering was examined to discriminate carob tree classes. Finally, the last step involved the comparison and validation of the results obtained against the ground truth.

Figure 1. Methodology flowchart

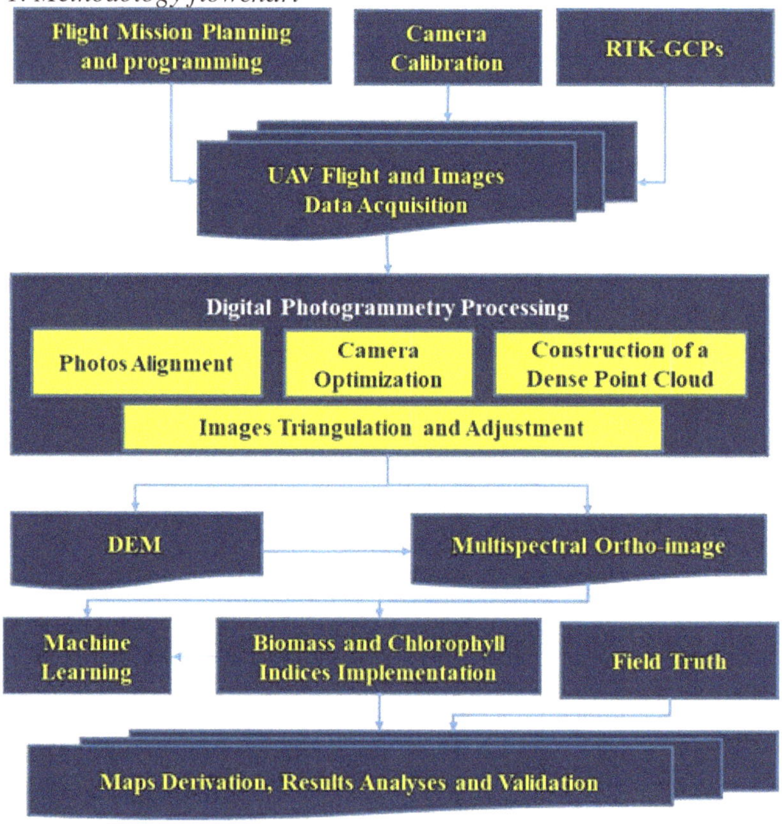

2.1. Study Area

The Taghzirt Valley is located approximately 15 km northeast of the city of Beni-Mellal in Morocco (Figures 2a-2c) in a watershed that forms a transitional zone between the mountain and the plain. It is a narrow, deep, and well-preserved valley (Figure 2b). The climate is sub-humid Mediterranean, varying from humid conditions in the mountainous areas to sub-arid conditions at lower elevations. It is characterized by a hot and dry summer and a cold and wet winter. Over the past 50 years, climate change-related disruptions in this region have caused cycles of intense droughts and catastrophic floods. Indeed, during this period, rainfall varied greatly between 260 and 1033 mm, with an annual average of about 580 mm, while the annual temperature ranged between 14 and 46°C for winter (January) and summer (July), respectively.

The topography is highly rugged characterized by very steep gradients, slopes exceeding 30°, and ridgelines reaching elevations of approximately 2,300 m. The morphological characteristics of the watershed largely determine the valley's configuration, with northeast-facing slopes contributing to an intense hydraulic regime along the valley floor. The geological structure shows a succession of anticlines and synclines with brittle tectonics. Limestone, carbonate, sandstone, and dolomite massifs with fracturing and karstic formations dominate the area. Along the valley and on the slopes, there are continental formations, scree, and alluvial deposits (Sallak, 2019).

In term of vegetation cover, the site is characterized by mixed and relatively dense vegetation. Carob trees occur in association with various other species including Holm Oak (Quercus sect. Quercus), Zean Oak (Quercus canariensis), Kermes Oak (Quercus coccifera), Thuyas maquis with common Cypress (Cupressus sempervirens), Phoenician Juniper (Juniperus phoenicea), Mediterranean Olive (Olea europaea), Aleppo Pine (Pinus halepensis), and Pistacia lentiscus (a dioecious evergreen shrub or small tree). In addition, some carob stands grow alongside other aromatic and medicinal dwarf shrubs, such as Greek sage (Salvia fruticosa), germander (Teucrium divaricatum and T. polium), savory (Thymbra spicata), and resinous spurge (Euphorbia resinifera) (Sallak, 2019). This lush environment, with a permanent stream at the valley floor, provides favorable ecological conditions that have enabled the Barbary macaque (*Macaca sylvanus*) to establish itself successfully (Figure 2d).

Figure 2. Study area, Taghzirt Valley location in Morocco (a, b, and c), and the Magot monkey sitting on a carob branch (d)

2.2. Data Acquisition

The data acquisition was conducted using a DJI Phantom-4 Pro+ quad-copter equipped with six digital cameras RGB and multispectral (Figure 3a). The UAV system is composed of four fundamental elements including aircraft, flight control system, cameras, and ground control segment. The aircraft was connected to the ground control segment via telemetry allowing the operator all commands and operations for flight mission control. Thanks to an inertial navigation system (INS) and RTK-GNSS devices onboard, the aerial system performs an automated survey, navigation between obstacles, and records coordinates of the photo perspective-center when each time the camera's shutter was activated. Moreover, other mission parameters were measured such as GPS location, acquisition time, and elevation that are required by the photogrammetric software to perform feature matching and block adjustment (Bannari et al., 2021). The six-camera bloc was mounted on a three-axis gimbal (gyroscope) for its stability, and the rotation angles (roll, pitch, and yaw) were measured by the INS; thus, allowing images to be acquired with the optical axis of the cameras pointing in the nadir direction towards the ground. The imaging system used consists of one RGB and five multispectral cameras (Figure 3b) covering the blue (450 ± 16 nm), green (560 ± 16 nm), red (650 ± 16 nm), red edge (730 ± 16 nm), and near infrared bands (840 ± 26 nm), all with 2 MP resolution (Figure 3c). These cameras employ CMOS sensors offering an excellent image clarity, reduced noise, and enhanced contrast with a field of view (FOV) of 62.7°, and a focal length of 5.74 mm. The camera system supports microSD cards with capacities up to 128 GB (DJI, 2020).

Figure 3. UAV DJI Phantom-4 Pro+ with the cameras onboard (a), P4 multispectral imaging system (b), RGB image and five spectral bands acquired with the six cameras (c), GCP measurements using DGPS rover station (d), and Real-time kinematic (RTK) GNSS module (e)

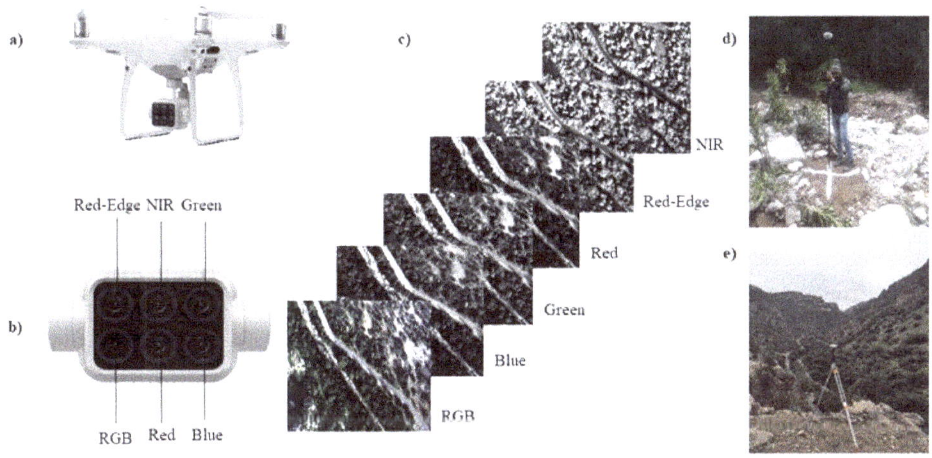

The flight mission was planned based on the coordinate corners of the study area and the predetermined camera parameters for the flight (i.e., flight altitude, image scale, overlaps, base, flight speed, etc.). To avoid discontinuities or gaps between photos and to ensure appropriate stereoscopic coverage, lateral and longitudinal overlaps between photos were set at 80% and 60%, respectively. During the field mission, all this information was uploaded to the UAV flight control system, which communicated with the ground remote control system to automatically manage flight execution and digital image acquisition. The flight mission was carried out vertically in a single day (December 28th, 2022) under excellent atmospheric conditions around noon, when the solar angle was minimal, generating weak shadow effects in the images. The UAV surveyed the study area at a horizontal speed of 4 m/s and at an altitude of 450 m above the valley floor, resulting in a pixel size of 25 cm. The mission was conducted along eight straight flight lines at the nadir, allowing the acquisition of 463 photos with the six cameras (77 photos for the RGB camera and 77 photos for each spectral band).

Prior to photos acquisition, 35 GCPs were distributed uniformly over the study site, considering all existing topographic variability (slopes, orientations, elevations, and roughness). Twenty-six (26) GCPs were used in the calibration and adjustment process, while the remaining nine (9) served as checkpoints. Since neither geodetic benchmarks nor fixed physical markers existed within the study area, all GCPs were marked before the flights using white-painted benchmarks of 50 × 50 cm, and their coordinates were measured with a rover station (Figure 3d) in Conic Conformal Lambert (CCL) projection using the Moroccan geodetic reference (Merchiche). These points are essential for photogrammetric operations, including stereo-preparation, geo-positioning, camera calibration, aero-triangulation; dense point cloud generation, 3D stereo-model construction, and projection of derived products into a specific ground coordinate system (Bannari et al., 1997). To measure the perspective center of each acquired photo, the DGPS Leica 500-RTK (Figure 3e) recorded the relative positions of the six camera centers and the phase center of the onboard D-RTK antenna. Measurements were corrected in real time by integrating differential corrections based on known geodetic benchmarks and GNSS satellite signals updated by the ground control segment (RTK base station, Figure 3e), ensuring accuracies of $\leq \pm$ 1 cm for planimetric coordinates and $\leq \pm$ 1.5 cm for altimetric coordinates (DJI, 2020).

2.3. Data Preprocessing

2.3.1 Geometric Rectifications

To perform photogrammetric preprocessing (i.e., geometric corrections and ortho-rectification) of digital photos acquired by UAV cameras, the "Agisoft PhotoScan" software was used (Pix4D-Mapper, 2017). The geometric correction and rectification workflow includes photo alignment, dense point cloud generation, polygonal model (mesh) construction, digital surface model (DSM) and digital elevation model (DEM) generation, mosaicking, and ortho-rectification (Bannari et al., 2021; 2024). Initially, the software automatically reads the metadata of each photo recorded onboard the UAV, including camera calibration parameters, approximate coordinates of the photo centers, flight altitude, and timestamps. The first task consists of checking the quality of each acquired photo individually and identifying any irregularities during the flight mission (e.g., blurred or distorted images). Once this step is completed and all photos are verified for quality, the photos alignment process begins. At this stage, the software identifies common points in the overlap zones and matches the photos for auto-calibration (Figure 4a) and correction of distortions (focal length, principal point coordinates, radial and tangential distortions), and subsequently builds a sparse point cloud. In other words, this stage includes aero-triangulation and bundle block adjustment. It searches for feature points in the photos and matches them across overlap zones using tie points. The software then determines the camera position for each photo and computes adjusted camera calibration parameters (internal and external orientations) with increased refinement and precision. Accordingly, the sparse point cloud is adjusted and geo-referenced to the ground coordinate system. The relationships between images coordinates (Figure 4c) and ground coordinates (Figure 4b) are computed using a stereo-matching algorithm based on the collinearity and coplanarity equations (Bannari et al., 1995a; 1997). This operation also allows the calculation of coordinate errors. Once these errors are within acceptable limits, the dense point cloud model is generated, followed by filtering operations (median filter with 7 × 7 windows) to remove outliers caused by noisy or poorly focused photos, or points not located on the ground surface. Finally, the adjusted dense point cloud is interpolated using triangulation to generate the polygonal model (mesh), DSM, DEM, and the ortho-rectified image.

Figure 4. Relationship between camera positions and GCP for self-calibration (a), GCP materialisation in the field (b), GCP appearing on the image acquired at the UAV altitude (c), and generate DEM (d)

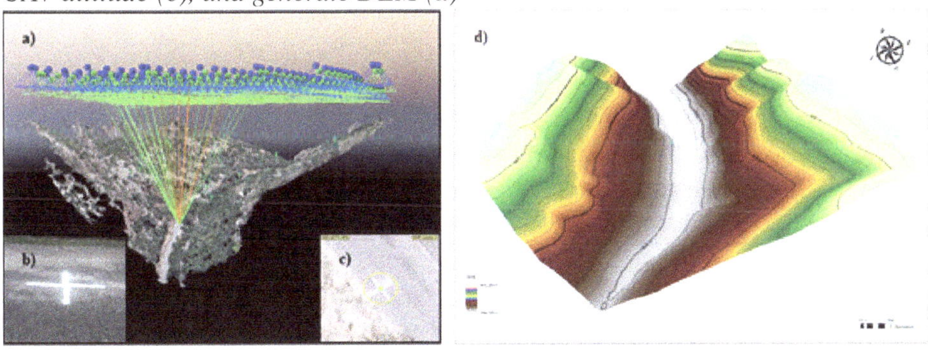

2.3.1 Radiometric Calibration

In the present study, a ground-based vicarious calibration was applied using *in-situ* spectroradiometric measurements and a semi-empirical line approach (SELA) based on pseudo-invariant targets (PITs) to convert the raw data (DN(λ)) acquired with a UAV-multispectral imaging system to the ground-surface reflectance factors ($\rho_G(\lambda)$) in the visible, red-edge and near-infrared (NIR) wavebands. This approach is an appropriate alternative to accurately establish the link between $\rho_G(\lambda)$ and DN(λ). According to Morin et al. (2001), to take into account the atmospheric path radiance, it is necessary to have at least two spectrally different targets within the scene (i.e. bright and dark targets). Whereas, the consideration of more than two targets allows greater confidence of the transformation equations (Bannari and Al-ali, 2020; Karpouzli and Malthus, 2003). For this calibration process, 16 samples of four types of homogeneous PITs (four samples of each target) were considered with a range of bright and dark surface cover types (i.e. covering a full range of brightness from high to low reflectance). These include white ceramic tiles, concrete surface, asphalt, and very dense carob trees (evergreen trees) all marked and accurately located using a DGPS before the flight (Figure 5). Their sizes were directly related to the camera's spatial resolution (larger than 2 to 4 times that of the pixel size) which facilitates their identification. Although these targets are non-Lambertian, they are potentially used in vicarious calibration operations (Slater et al., 1987; Bannari et al., 1999; Wang and Myint, 2015).

Figure 5. Calibration targets on the left [white ceramic tiles (1), concrete surface (2), asphalt (3), and dense vegetation cover (4)]. On the right [relative spectral response functions of the DJI Phantom-4 Pro+ cameras (visible, red-edge, and NIR) and field-measured spectra of the PITs used]

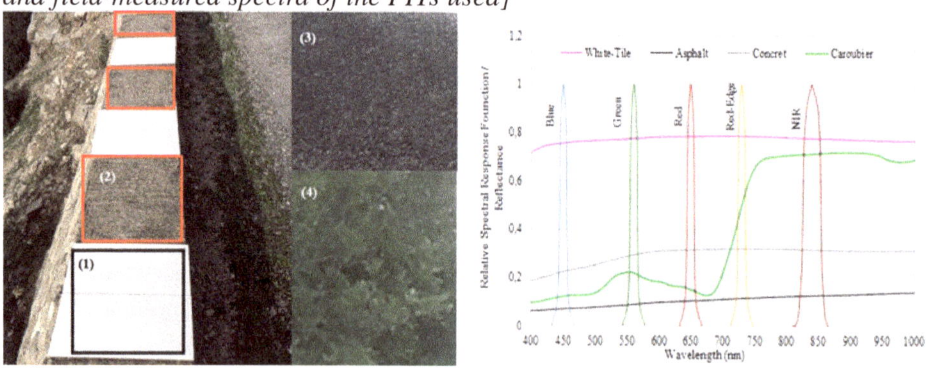

Concurrent with the UAV overflight, spectroradiometric measurements were acquired above each PIT using an ASD spectroradiometer (ASD, 2015) in the visible, red-edge, and NIR wavelengths. This instrument acquires a continuous spectrum with a 1.4 nm sampling interval from 350 to 1000 nm and resamples the measurements at 1-nm intervals. The data were collected at nadir to minimize angular dependence, with a field of view (FOV) of 25° and a solar zenith angle at nadir, averaging 40 measurements per target. The ASD was positioned over each PIT, allowing observation of a surface area of approximately 900 cm^2. The reflectance factor of each sample was calculated by taking the ratio of the target radiance to the radiance obtained from a calibrated "Spectralon panel . Moreover, the corrections were applied for the wavelength dependence and non-Lambertian behavior of the panel (ASD, 2015). The measured spectra were then resampled and convolved using the radiative transfer code (RTC) CAM5S (Teillet and Santer, 1991) and the relative spectral response functions specific to the Phantom DJI Pro 4+ sensor bands (Figure 5). Accordingly, the SELA was implemented, linear regression fits ($p < 0.05$) were established, and the calibration equations were developed (Equation 1). An accuracy assessment was performed by comparing the calibrated reflectances to those measured in the field.

$$\rho_G(\lambda)_i = A_i \,.\, DN(\lambda)_i + B_i \tag{1}$$

Where: $\rho_G(\lambda)_i$ is the ground-surface reflectance factor (PITs) in each spectral band "i". A_i and B_i are the slope and the intercept of the radiometric calibration equation established for each spectral band "i".

2.4. Data Processing

2.4.1. Biomass and Chlorophyll Indices

Since the emergence of remote sensing as a new scientific discipline in the early 1970s, vegetation indices (VI's) were involved as radiometric measurements of the spatial and temporal distribution of vegetation photo-synthetically active (Bannari et al., 1995b). Based on the strong absorption of chlorophyll in the red and intense reflectivity of canopy biomass in the NIR, these indices play an important role in deriving various biophysical parameters, including percentage of vegetation cover, leaf area index (LAI), absorbed photo-synthetically active radiation (APAR), production rate of the biomass, etc. Using the red and NIR bands, the *Normalized Difference Vegetation Index* (NDVI) was proposed by Rouse et al. (1974) at the dawn of remote sensing. To minimize the effects caused by soil background (color and brightness) on the NDVI, the *Soil Adjusted Vegetation Index* (SAVI) was suggested by Huete (1988). Moreover, to overcome the limitations of linearity and saturation, reduce the noise of atmospheric effects, and remove the artefacts of soil optical properties the *Enhanced Vegetation Index* (EVI) was developed also by Huete et al. (2002). Furthermore, to describe the vegetation cover fraction independently to the soil background, to reduce the saturation problem, and to enhance the vegetation dynamic range linearly, Bannari et al. (2002) established the *Transformed Difference Vegetation Index* (TDVI). Likewise, the *Visible Atmospherically Resistant Index* (VARI) was proposed by Gitelson et al. (2002a) to estimate the green vegetation fraction. These vegetation or biomass indices were developed and used to establish a close relationship between radiometric responses and vegetative cover density, health and stress conditions.

Likewise, the chlorophyll (Chl) content is an essential biochemical parameter to track the main developmental stages of vegetation cover in forestry and agriculture. In this perspective, several chlorophyll indices were developed to estimate chlorophyll content both at the leaf and canopy level (Bannari et al., 2007a and 2008) and to discriminate between healthy, stressed, and senescent vegetation (Ustin et al., 1988; Coops et al., 2009). For instance, the *Modified Red Edge Simple Ratio* (MRESR) index, which is a modification of the broadband simple ratio using the blue and red-edge bands for vegetation stress detection in forestry and precision agriculture (Datt, 1999; Sims and Gamon, 2002). While, the *Chlorophyll Index Green* (CI_G) or *Chlorophyll Index Red-Edge* ($CI_{Red\text{-}Edge}$) was developed for estimating the chlorophyll content using the ratio of reflectivity in the NIR and green or red-edge bands (Gitelson et al., 2006). Additionally, the *Green Normalized Difference Vegetation Index* (GNDVI) was developed by Gitelson et al. (1996) using the green band instead of the red one for the estimation of chlorophyll only without being influenced by the soil background and the atmospheric effects. Penuelas et al. (1993; 1994) proposed the *Simple Ratio Pigment Index* (SRPI) based on the ratio of the carotenoid (Car) and Chl-a content. They found that this index correlates very well with different levels of mite attacks in apple trees, as the carotenoid / Chl-a ratio increases with increasing levels of mite attack. Similar performances were observed from a wide range of leaves from different species (maize, wheat, tomato, soybean, sunflower, sugar beet, and maple) when the SRPI was highly correlated with the carotenoid / Chl-a ratio (Penuelas *et al.,* 1995). Developed by Kim et al. (1994), the *Chlorophyll Absorption in Reflectance Index* (CARI) was designed to reduce the variability of photo-synthetically active radiance due to the presence of diverse non-photosynthetic materials.

Proposed by Merzlyak et al. (1999), the *Plant Senescence Reflectance Index* (PSRI) is an indicator of the senescence process of tree crown leaves, as color changes in senescent plant tissues are associated with the degradation of chlorophyll and carotenoids. It estimates the characteristic features and the evolution over time of the changes accompanying natural leaf senescence and possibly induced by stress, dieback, or outbreak. Obviously, the quantitative assessment of these physiological disturbances was based on the difference in the transformation of chlorophyll and carotenoid. All these biomass and chlorophyll indices discussed above were tested in this research, and their mathematical formulations are summarized below. They were selected based on their performance and they integrate in their equations the five spectral bands collected by UAV DJI Phantom-4 multispectral camera.

$$\text{NDVI} = (\rho_{\text{NIR}} - \rho_{Red}) \,/\, (\rho_{NIR} + \rho_{Red}) \text{ (Rouse et al., 1974)} \quad (2)$$

$$\text{SAVI} = 1.5 * (\rho_{\text{NIR}} - \rho_{Red}) \,/\, (\rho_{NIR} + \rho_{Red} + 0.5) \text{ (Huete, 1988)} \quad (3)$$

$$\text{TDVI} = 1.5 * (\rho_{\text{NIR}} - \rho_{Red}) / \left(\sqrt{(\rho_{NIR}^2 + \rho_{Red} + 0.5)} \right) \text{ (Bannari et al., 2002)} \quad (4)$$

$$\text{EVI} = 2.5 * (\rho_{\text{NIR}} - \rho_{Red}) / (\rho_{NIR} + 6 * \rho_{Red} - 7.5 * \rho_{Blue} + 1) \text{ (Huete et al., 2002)} \quad (5)$$

$$\text{VARI} = (\rho_{\text{Green}} - \rho_{Red}) \,/\, (\rho_{Green} + \rho_{Red} - \rho_{Blue}) \text{ (Gitelson et al., 2002a)} \quad (6)$$

MRESR = (NIR – Blue) / (Red-Edge – Blue) (Datt, 1999) (7)

CI_G = (NIR/Green) -1 (Gitelson et al., 2006) (8)

$CI_{\text{Red-Edge}}$ = (NIR/Red-Edge) -1 (Gitelson et al., 2006) (9)

GNDVI = (NIR – Green) / (NIR + Green) (Gitelson et al., 1996) (10)

SRPI = Blue / Read (Penuelas *et al.,* 1993) (11)

CARI= [(Red-Edge – Red) - 0.2 * (Red-Edge - Green)] (Kim *et al.,* 1994) (12)

PSRI = (Red - Green)/Red-Edge (Merzlyak et al., 1999) (13)

2.4.1. Machine Learning

Machine learning (ML) approaches, as a subset of artificial intelligence, combined with robotic platforms, have increasingly been incorporated with UAV-based communications and image processing to improve various scientific domains, especially natural resource management (Bithas et al., 2019) and remote monitoring of forest ecosystems and health (Shi, 2020; Estrada et al., 2023). When integrated into supervised and unsupervised classifiers (i.e., decision trees, k-nearest neighbors, K-means clustering, neural networks, support vector machines, random forests, etc.), these innovative methods offer powerful capabilities for processing large remote sensing datasets, managing high spectral dimensionality, and reducing data redundancy and variability (Liang and Gamarra, 2020). MLC algorithms have achieved significant advancements in forestry, supporting applications such as tree species characterisation, diseases and dieback detection, degradation and mortality assessment, evaluation of water and nitrogen stress, and tree-crown segmentation, etc. (Shi, 2020). For instance, Su et al. (2024) demonstrated the appropriateness of ML for understanding the complex phenomenon of Oak forest dieback. Moreover, the synergy between ML and UAV data has enabled accurate detection of Pine wilt (Yu et al., 2021), Citrus canker disease (Abdulridha et al., 2019), Ash dieback (Bates et al., 2025), and identification of Spruce bark beetle outbreaks (Bozzini et al., 2023). Furthermore, Murugesan and Nanjappan (2025) proposed a hybrid approach combining two ML methods (i.e., spatial knowledge from K-means clustering and temporal modeling using recurrent neural networks) for forest dynamics prediction. For forest fuels and fire management, Chávez-Durán et al. (2022) demonstrated the potential of a hybrid ML method combining K-means cluster analysis with random-forest classification.

The promising interaction between UAVs and satellite systems has enabled the assessment of forest health on a larger scale. Catal et al. (2022) used K-means clustering and vegetation indices to evaluate the potential of UAV and Pleiades data for detecting and mapping Cork Oak mortality. Similarly, Navarro et al. (2019) showed the potential of biomass-chlorophyll indices derived from Sentinel-MSI and UAV data processed with ML classifiers for monitoring Cork Oak decline. In this context, Ni et al. (2025) provided a comprehensive literature review on the importance and role of integrating drone and satellite data with ML classification methods to monitor forest dynamics and tree mortality. In the present study, two unsupervised ML K-means clustering scenarios were considered. The first scenario incorporate only the five preprocessed and standardized spectral bands (visible, red-edge, and NIR). The second scenario integrated the most relevant biomass-chlorophyll indices mentioned above in addition to the five original bands. This unsupervised ML (UML) scheme does not require ground truth for initialization or training parameter selection, relying solely on spectral similarity to determine the optimal number of class clusters (Sinaga and Yang, 2020).

3. RESULTS ANALYIS

3.1. Preprocessing

Within the framework of UAV data preprocessing, the acquired photos were preprocessed geometrically, topographically and grouped into a mosaic image over 2 h 40 min using the “Agisoft PhotoScan software. A workstation equipped with Windows 10 Professional, 64 GB of RAM, an Intel(R) Xeon(R) processor, and a CPU E5-2670 running at 2.5 GHz was used. The output file of the adjusted dense point cloud was interpolated using triangulation to generate a polygon model (mesh), DSM, and DEM. The ortho-image (mosaic) covering the entire study area was generated in five spectral bands and geo-referenced automatically using the WGS84 ellipsoid and CCL map projection based on TRK and GNSS constellation. According to the results, absolute camera position uncertainties after calibration were 0.4 cm and 1.0 cm in planimetry and altimetry, respectively. While, external orientation uncertainties were 0.011° for phi (rotation about Y-axis) and 0.008° for and kappa (rotation about Z-axis). The absolute ground geolocation errors were 0.9 cm for X and Y coordinates and 0.55 cm for altitude (Z). These negligible differences demonstrate the excellent technological performance of the camera, as well as the high accuracy of the RTK-GNSS, D-RTK antenna, and INS devices integrated onboard the UAV.

Additionally, variations in solar irradiance induced by topographic changes in the valley were corrected through the ortho-rectification procedure, which integrates the DEM (Figure 4d) derived throughout the geometric preprocessing chain. This process allows correction of relief distortions and parallax effects according to the spatial arrangement of pixels along the slopes (Bannari et al., 1995a, 1997), while minimizing atmospheric and shadow effects and normalizing the bidirectional reflectance distribution factor (Song et al., 2001; Roy et al., 2016; Schläpfer et al., 2020). Regarding vicarious radiometric calibration, the results showed that the established calibration equations were achieved with excellent coefficients of determination ($0.94 \leq R^2 \leq 0.97$) for all bands. Ground reflectances ($\rho G(\lambda)$) were retrieved with RMSE ≤ 5.4% for the visible bands and around 3% for the red-edge and NIR bands (Bannari et al., 2024b). These RMSE values are consistent with the expected absolute radiometric calibration accuracies reported in the literature, ensuring reliable retrieval of biophysical and physiological parameters comparable over time and consistent with other calibrated datasets (Thome, 2001; Thome et al., 2004; Bannari et al., 1999; Bannari and Al-Ali, 2020).

3.2. Spectral Analysis

It is well known that each vegetation pigment exhibits distinct spectral absorption features, allowing remote sensing to assess tree conditions and, consequently, indicate overall physiological state. To corroborate the results of this research, prior to the implementation and derivation of the tested indices, additional analyses concerning the spectral capability of the UAV camera were conducted to examine its discriminatory power among the considered carob forest classes. For this purpose, analyses were performed on five selected classes: tree crowns (healthy, dieback, and dead), bare soil, and asphalt (Figure 6). Tree crown segmentation and spectral information extraction were implemented using PCI-Geomatica (PCI, 2018). Considering a 3×3 pixel window, the average of the nine pixels provided the most representative spectral signature for each class. The positions of the spectral bands of the DJI Phantom-4 camera are illustrated in Figure 7a.

The spectra of healthy carob tree crowns (bright red in Figure 7a) show strong absorption in the blue and red bands due to high chlorophyll content, and an inflection point in the green wavelength (called the green peak) caused by carotenoids. The red-edge is the inflection point in the reflectance slope located in the transition region between the red and NIR. It exhibits an increasing asymptotic pattern between 0.15 and 0.75 (reflectance units), indicating very good tree health. In the NIR wavelength, a large fraction of incoming radiation is reflected toward space, according to biomass density and internal leaf structure (Figure 7a). While, visual analysis of a dieback tree shows a large dark-brown area in the center of the canopy (Figure 6), indicating the onset of dieback and damage. Dieback progresses from the center toward the outer crown (i.e., reduction of carob foliage and branch mortality). The derived spectral signature of the affected tree shows a drastic change caused by internal alterations of the bio-physiological system (carotenoids, chlorophyll, and water) due to nutrient disturbances. Specifically, a decrease in absorption in the blue and red bands is observed, along with a shift in the green inflection point and a substantial decrease in NIR reflectance due to biomass loss in branches and crown. Moreover, the asymptotic shape of the red-edge transition is bent, indicating that the tree is under intense dieback stress (Figure 7a).

Furthermore, the completely dead tree exhibits total defoliation and biomass loss, as well as destruction of the outer bark layer, which is distinguishable in the radiometry of the image as a dark-brown spot (Figure 6). This decline is reflected spectrally and radiometrically by very low reflectance in the NIR, the disappearance of the red-edge transition zone, and a total absence of absorption by pigments in the visible wavelengths. Consequently, reflectance amplitudes are almost identical across the visible, red-edge, and NIR bands (Figure 7a). Finally, bare soil and asphalt (Figure 6) exhibit spectral signatures completely different from those of the examined trees (Figure 7a). Moreover, the analysis of spectral differences between dieback and dead trees, compared to healthy trees, indicates that the NIR, red-edge, and red bands are the most relevant for calculating the selected indices and for discriminating between the considered classes (Figure 7b). These analyses clearly demonstrate the radiometric and spectral potential of the DJI Phantom-4 camera for effective discrimination between healthy, dieback, and damaged trees, as well as other land-use classes. Thus, as reported in numerous studies (Rouse et al., 1974; Huete, 1988; Bannari et al., 1995b; Thenkabail et al., 2018; Näsi et al., 2015), the VNIR spectral domains are the most prominent spectral regions used in biomass and chlorophyll indices for vegetation cover assessment, discrimination, and management in forests and agricultural areas.

Figure 6. False color composite (NIR, Red-Edge, Red) showing the selected classes for spectral analysis: healthy tree, dieback tree, dead tree, bare soil, and asphalt

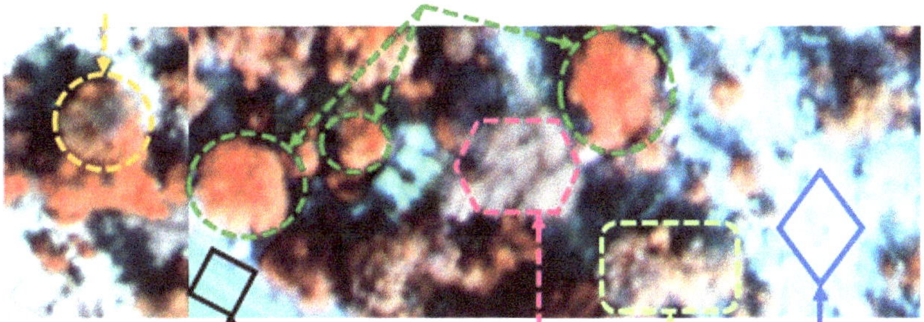

Figure 7. Positions of spectral bands of the DJI Phantom-4 camera on the average of extracted spectral signatures from the image of the healthy, dieback and dead trees, bare soil, and asphalt (a); as well as, the difference of spectrums of dieback and dead trees, by reference to healthy carob tree (b)

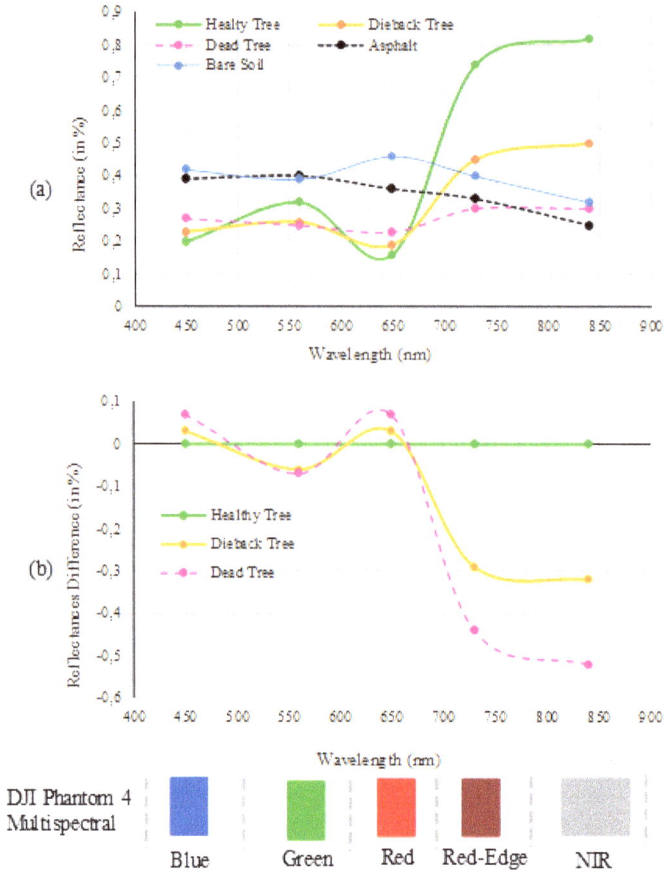

3.3. Biomass and Chlorophyll Indices

Four biomass indices and eight chlorophyll indices, which are the most widely used in the literature, were implemented. Maps were derived, and their potential was analyzed for detecting and discriminating among carob tree classes. Similar to the spectral analyses, the five considered classes are healthy, dieback, and dead trees, bare soil, and asphalt. The results obtained are very contrasting, showing two groups of indices with different discriminating power. The first group includes indices with poor discriminatory power, such as the EVI biomass index and five chlorophyll indices (VARI, MRESR, $CIR_{ed\text{-}Edge}$, PSRI, and SRPI) (Figure 8). This group is not sensitive to changes in land-cover types or physiological variations in the carob forest. For example, EVI is shade-sensitive and does not separate healthy trees from the underlying soil; some confusion is observed between healthy and dieback trees.

Otherwise, the senescence index PSRI shows no sensitivity to the considered land-cover classes; its value is always null regardless of the target under analysis. However, it shows some sensitivity to senescent herbaceous vegetation mixed with bare soil under and around carob trees (Figure 8). The pigment index SRPI groups all land-use categories into a single class, including healthy trees, bare soil, and asphalt. These poor performances are explained by the fact that PSRI and SRPI rely solely on the visible spectrum and do not consider the NIR band, which is most informative about biomass density. However, although the MRESR index uses the NIR and red-edge bands, it fails to distinguish between different land-use classes, grouping all classes undeservedly into a single category of healthy carob trees (Figure 8). Furthermore, even though the VARI index shows some sensitivity to healthy trees with high biomass density, its potential to mitigate the influence of underlying bare soil remains very limited. Likewise, it is difficult to distinguish between dieback and dead trees, bare soil, and asphalt. This index was developed specifically for very dense crops (100%) using coarse data acquired by SeaWiFS, MODIS, MISR, and MERIS sensors. According to Gitelson et al. (2002b), factors such as vegetation species, canopy architecture, and sun illumination geometry can reduce VARI's accuracy. Like PSRI and SRPI, VARI does not include the NIR band. Finally, despite incorporating the NIR and red-edge bands, the $CI_{Red\text{-}Edge}$ index does not respond effectively to dense biomass; its range does not exceed 0.3. Consequently, it cannot clearly discriminate between healthy and dieback trees, or between damaged and dead trees. Overall, it is likely that these chlorophyll indices are not efficient because they were developed for narrow spectral bands, whereas the DJI P4-multispectral imaging camera lacks the specific wavelengths needed to accurately measure pigment contents.

Figure 8. Pseudo-color representation of powerless biomass-chlorophyll indices for discrimination among carob tree classes, bare soil and asphalt

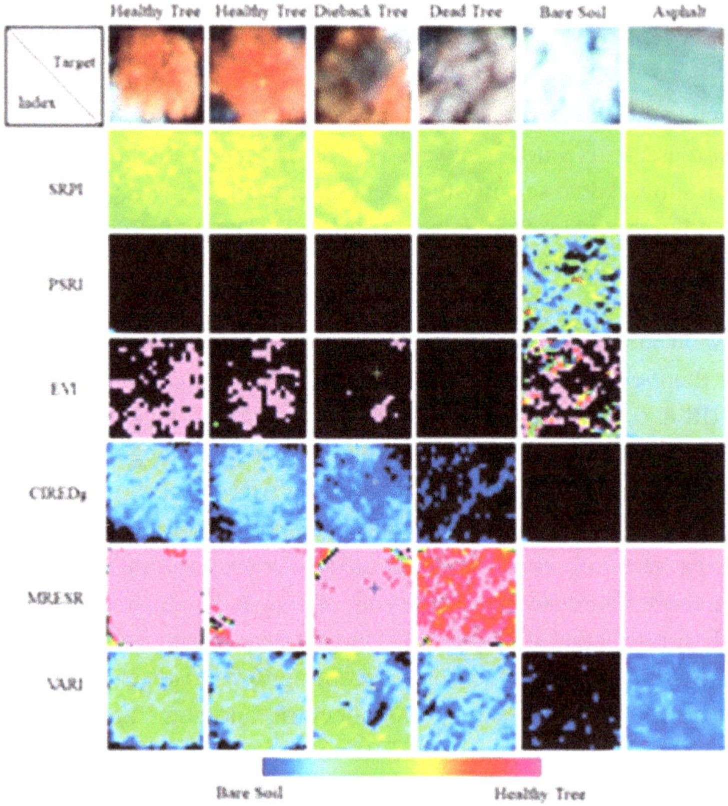

Furthermore, Figure 9 illustrates the potential of powerful biomass-chlorophyll indices for discriminating among the considered carob forest classes. This group effectively removes the impact of bare soil underlying the canopies by highlighting only the target tree classes. They exhibit different sensitivities depending on canopy density and internal conditions, which obviously influence the ability of NIR wavelengths to penetrate the carob canopy. Indeed, it is observed that indices with higher NIR weighting coefficients offer better performance in discriminating between different states of the carob canopies. Moreover, they respond differently to linearity and saturation issues. For instance, although CARI satisfactorily describes the chlorophyll content of healthy trees, it remains sensitive to soil artefacts and exhibits non-linearity (Figure 9) due to the absence of the NIR band in its equation. Consequently, its dynamic range varies between 0.0 and 0.35 for bare soils and dense, healthy carob trees, respectively. This limitation leads to a strong overlap of class values during the threshold process.

Similarly, NDVI and GNDVI show almost the same behavior toward the analyzed classes. However, NDVI highlights healthy and dieback trees more clearly due to the strong absorption of chlorophyll in the red band. In contrast, GNDVI substitutes the red band with the green band, reflecting carotenoid content rather than chlorophyll. Therefore, incorporating the green band into GNDVI does not improve carob tree mapping. Nevertheless, both indices are biased toward higher values because they saturate over very dense carob biomass and healthy trees; their maximum values do not exceed 0.72 for NDVI and 0.60 for GNDVI. While, the SAVI mimics the behavior of NDVI but with higher values and saturates less on healthy and dieback trees due to variations in NIR reflectance (Figures 9 and 10). Its dynamic range values are approximately null for bare soils, 0.20 for dead trees, 0.45 for withered trees, and 0.85 for healthy trees. These results highlight the strong discriminating power of SAVI for distinguishing among carob classes.

Figure 9. Pseudo-color representation of powerful biomass-chlorophyll indices for discrimination of carob tree classes, bare soil and asphalt

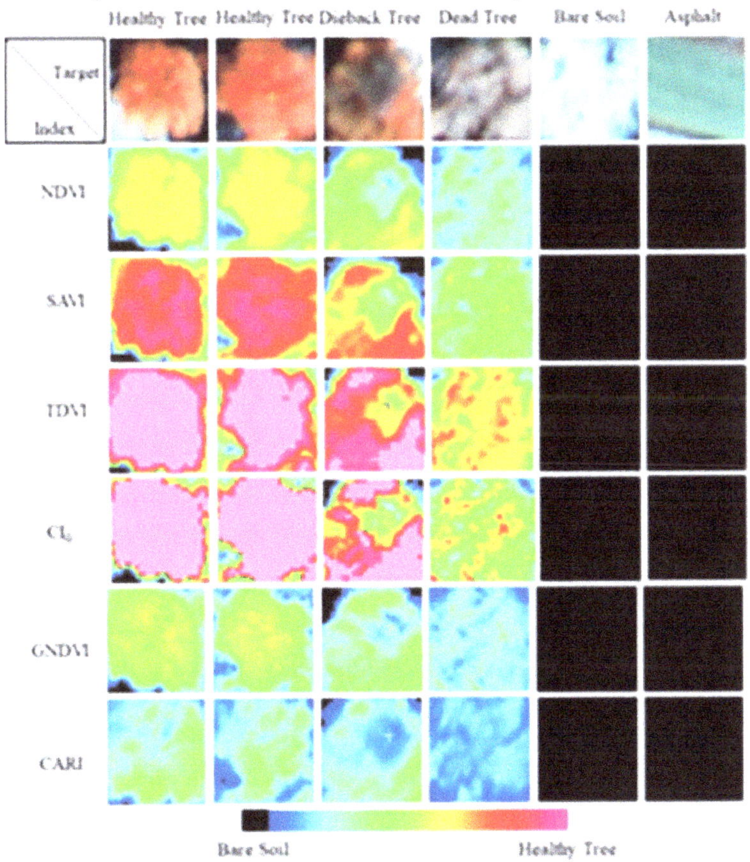

Ultimately, TDVI and CI_G are the most sensitive indices, yielding comparable results with excellent discriminating power among all classes considered, particularly for the different stages of carob tree dieback (early, intermediate, and advanced). They demonstrate a good dynamic range and sensitivity for monitoring and assessing the spatial distribution of carob, as well as detecting bio-physiological variations under stress and dieback conditions (Figures 9 and 10). These indices are completely independent of underlying soil artefacts, and they avoid the linearity and saturation issues encountered in other indices (NDVI, GNDVI, and CARI). They show a curvilinear relationship between their computed values and the classes considered, whit range values generally varying between 0.0 and 1.0: zero for bare soil, 0.20 for dead trees, 0.60 for dieback trees, and approximately the maximum (1.0) for healthy trees with very dense biomass, as illustrated in Figure 10. Accordingly, from both the spectral and radiometric perspectives of the camera used, TDVI and CI_G are sensitive to major structural and bio-physiological changes in carob canopies. Therefore, they are the most relevant and useful indices for monitoring carob trees, discriminating bio-physiological classes, and mapping these characteristics.

Figure 10. Curvilinear relationship between the calculated values of the powerful indices and the classes considered (i.e., soil, dead, dieback and healthy carob trees)

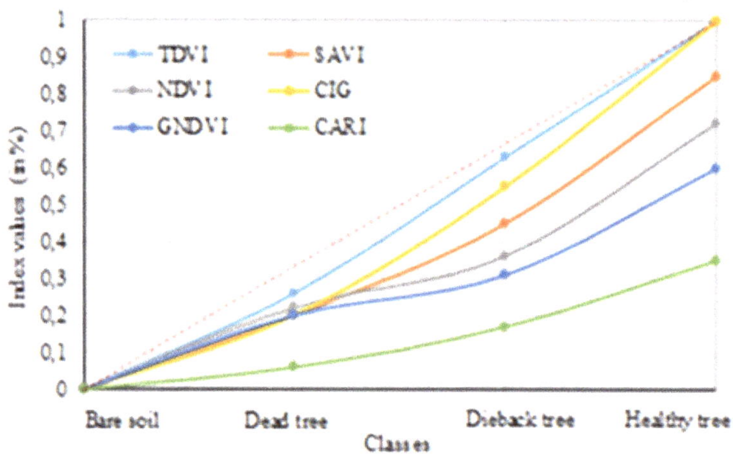

Since these two indices show similar results, only TDVI was thresholded. A map of the spatial distribution and conditions of carob trees was then derived (Figure 11a) using its histogram and the values of the centroids and standard deviations of the considered classes. It was then validated by reference to the ground truth using an approach commonly applied in forest sciences, considering dieback symptoms and tree-crown density (Dash et al., 2017). To reduce the variance associated with subjective evaluation, two inspectors were involved simultaneously in sample selection and assessment during the same days as the UAV data acquisition. Each individual tree was evaluated from multiple angles to ensure a complete assessment of the canopy (Dash et al., 2018). Subsequently, each sampled tree was photographed and its location recorded using an accurate GPS ($\sigma \leq 30$ cm).

First, it is observed that the road, the rocky bottom of the valley, the underlying bare soils, and the mountainous rocks are displayed in black (Figure 11a), indicating a clear differentiation of these subclasses from the carob tree classes. The dieback trees class, illustrated in light green (Figure 11a), shows some fragmentation and a mixture of sub-class fractions, such as bark-less branches with yellowish leaves (indicative of reduced nutrients and chlorophyll), defoliation, low biomass, or completely dead branches due to monkey activity. While, the dead trees class (red in Figure 11a), characterized by small and medium crown sizes, is also identifiable, showing that monkeys primarily attack young trees. In contrast, the crowns of healthy trees are well-foliated and clearly demarcated, with almost circular shapes due to their high biomass density and bio-physiological activities (dark green in Figure 11a).

However, in dense areas with young, small healthy trees, some mixture is observed between carob, Oak (green and Zean), and Pistacia lentiscus, particularly in areas where shadows were not fully corrected. This occasionally limits the ability to separate clearly these young tree species. To separate successfully among these classes, more narrow spectral bands are needed as reported by Näsi et al. (2015). Nevertheless, despite this minor limitation, the validation confirms the efficiency and discriminative power of the proposed method for mapping and classifying carob trees. This simple and rapid approach, integrating UAV multispectral data with the best-selected indices (TDVI, CI_G, and SAVI), offers a practical decision-support tool for the protection of carob trees and the promotion of sustainable development in mountainous areas. Its simplicity allows easy transfer between sensors and ensures reproducibility and applicability in other geographic regions using different multispectral or hyperspectral sensors.

Figure 11. TDVI Threshold (a) and unsupervised ML classification (b) maps

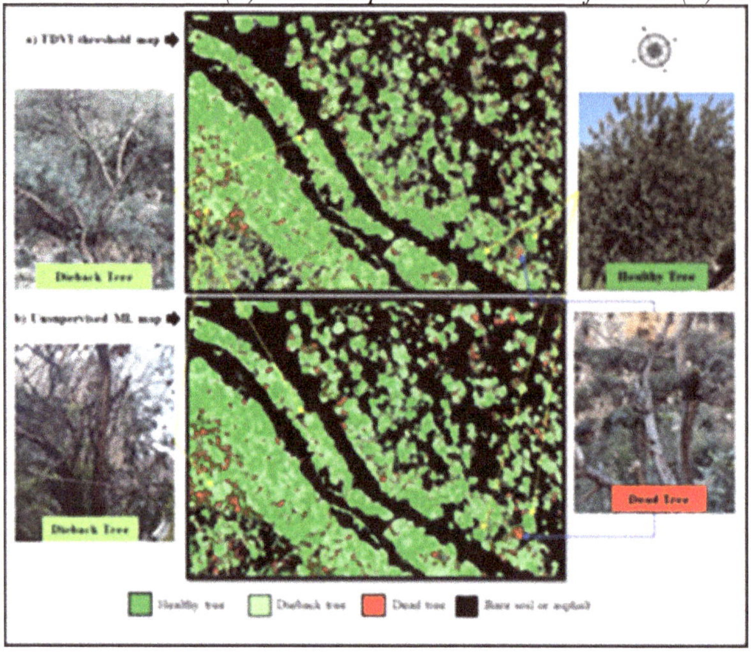

3.4. Machine Learning

In this study, two unsupervised machine learning classification (UMLC) scenarios were established. The first scenario uses only the five spectral bands, while the second incorporates these spectral bands along with six selected biomass-chlorophyll indices (NDVI, SAVI, TDVI, GNDVI, CARI, and CI_G) identified as the most relevant in the previous section. For both scenarios, major shadow areas and underlying herbaceous vegetation classes (senescent and photo-synthetically active) were thresholded and masked. The UMLCs were performed in a standard procedure in which all classes were assumed *a priori* to have equal probability. This assumption allows the evaluation of pixel membership to each during centroid calculation and pixel clustering. To ensure that all pixels were classified, the probability for the null hypothesis was set to 95% meaning that for each class; fewer than 5% of the pixels with the lowest membership probability were rejected. Under these conditions, the classifications converged from one iteration to the next, and the iterative process stopped when less than 0.05 of the pixels changed their cluster assignment compared to the previous iteration. To identify the optimal number of classes, classifications were performed considering a maximum of 30 iterations. The initial number of classes was set to ten and subsequently aggregated into the main classes (bare soil, asphalt, dead trees, dieback trees and healthy trees).

The results of the two scenarios were notably different. The first scenario, using only the five spectral bands, showed limited discrimination between the desired classes. In contrast, the second scenario, which combined the spectral bands with the selected biomass-chlorophyll indices, enhanced biomass detection, highlighted chlorophyll content, and effectively discriminated underlying soil optical properties. This integration has demonstrated an advantage by improving both of the probability calculation of centroids and thresholding process, ultimately achieving an excellent overall classification accuracy of 93%. Figure 11b shows the map from the second scenario, which closely resembles the TDVI-threshold map (Figure 11a) with only minor differences. Healthy trees, bare soil, and asphalt were accurately classified, reflecting similar density and spatial patterns as in the TDVI-threshold map. Whereas, dieback and dead carob trees exhibited irregular distributions and variable densities, ranging from isolated trees along slopes to clusters in higher altitudes where primates are less disturbed by human activity. Monkeys preferentially feed on young shoots rich in water and tender bark, explaining the relatively higher number of dieback and dead trees.

Although the two maps are largely similar (Figures 11a and b), slight overestimations of dead and dieback trees are observed particularly in the west-northeast direction. Theoretically, dieback and dead classes should have less absorption and reflection, respectively, in red and NIR bands due to chlorophyll and biomass loss. Therefore, they can be better distinguished from the other classes and increasing the overall classification exactitude. However, the slight overestimation of dead trees could be due to the shadow of the outer-ring of the crown of healthy or dieback trees. The spectral feature of dieback trees and shadow are similar to some extent, so it was difficult to discriminate them. Undeniably, despite the correction and elimination of shadow effects during the ortho-rectification and threshold processes, spectral confusion can occur due to the residual shadow. While, the overestimation of dieback trees class is due to misclassification caused by several source of errors. For instance, the mixt pixels between the branches of dead carob trees and those of Oaks (green and zean) or Pistacia lentiscus trees, particularly in areas where the shadow was not rigorously eliminated. Moreover, spectral confusion may occur, particularly when the ten derived classes were aggregated into only five main classes. Indeed, there is greater heterogeneity of vegetation species that probably shared similar spectral characteristics and, therefore, higher spectral confusion can occur during the aggregation process. Likewise, sometimes dieback trees can be assigned to other classes due to the spectral overlap caused by varying proportion of healthy and dieback or dead branches in the trees crown. As well as, some trees with early dieback can be classified as healthy due to the spectral transition between these classes, as reported by Naseri et al. (2023). Other issues such as drought and nutrient deficiency, characterized by the death of tree foliage, could lead to spectral confusion and, therefore, inaccurate discrimination of dieback trees. Although we discussed the potential sources of some errors that may be responsible for this slight overestimation of the dead and dieback tree classes, the UMLC method performed with an accuracy of 93%, producing a map consistent with the ground truth. The validation of the UMLC map performed using accurate GPS and photographs by reference to ground truth and field observations (following the same validation procedure described above). The results showed errors of approximately 4.2% and 6.5% for dieback and dead tree classes, respectively, particularly affecting small-crown trees surrounded by other species with a dominant broad crown. This confirms the efficiency and separating power of this method for mapping carob forest dieback.

4. DISCUSSION

In several forest studies, biomass and chlorophyll indices showed sufficient robustness to detect bio-physiological stress, defoliation, outbreaks, etc. The behavioral differences among these indices were observed due to the variations and dissimilarity between the spectral ranges covered by the bands of examined sensors. In the present study about carob trees dieback using DJI Phantom-4 multispectral camera, the red waveband provided the best results compared to the integration of red-edge and green bands. This relative importance is probably due to the red-edge band position of this used camera, resulting in an inability to detect spectral changes associated with physiological stress due to dieback of carob trees. This conclusion is consistent with Otsu et al. (2019) who demonstrated no significant difference between NDVI and RE-NDVI because the red-edge band was not particularly sensitive to defoliation. However, previous studies have related changes in foliage density and chlorophyll content to shifts in the red-edge, the inflection point that occurs in the rapid transition between red and NIR reflectance (Curran, 1983), and these shifts have been associated with vegetation stress, defoliation and forest decline (Collins, 1978; Ustin et al., 1988). Moreover, some studies demonstrated that indices integrating the red-edge band have the potential to improve considerably the detection of forest stress (Eitel et al., 2011) and defoliation resulting from insect attack on conifers (Dennison et al., 2010), compared to indices incorporating the green (GNDVI) or the red (NDVI) bands. Conversely, Dash et al. (2018) showed that for physiological stress in a mature *Pinus radiate*, the NDVI derived from the UAV data was more predictor than RE-NDVI. Whereas, Otsu et al. (2019) reported that NDVI and RE-NDVI performed similarly and powerfully for conifer defoliation detection and mapping.

Theoretically, as discussed above, the spectral bands located in red-edge spectral transition are well suited for estimating the chlorophyll content, which decreases significantly as the dieback or infection progresses (Ustin et al., 1988). However, the examined chlorophyll indices integrating red-edge bands in the present study such as $CI_{Red\text{-}Edge}$, MRESR, PSRI and CARI have not achieved any relevant results. They had high misjudgment pigmentation rates and were unable to detect the presence of biological or physiological variables. These observations confirmed the finding of Ye et al. (2020) that the CARI was unable to discriminate among banana-tree infestation classes since it was insensitive to variations in chlorophyll content. Among several biomass-chlorophyll indices examined for kiwifruit trees stress, the CARI obtained weak discrimination results (Jopia et al., 2020).

Furthermore, EVI, VARI, and SRPI were not conclusive for physiological carob dieback detection. These results are in agreement with those of Hornero et al. (2021) who tested 95 biomass-chlorophyll indices, including NDVI, EVI, VARI, SRPI and PSRI, for the early detection of Phytophthora-induced symptoms in Holm Oak dieback and decline. They showed that these indices are unfortunately irrelevant and inconclusive for discriminating and classifying physiological conditions of this tree species. As well as, Van-Beek et al. (2016) demonstrated that SRPI index is insensitive to orchard tree stress, inaccurate for estimating biophysical variables, and highly dependent on the sensor acquisition geometry. Moreover, to monitor heavy metal stress levels in the rice canopy, Zhang et al. (2018) concluded that neither PSRI nor $CI_{Red\text{-}Edge}$ could accurately distinguish between different stress levels. In addition, in the context of assessing and estimating defoliation of Scots Pine stands, the PSRI proved to be a poor predictor (Hawryło et al., 2018). Similarly to the results obtained in the present study, Tilly et al. (2020) showed that for Oak tree decline and vitality classes' separation (healthy, dieback, and dead), the maps derived from PSRI, GNDVI, $CI_{Red\text{-}Edgre}$ and EVI indices were neither accurate nor reliable compared to the ground truth. As also pointed out by Bilintoh (2019), the EVI index was inappropriate for detecting dieback trees in Pinus pinaster forest (Moreno-Fernández et al., 2021), and insensitive to armyworm infestations in agricultural environments.

Otherwise, some studies suggest that the integration of the green band in NDVI (GNDVI) allows its potential for estimating chlorophyll pigmentation and a significant correlation with LAI (Yoder & Waring, 1994; Gitelson et al., 1996). To discriminate between deciduous forests and grassland, Motohka et al. (2010) reported that GNDVI is better for detecting subtle disturbances in the middle of the growing period. Contrariwise, for conifer defoliation detection the GNDVI showed poor performance (Otsu et al., 2019). Like the present study, GNDVI shows a very weak sensitivity to the chlorophyll content in carob trees and a severe saturation confusing among dieback and healthy carob trees. Moreover, Spruce et al. (2011) noted that NDVI was more effective than GNDVI in distinguishing between medium and low damage within deciduous forests, as well as in mapping insect pest defoliation in forest areas. Additionally, Jin and Sader (2005) reported that NDVI products provide better detection of forest disturbances in a mixed environment (deciduous and coniferous).

Among the twelve indices examined in the present study, the results achieved with NDVI, SAVI, CI_G and TDVI are the best predictors for carob trees dieback and are consistent with the results of other studies. However, their power of detection and discrimination is relatively variable from one index to another. For instance, for healthy Spruce stands and those infested by the European Spruce beetle, SAVI showed a good ability to detect infested and physiologically stressed trees (Kampen et al., 2019). While other studies have shown that NDVI and SAVI have relatively acceptable potential as they tend to saturate when the LAI increases due to the saturation and linearity issues (Curran, 1983; Asrar et al., 1984; Huete, 1988; Bannari et al., 2002). Obviously, this saturation could be a major obstacle when sensing between healthy and dieback of carob trees, thus introducing significant errors during the classification and mapping process. Yet, to make these two indices (NDVI and SAVI) operational for recognizing carob tree conditions and discrimination, further investigation and field survey for calibration are required.

Whereas, CI_G and TDVI are more representative of ground truth, they offer excellent dynamic values between 0% and 100% allowing very good distinction among carob classes (i.e., zero for bare soil, 0.20 for dead trees, 0.60 for dieback trees, and around 1.0 for healthy trees and dense crown-trees). Similarly, for the identification and mapping of infected and non-infected banana trees (*Fusarium wilt*) using UAV-multispectral data, CI_G was the optimal chlorophyll index exhibiting the best performance due to its high sensitivity to small variations in chlorophyll content as reported by Ye et al. (2020). The performance of these two indices (TDVI and CI_G) is not only due to the strong contrast of reflectances in the red or green and NIR but also to their mathematical principles integrated into their equations. As in dieback or dead trees under physiological stress, the water content declines, chlorophyll and carotenoid molecules break down and declines completely, senescence and defoliation set in, thus the reflectance increase in the visible bands and decrease in NIR especially for broad leaves (as carob) than needle leaves (Sims and Gamon, 2002; Motohka et al., 2010).

The analysis of spectral signatures allowed a clear separation among the considered land-use classes. As well as the approach of tree classes identification based on TDVI threshold provided satisfactory results that met our expectations. Of course, the threshold accuracy relies on the operator's *a priori* knowledge of the study area, excellent analysis of the index histogram, optimal statistical definition of the centroid of each class, and the availability of *in situ* data (ground truth) for verification and validation. The classes of bare soil and asphalt (black color in Figure 11a) are clearly distinguishable from the carob tree classes as presented in Figure 11a. Obviously, the very high spatial resolution of UAV data (25 cm) allowed the possibilities for various class feature identification and verification by reference to the ground truth. Overall, the TDVI threshold classification is robust enough to demonstrate the potential of this index or that of CI_G derived from UAV-multispectral imagery. These results are consistent with other relevant studies that confirmed the ability of these indices to separate among different categories of tree decline (healthy, infested and dead) using UAV-multispectral data acquired above Spruce forest with a heavy beetle infestation (Minarik and Langhammer, 2016; Brovkina et al., 2018).

Furthermore, K-means clustering is an efficient UMLC in pattern recognition. Initially, it selects groups of pixels based only on their spectral similarity, which some time generates spectral classes that are not matching those of the searched themes. For instance, spectral confusion between healthy and dieback trees occurs because sick trees have altered leaf chemistry and canopy structure, which results in a change in their spectral signatures making them spectrally similar. Moreover, it can group pixels that share common spectral features, which may include healthy and dieback trees, thereby highlighting one class at the expense of another. Likewise, mixed pixels might contain both healthy and dieback branches, or even other underlying cover, further complicating the spectral signature composition and identification. It my also be due to disturbance effects corresponding to spectral confusion of shadow with dead trees, bare soils with senescent vegetation, and healthy trees with underlying herbaceous vegetation. In addition, tree mortality reduces water, chlorophyll and organic matter contents, significantly influencing the reflectance in the VNIR bands and, consequently, further modifying their spectral characteristics. In this study, we demonstrate that detecting carob crown dieback using biomass-chlorophyll indices, either separately or integrated with spectral bands in UMLC, is feasible; this corroborates the results reported by other scientists for other forest species (Şandric et al., 2022; Naseri et al., 2023; Allen et al., 2024). Certainly, the methodological framework proposed to detect the carob trees dieback based on TDVI or CI_G indices aligned well with field surveys and observations during validating. It relies exclusively on the knowledge of spectral absorption features that characterize specifically the target under investigation, easily transferable between sensors and constitutes a robust alternative to the revolutionary and sophisticated modeling approaches. Of course, issues regarding different acquisition geometries, different lighting conditions, radiometric calibration, geometric corrections, and ortho-rectification are particularly relevant and need to be addressed correctly.

Therefore, since biomass-chlorophyll indices performance can vary from site to site, it is recommendable to explore several indices to select the best index for a particular application. Furthermore, although many strategies exist in the literature to set the right threshold to optimize the selection of dominant classes, the threshold process must consider the environmental and biogeographic conditions of the studied area, the climatic conditions, the forest species studied and the conditions (internal and external) of data acquisition (Heiden et al., 2022). In addition, specific work on the classification models could improve dieback detection, especially testing other advances ML models. Likewise, we demonstrated that the DJI camera is a promising technology for carob trees dieback detection and discrimination, thus facilitating proactive interventions with rapid mapping of forest health. However, the distinction between healthy trees and the different stages of dieback (early, intermediate and advanced) remains a challenge to achieve with high accuracy using limited multispectral bands. Obviously, it is necessary to conduct further studies using UAV-hyperspectral cameras (VNIR, SWIR and thermal) with very narrow wavelengths as reported by Näsi et al. (2015).

Finally, as this research was conducted in a mountainous area with strong undulation and deep valleys, it is important to highlight some of the difficulties encountered. Among these, the complexity of the drone flight procedure, the elimination of existing risks when maneuvering between narrow rock walls and the successful landing of the drone in a safe place are conditioned by the operator's experience, the reliability of telemetry control, and the availability of GPS (or GNSS) signals in this harsh environment.

5. CONCLUSIONS

The use of biomass-chlorophyll indices for biological and/or physiological parameter retrievals and analysis is a challenging task. This is because they represent a simple and integrative measure of vegetation photosynthetic activities and canopy structural variations that are useful in spatiotemporal monitoring and land-use change detection studies. In this preliminary performance study for discriminating and mapping the dieback and damage of carob trees caused by the Magot monkey, we demonstrated the usefulness of a geomatics approach based on the synergy among the very high spatial resolution of UAV-multispectral data, analytical photogrammetry, DGPS, biomass-chlorophyll indices and UML classifier. To achieve so, the UAV data were rigorously preprocessed (i.e., geometry, topography, and radiometry), biomass (NDVI, GNDVI, SAVI, EVI, and TDVI) and chlorophyll (CARI, CI_G, VARI, SRPI, PSRI, $CI_{Red\text{-}Edge}$, and MRESR) indices were derived and analyzed, UML K-Means clustering was implemented, and the obtained results were validated against ground truth.

The results obtained highlighted the radiometric and spectral performance of the DJI Phantom-4 camera for effective discrimination between healthy and dieback trees, as well as other land-use classes. Among the indices tested, biomass indices such as TDVI and SAVI, and the chlorophyll index CI_G performed better than the other indices tested. They are completely independent of the optical properties of the bare soils (color and brightness) underlying the carob trees, and they avoid the linearity and saturation problems encountered in the other indices tested. Particularly TDVI and CI_G showed a curvilinear relationship between their computed values and the considered classes, discriminating powerfully among the carob tree classes (healthy, dieback, and dead). Both indices demonstrated good and similar dynamic range values (0.0 to 1.0), good sensitivity to spatial variation of different classes considered, and provided useful biological information about dieback and damage of carob trees caused by Magot monkeys. By reference to the ground truth, they show an excellent separating power between the different carob tree classes. In addition, the TDVI threshold and classification allowed the identification of carob tree classes (healthy, dieback, and dead) with satisfactory results. Overall, these two indices highlight their capacity for separating the different classes considered and their high conformity with the ground truth.

Similarly, UMLC demonstrated its potential to classify trees according to their degree of dieback in a complex environment, achieving excellent overall classification accuracy of 93% with errors of 4.2% and 6.5%, respectively, for the "dieback" and "dead" tree classes. Nevertheless, although this classification is a revolutionary approach providing satisfactory results for carob dieback mapping, the methodological framework based on TDVI or CI_G indices is recommended for several benefits. Indeed, this method relies solely on the knowledge of spectral absorption features that characterize specifically the target under investigation. It faithfully reflects the reality on the ground in terms of detecting and discriminating the carob forest dieback. It is simple and fast, easily transferable between sensors, and constitutes a robust alternative to sophisticated modeling approaches. Undoubtedly, it is a useful decision-making tool for monitoring and protecting carob forest on a large scale, and relevant for stakeholders such as forest management authorities, environmental policy makers, and forest health researchers to promote sustainable development. However, it is desirable to analyze the power of UAV hyperspectral camera in this field of application because this new technology consists of the simultaneous acquisition of images in many narrow and contiguous spectral bands. In this type of imaging system, each pixel provides a detailed spectral signature between 350 and 2500 nm, allowing much deeper examination and discrimination than multispectral data.

REFERENCES

Abdulridha, J., Batuman, O., & Ampatzidis, Y. (2019). UAV-Based Remote Sensing Technique to Detect Citrus Canker Disease Utilizing Hyperspectral Imaging and Machine Learning. *Remote Sensing (Basel), 11*(11), 1373. DOI: 10.3390/rs11111373

Allen, M. J., Moreno-Fernández, D., Ruiz-Benito, P., Grieve, S. W. D., & Lines, E. R. (2024). Low-cost tree crown dieback estimation using deep learning-based segmentation. *Environmental Data Science, 3*(e18), 1–18. DOI: 10.1017/eds.2024.16

ASD. (2015). *Analytical Spectral Devices. Technical Guide, 4th ed.; ASD Inc.: Boulder, CO, USA, 2015.* http://www.asdi.com/products-spectroradiometers.asp

Asrar, G., Fuchs, M., Kanemasu, E. T., & Hatfield, J. L. (1984). Estimating absorbed photosynthetic radiation and leaf area index from spectral reflectance in wheat. *Agronomy Journal, 76*(2), 300–306. DOI: 10.2134/agronj1984.00021962007600020029x

Bahri, M., Haboudane, D., Bannari, A., & Bonn, F. (2008). Mapping the dominant forest species of the Middle Atlas cedar ecosystem in Morocco: Comparison between methods based on spectral similarity. *Revue Télédétection*, vol. 7, no. 1-2-3-4, pp. 283-301.

Bannari, A., & Al-ali, Z. M. (2020). Ground Reflectance Factor Retrieval from Landsat (MSS, TM, ETM+, and OLI) Time Series Data based on Semi-empirical Line Approach and Pseudo-invariant Targets in Arid Landscape. *Proceedings of IEEE-IGARSS* (pp. 5990-5993). Waikoloa, Hawaii, USA, 26 September to 02 October 2020.

Bannari, A., Almansoori, T. A., Mohamed, A. M. A., El-Battay, A., & Hameid, N. (2018). Modeling Water Stress as an Indicator of Red Palm Weevil Infestation using Field Sampling, WorldView-3 Reflectance, and Laboratory Analysis. *Proceedings of IEEE-IGARSS* (pp. 9056-9059). Valencia, Spain, 23-27 July 2018. DOI: 10.1109/IGARSS.2018.8518687

Bannari, A., Asalhi, H., & Teillet, P. M. (2002). Transformed Difference Vegetation Index (TDVI) for Vegetation Cover Mapping. *Proceedings of IEEE-IGARSS'* (pp. 3053-3055), Toronto (Ontario), Canada, 9-13 July 2002. DOI: 10.1109/IGARSS.2002.1026867

Bannari, A., El-Ghmari, A., Rhinane, H., Selouani, A., & Loulad, S. (2024b). Vicarious Radiometric Calibration of UAV-Data in Mountainous area using Semi-Empirical Line Approach and *in-situ* Spectroradiometric Measurements. *Proceedings of IEEE-IGARSS* (pp. 11449-11452). Athens, Greece, 7-12 July 2024. DOI: 10.1109/IGARSS53475.2024.10642636

Bannari, A., Gibson, J. R., & Morin, D. (1997). The Necessity of Exterior Orientation Parameters for the Rigorous Geometric Correction of MEIS-II Airborne Digital Image. *Remote Sensing Reviews*, *16*(1-2), 135–156. DOI: 10.1080/02757259709532349

Bannari, A., Khurshid, K. S., Staenz, K., & Schwarz, J. (2007a). A Comparison of Hyperspectral Chlorophyll Indices for Wheat Crop Chlorophyll Content Estimation Using Laboratory Reflectance Measurements. *IEEE Transactions on Geoscience and Remote Sensing*, *45*(10), 3063–3073. DOI: 10.1109/TGRS.2007.897429

Bannari, A., Khurshid, S. K., Staenz, K., & Schwarz, J. (2008). Potential of Hyperion EO-1 Hyperspectral Data for Wheat Crop Chlorophyll Content Extraction in Precision Agriculture. *Canadian Journal of Remote Sensing*, *34*(1), 139–157. DOI: 10.5589/m08-001

Bannari, A., Mohamed, A. K., & Peddle, D. R. (2016). Bio-Physiological Parameters Retrieval and Statistical Analysis for Red Palm Weevils Stress-Attack Prediction Using WorldView-3 data. *Proceedings of IGARSS* (pp. 7129-7132). Beijing, China, 10-15 July 2016.

Bannari, A., Mohamed, A. M. A., & El-Battay, A. (2017). Water Stress Detection in Palm Threes Canopies as an Indicator for Red Palm Weevil Stress-Attack using WorldView-3 data. *Proceedings of IGARSS* (pp. 4000-4003). Texas, USA, 23-28 July 2017.

Bannari, A., Morin, D., Bénié, G. B., & Bonn, F. (1995b). A Theoretical Review of Different Mathematical Models of Geometric Corrections Applied to Remote Sensing Images. *Remote Sensing Reviews*, *13*(1-2), 27–47. DOI: 10.1080/02757259509532295

Bannari, A., Morin, D., Huete, A. R., & Bonn, F. (1995a). A Review of Vegetation indices. *Remote Sensing Reviews*, *13*(1-2), 95–120. DOI: 10.1080/02757259509532298

Bannari, A., Selouani, A., El-basri, M., Rhinane, H., El-Harti, A., & El-Ghmari, A. (2021). Multi-scales Analysis of DEMs Derived from Unmanned Aerial Vehicle in Precision Agriculture Context. *Proceedings of IEEE-IGARSS* (pp. 8285-8188). Brussels, Belgium, 12-14 July 2021. DOI: 10.1109/IGARSS47720.2021.9554020

Bannari, A., Selouani, A., Nehari, W., Rhinane, H., El-Ghmari, A., & Oulbacha, S. (2024). Accuracy Evaluation of DEMs Derived from UAV data acquired over a Narrow-Deep Valley in the Middle-Atlas Mountain. *Proceedings of IEEE-IGARSS* (pp. 11449-11452). Athens, Greece, 17-12 July 2024. DOI: 10.1109/IGARSS53475.2024.10641039

Bannari, A., Teillet, P. M., & Richardson, G. (1999). The necessity of radiometric calibration and standardization of remote sensing data. *Canadian Journal of Remote Sensing*, *25*(1), 45–59. DOI: 10.1080/07038992.1999.10855262

Bates, E., Popović, M., Marsh, C., Clark, R., Kovac, M., & Bahadir-Kocer, B. (2025). Leaf Level Ash Dieback Disease Detection and Online Severity Estimation with UAVs. *In IEEE-Access*, 13, 55499-55511; DOI: 10.1109/ACCESS.2025.3541980

Batlle, I., & Tous, J. (1997). Carob tree. *Ceratonia siliqua* L. Promoting the conservation and use of underutilized and neglected crops. 17 Institute of Plant Genetics and Crop Plant Research, Gatersleben. International Plant Genetic Resources Institute, Rome, Italy.

Bilintoh, T. (2019). *Remote Sensing for Mapping and Modeling Presence of Armyworm Infestation on Maize in Ejura, Ghana*. Master Thesis, Department of Geography, Michigan State University, United States, 50 pages.

Bithas, P. S., Michailidis, E. T., Nomikos, N., Vouyioukas, D., & Kanatas, A. G. (2019). A Survey on Machine-Learning Techniques for UAV-Based Communications. *Sensors (Basel)*, *19*(23), 5170. DOI: 10.3390/s19235170

Bonsignore, C. P. (2012). Apate monachus (Fabricius, 1775), a bostrichid pest of pomegranate and carob trees in nurseries. *Plant Protection Science*, *48*(2), 94–97. DOI: 10.17221/53/2011-PPS

Bozzini, A., Francini, S., Chirici, G., Battisti, A., & Faccoli, M. (2023). Spruce Bark Beetle Outbreak Prediction through Automatic Classification of Sentinel-2 Imagery. *Forests*, *14*(6), 1116. DOI: 10.3390/f14061116

Brovkina, O., Cienciala, E., Surový, P., & Janata, P. (2018). UAV for assessment of qualitative classification of Norway spruce in temperate forest stands. *Geo-Spatial Information Science*, *21*(1), 12–20. DOI: 10.1080/10095020.2017.1416994

Candiago, S., Remondino, F., De Giglio, M., Dubbini, M., & Gattelli, M. (2015). Evaluating Multispectral Images and Vegetation Indices for Precision Farming Applications from UAV Images. *Remote Sensing (Basel)*, *7*(4), 4026–4047. DOI: 10.3390/rs70404026

Cardil, A., Vepakomma, U., & Brotons, L. (2017). Assessing Pine Processionary Moth Defoliation Using Unmanned Aerial Systems. *Forests*, *8*(10), 402. DOI: 10.3390/f8100402

Catal, J., Navarro, A., & Calv.o, J. (2022). Mapping Cork Oak Mortality Using Multitemporal High-Resolution Satellite Imagery. *Remote Sens.*, 14, 2750.

Catarino, F. (1996). Carob - an old and sustainable Mediterranean crop: Its potential for the future. In: *Proc. Third International Carob Symposium* (Tavira, Portugal), University of Lisbon: Portugal.

Chaponniere, A., & Smakhtin, V. (2006). A review of climate change scenarios and preliminary rainfall trend analysis in the Oum er Rbia Basin, Morocco. *Working Paper 110 (Drought Series: Paper 8) Colombo, Sri Lanka: International Water Management Institute (IWMI)*, 15 pages.

Chávez-Durán, Á. A., Olvera-Vargas, M., Figueroa-Rangel, B., García, M., Aguado, I., & Ruiz-Corral, J. A. (2022). Mapping Homogeneous Response Areas for Forest Fuel Management Using Geospatial Data, K-Means, and Random Forest Classification. *Forests*, *13*(12), 1970. DOI: 10.3390/f13121970

Collins, W. (1978). Remote sensing of crop type and maturity. *Photogrammetric Engineering and Remote Sensing*, *44*, 43–55.

Coops, N., Waring, R., Wulder, M., & White, J. (2009). Prediction and assessment of bark beetle-induced mortality of lodge pole pine using estimates of stand vigor derived from remotely sensed data. *Remote Sensing of Environment*, *113*(5), 1058–1066. DOI: 10.1016/j.rse.2009.01.013

Curran, P. J. (1983). Multispectral remote sensing for the estimation of green leaf area index. *Philosophical Transactions of the Royal Society of London. Series A, Mathematical and Physical Sciences*, *309*(1508), 257–270. DOI: 10.1098/rsta.1983.0039

Dash, J. P., Pearse, G. D., & Wat, M. S. (2018). UAV Multispectral Imagery Can Complement Satellite Data for Monitoring Forest Health. *Remote Sensing (Basel)*, *10*(8), 1216. DOI: 10.3390/rs10081216

Dash, J. P., Watt, M. S., Pearse, G. D., Heaphy, M., & Dungey, H. S. (2017). Assessing Very High Resolution UAV Imagery for Monitoring Forest Health during a Simulated Disease Outbreak. *ISPRS Journal of Photogrammetry and Remote Sensing*, *131*, 1–14. DOI: 10.1016/j.isprsjprs.2017.07.007

Datt, B. (1999). A New Reflectance Index for Remote Sensing of Chlorophyll Content in Higher Plants: Tests Using Eucalyptus Leaves. *Journal of Plant Physiology*, *154*(1), 30–36. DOI: 10.1016/S0176-1617(99)80314-9

Dennison, P. E., Brunelle, A. R., & Carter, V. A. (2010). Assessing canopy mortality during a mountain pine beetle outbreak using GeoEye-1 high spatial resolution satellite data. *Remote Sensing of Environment*, *114*(11), 2431–2435. DOI: 10.1016/j.rse.2010.05.018

Diamantoglou, S., & Mitrakos, K. (1981). Leaf longevity in Mediterranean evergreen sclerophylls. Chapter. In Margaris, N. S., & Mooney, H. A. (Eds.), *Components of Productivity of Mediterranean-climate Regions - Basic and Applied Aspects. Tasks for Vegetation Science* (Vol. 4, pp. 17–19). Springer. DOI: 10.1007/978-94-009-8683-1_4

DJI. (2020). *P4 Multi-spectral, User Manual*. 55 pages. https://www.dji.com/p4.multispectral

Eitel, J. U., Vierling, L. A., Litvak, M. E., Long, D. S., Schulthess, U., Ager, A. A., Krofcheck, D. J., & Stoscheck, L. (2011). Broadband, red-edge information from satellites improves early stress detection in a New Mexico conifer woodland. *Remote Sensing of Environment*, *115*(12), 3640–3646. DOI: 10.1016/j.rse.2011.09.002

Estrada, J. S., Fuentes, A., Reszka, P., & Auat-Cheein, F. (2023). Machine learning assisted remote forestry health assessment: A comprehensive state of the art review. *Frontiers in Plant Science*, *14*(1139232), 1–25. DOI: 10.3389/fpls.2023.1139232

Giannakopoulos, C., Bindi, M., Moriondo, M., LeSager, P., & Tin, T. (2005). Climate change impacts in the Mediterranean resulting from a _oC global temperature rise. A report for WWF, Published in July 2005 by WWF, the global conservation organization, Gland, Switzerland.

Gitelson, A., Kaufman, Y. J., Stark, R., & Rundquist, D. (2002a). Novel algorithms for remote estimation of vegetation fraction. *Remote Sensing of Environment*, *80*(1), 76–87. DOI: 10.1016/S0034-4257(01)00289-9

Gitelson, A. A., Kaufman, Y. J., & Merzlyak, M. N. (1996). Use of a green channel in remote sensing of global vegetation from EOS-MODIS. *Remote Sensing of Environment*, *58*(3), 289–298. DOI: 10.1016/S0034-4257(96)00072-7

Gitelson, A.A., Keydan, G.P., & Merzlyak, M.N. (2006). Three-band model for non-invasive estimation of chlorophyll, carotenoids, and anthocyanin contents in higher plant leaves. *Geophys. Res. Lett.*, 33(11), L11402, 1-5.

Gitelson, A. A., Stark, R., Grits, U., Rundquist, D., Kaufman, Y., & Derry, D. (2002b). Vegetation and soil lines in visible spectral space: A concept and technique for remote estimation of vegetation fraction. *International Journal of Remote Sensing*, *23*(13), 2537–2562. DOI: 10.1080/01431160110107806

Gomez, D. F., Ritger, H. M. W., Pearce, C., Eickwort, J., & Hulcr, J. (2020). Ability of Remote Sensing Systems to Detect Bark Beetle Spots in the Southeastern US. *Forests*, *11*(11), 1167. DOI: 10.3390/f11111167

Gugliuzzo, A., Mazzeo, G., Mansour, R., & Tropea-Garzia, G. (2019). Carob pests in the Mediterranean region: Bio-ecology, natural enemies and management options. *Phytoparasitica*, *47*(5), 605–628. DOI: 10.1007/s12600-019-00766-7

Hawryło, P., Bednarz, B., Wężyk, P., & Szostak, M. (2018). Estimating defoliation of Scots pine stands using machine learning methods and vegetation indices of Sentinel-2. *European Journal of Remote Sensing*, *51*(1), 194–204. DOI: 10.1080/22797254.2017.1417745

Heiden, U., d'Angelo, P., Schwind, P., Karlshöfer, P., Müller, R., Zepp, S., Wiesmeier, M., & Reinartz, P. (2022). Soil Reflectance Composites: Improved Thresholding and Performance Evaluation. *Remote Sensing (Basel)*, *14*(18), 4526. DOI: 10.3390/rs14184526

Hornero, A., Zarco-Tejada, P. J., Quero, J. L., North, P. R. J., Ruiz-Gómez, F. J., Sánchez-Cuesta, R., & Hernandez-Clemente, R. (2021). Modelling hyperspectral- and thermal-based plant traits for the early detection of Phytophthora-induced symptoms in Oak decline. *Remote Sensing of Environment*, *263*, 112570. DOI: 10.1016/j.rse.2021.112570

Huete, A. R. (1988). A soil-adjusted vegetation index (SAVI). *Remote Sensing of Environment*, *25*(3), 295–309. DOI: 10.1016/0034-4257(88)90106-X

Huete, A. R., Didan, K., Miura, T., Rodriguez, E. P., Gao, X., & Ferreira, L. G. (2002). Overview of the radiometric and biophysical performance of the MODIS vegetation indices. *Remote Sensing of Environment*, *83*(1), 195–213. DOI: 10.1016/S0034-4257(02)00096-2

Innes, J. (1993). *Forest health: its assessment and status.* Published by CAB International.

Jin, S., & Sader, S. A. (2005). MODIS time-series imagery for forest disturbance detection and quantification of patch size effects. *Remote Sensing of Environment*, *99*(4), 462–470. DOI: 10.1016/j.rse.2005.09.017

Jopia, A., Zambrano, F., Pérez-Martínez, W., Vidal-Páez, P., Molina, J., & Mardones, F. (2020). Time-Series of Vegetation Indices (VNIR/SWIR) Derived from Sentinel-2 (A/B) to Assess Turgor Pressure in Kiwifruit. *ISPRS International Journal of Geo-Information*, *9*(11), 641. DOI: 10.3390/ijgi9110641

Kampen, M., Lederbauer, S., Mund, J.P., & Immitzer, M. (2019). UAV-Based Multispectral Data for Tree Species Classification and Tree Vitality Analysis. Dreiländertagung der DGPF, der OVG und der SGPF in Wien, Österreich - Publikationen der DGPF, Band 28, pp. 623-639.

Karpouzli, E., & Malthus, T. (2003). The empirical line method for the atmospheric correction of IKONOS imagery. *International Journal of Remote Sensing*, *24*(5), 1143–1150. DOI: 10.1080/0143116021000026779

Kim, M. S., Daughtry, C. S. T., Chappelle, E. W., McMurtrey, J. E., III, & Walthall, C. L. (1994). The use of high spectral resolution bands for estimating absorbed photosynthetically active radiation (APAR). *Proceedings of the 6th Symposium on Physical Measurements and Signatures in Remote Sensing* (pp. 299-306). Val D'Isère, France.

Lehmann, J. R. K., Nieberding, F., Prinz, T., & Knoth, C. (2015). Analysis of Unmanned Aerial System–Based CIR Images in Forestry–a New Perspective to Monitor Pest Infestation Levels. *Forests*, *6*(3), 594–612. DOI: 10.3390/f6030594

Liang, J., & Gamarra, J. G. P. (2020). The importance of sharing global forest data in a world of crises. *Scientific Data*, *7*(1), 424. DOI: 10.1038/s41597-020-00766-x

Mahdad, M. Y. (2013). *Situation et perspectives d'amélioration du caroubier (Ceratonia siliqua L.) dans la wilaya de Tissemsilt.* Thèse de Maîtrise, Faculté des Sciences de la Technologie, Département des Sciences et de la Nature et de la Vie, Université de Tissemsilt, Algérie, 71 pages.

Manfreda, S., McCabe, M. F., Miller, P. E., Lucas, R., Madrigal, V. P., Mallinis, G., Dor, E. B., Helman, D., Estes, L., & Ciraolo, G.. (2018). On the Use of Unmanned Aerial Systems for Environmental Monitoring. *Remote Sensing (Basel)*, *10*(4), 641. DOI: 10.3390/rs10040641

Merzlyak, M. N., Gitelson, A. A., Chivkunova, O. B., & Rakitin, V. Y. (1999). Non-destructive optical detection of pigment changes during leaf senescence and fruit ripening. *Physiologia Plantarum*, *106*(1), 135–141. DOI: 10.1034/j.1399-3054.1999.106119.x

Michez, A., Piégay, H., Lisein, J., Claessens, H., & Lejeune, P. (2016). Classification of Riparian Forest Species and Health Condition Using Multi–Temporal and Hyperspatial Imagery from Unmanned Aerial System. *Environmental Monitoring and Assessment*, *188*(3), 1–19. DOI: 10.1007/s10661-015-4996-2

Minarik, R., & Langhammer, J. (2016). Use of a Multispectral UAV Photogrammetry for Detection and Tracking of Forest Disturbance Dynamics. Proceedings of *International Archives of Photogrammetry Remote Sens. and Spatial Informational Science* (Vol. XLI-B8, pp. 1-8), XXIII ISPRS Congress, Prague, 23 June 2026.

Moreno-Fernández, D., Viana-Soto, A., Camarero, J. J., Zavala, M. A., Tijerín, J., & García, M. (2021). Using spectral indices as early warning signals of forest dieback: The case of drought-prone Pinus pinaster forests. *The Science of the Total Environment*, *793*, 148578. DOI: 10.1016/j.scitotenv.2021.148578

Motohka, T., Nasahara, K. N., Oguma, H., & Tsuchida, S. (2010). Applicability of Green-Red Vegetation Index for Remote Sensing of Vegetation Phenology. *Remote Sensing (Basel)*, *2*(10), 2369–2387. DOI: 10.3390/rs2102369

Murugesan, K., & Nanjappan, R. (2025). Novel hybrid RNN K-Means algorithm for the forest cover type prediction. *Proceedings of AIP Conference* (vol. 3258, 1: 020015). Madurai, India, 6-17 April 2025. DOI: 10.1063/5.0271542

Naggar, M., & Lahssini, S. (2015). La filière caroubier au service du développement socioéconomique des territoires forestiers de montagne. *Actes du XIVe Congrès forestier mondial* (8 pages). Durban, Afrique du Sud, 7-11 Septembre 2025.

Naseri, M. H., Jouibary, S. S., & Habashi, H. (2023). Analysis of forest tree dieback using Ultra-Cam and UAV-imagery. *Scandinavian Journal of Forest Research*, *38*(6), 392–400. DOI: 10.1080/02827581.2023.2231349

Näsi, R., Honkavaara, E., Lyytikäinen-Saarenmaa, P., Blomqvist, M., Litkey, P., Hakala, T., Viljanen, N., Kantola, T., Tanhuanpää, T., & Holopainen, M. (2015). Using UAV–Based Photogrammetry and Hyperspectral Imaging for Mapping Bark Beetle Damage at Tree–Level. *Remote Sensing (Basel)*, *7*(11), 15467–15493. DOI: 10.3390/rs71115467

Navarro, A., Catalao, J., & Calvao, J. (2019). Assessing the Use of Sentinel-2 Time Series Data for Monitoring Cork Oak Decline in Portugal. *Remote Sensing (Basel)*, *11*(21), 2515. DOI: 10.3390/rs11212515

Ni, M., Wu, Q., Li, G., & Li, D. (2025). Remote Sensing Technology for Observing Tree Mortality and Its Influences on Carbon–Water Dynamics. *Forests*, *16*(2), 194. DOI: 10.3390/f16020194

Otsu, K., Pla, M., Duane, A., Cardil, A., & Brotons, L. (2019). Estimating the Threshold of Detection on Tree Crown Defoliation Using Vegetation Indices from UAS Multispectral Imagery. *Drones (Basel)*, *3*(4), 80. DOI: 10.3390/drones3040080

Otsu, K., Pla, M., Vayreda, J., & Brotons, L. (2018). Calibrating the Severity of Forest Defoliation by Pine Processionary Moth with Landsat and UAV Imagery. *Sensors (Basel), 18*(10), 3278. DOI: 10.3390/s18103278

Palacios-Rodríguez, G., Quinto, L., Lara-Gómez, M. A., Pérez-Romero, J., Recio, J. M., Álvarez-Romero, M., Cachinero-Vivar, A. M., Hernández-Navarro, S., & Navarro-Cerrillo, R. M. (2022). Carbon Sequestration in Carob (Ceratonia siliqua L.) Plantations under the EU Afforestation Program in Southern Spain Using Low-Density Aerial Laser Scanning (ALS) Data. *Forests*, *13*(2), 285. DOI: 10.3390/f13020285

PCI-Geomatics. (2018). *Using PCI Software*. PCI-Geomatics.

Penuelas, J., Baret, F., & Filella, I. (1995). Semi-empirical indices to assess carotenoids/chlorophyll a ratio from leaf spectral reflectance. *Photosynthetica*, *31*, 221–230.

Penuelas, J., Filella, I., Biel, C., Serrano, L., & Save, R. (1993). The reflectance at the 950-970 nm region as an indicator of plant water status. *International Journal of Remote Sensing*, *14*(10), 1887–1905. DOI: 10.1080/01431169308954010

Penuelas, J., Gamon, J., Freeden, A., Merino, J., & Field, C. (1994). Reflectance indices associated with physiological changes in nitrogen and water limited sunflower leaves. *Remote Sensing of Environment*, *48*(2), 135–146. DOI: 10.1016/0034-4257(94)90136-8

Pix4D-Mapper-V4.1. (2017). *Manual of Pix-4D Mapper*. https://support.pix4d.com/hc/en-us/articles/204272989-Offline-Getting-Started-and-Manual-pdf. Manual-Pix4D-Mapper.pdf

Rouse, J. W., Haas, R. W., Schell, J. A., Deering, D. W., & Harlan, J. C. (1974). *Monitoring the vernal advancement and retro-gradation (Green wave effect) of natural vegetation*. NASA/GSFCT Type III Final Report Number: NASA-CR-139243, Greenbelt, MD, USA, 250 pages.

Roy, D. P., Zhang, H. K., Ju, J., Gomez-Dans, J. L., Lewis, P. E., Schaaf, C. B., Sun, Q., Li, J., Huang, H., & Kovalskyy, V. (2016). A general method to normalize Landsat reflectance data to nadir BRDF adjusted reflectance. *Remote Sensing of Environment*, *176*, 255–271. DOI: 10.1016/j.rse.2016.01.023

Rullan-Silva, C. D., Oltho, A. E., Delgado de la Mata, J. A., & Pajares-Alonso, J. A. (2013). Remote Monitoring of Forest Insect Defoliation - A Review. *Forest Systems*, *22*(3), 377–391. DOI: 10.5424/fs/2013223-04417

Salamí, E., Barrado, C., & Pastor, E. (2014). UAV flight experiments applied to the remote sensing of vegetated areas. *Remote Sensing (Basel)*, *6*(11), 11051–11081. DOI: 10.3390/rs61111051

Sallak, B. (2019). *Les risques hydrologiques d'inondations et la problématique d'aménagement des territoires de piémont: Cas de Taghzirt*. Thèse de Doctorat, Université Sultan Moulay Slimane, Béni-Mellal, Maroc, 289 pages.

Şandric, I., Irimia, R., Petropoulos, G. P., Anand, A., Srivastava, P. K., Pleşoianu, A., Faraslis, I., Stateras, D., & Kalivas, D. (2022). Tree's Detection and Health's assessment from ultra-high resolution UAV imagery and deep learning. *Geocarto International*, *37*(25), 10459–10479. DOI: 10.1080/10106049.2022.2036824

Schläpfer, D., Popp, C., & Richter, R. (2020). Drone data Atmospheric Correction Concept for Multi – and Hyperspectral Imagery - The DROACOR Model. *The Int. Arch. of Photogram. Remote Sens. and Spatial Inf. Sci.*, XLIII-B3-2020. *Commission III, WG*, *III*(4), 473–478.

Shi, Y. (2020). Remotely sensing the species of individual trees. *International J. of Applied Earth Observation and Geo-information*, *73*, 207–219. DOI: 10.1016/j.jag.2018.06.018

Sims, D. A., & Gamon, J. A. (2002). Relationships between Leaf Pigment Content and Spectral Reflectance Across a Wide Range of Species, Leaf Structures and Developmental Stages. *Remote Sensing of Environment*, *81*(2-3), 337–354. DOI: 10.1016/S0034-4257(02)00010-X

Sinaga, K. P., & Yang, M. S. (2020). Unsupervised K-Means Clustering Algorithm. *IEEE Access : Practical Innovations, Open Solutions*, *8*, 80716–80727. DOI: 10.1109/ACCESS.2020.2988796

Slater, P. N., Biggar, S. F., Holm, R. G., Jackson, R. D., Mao, Y., Moran, M. S., Palmer, J. M., & Yuan, B. (1987). Reflectance- and radiance-based methods for the in-flight absolute calibration of multispectral sensors. *Remote Sensing of Environment*, *22*(1), 11–37. DOI: 10.1016/0034-4257(87)90026-5

Song, C., Woodcock, C. E., Seto, K. C., Lenney, M. P., & Macomber, S. A. (2001). Classification and change detection using Landsat TM data: When and how to correct atmospheric effects. *Remote Sensing of Environment*, *75*(2), 230–244. DOI: 10.1016/S0034-4257(00)00169-3

Spruce, J., Sader, S., Ryan, R., Smoot, J., Kuper, P., Ross, K., Prados, D., Russell, J., Gasser, G., McKellip, R., & Hargrove, W. (2011). Assessment of MODIS NDVI time series data products for detecting forest defoliation by gypsy moth outbreaks. *Remote Sensing of Environment*, *115*(2), 427–437. DOI: 10.1016/j.rse.2010.09.013

Su, L., Heydari, M., Mousavi, S. R., & Prévosto, B. (2024). Role of environmental and stand factors on forest dieback: An approach using structural equation modelling and machine learning. *Forest ecology and Management*, 562, 121927, ISSN 0378-1127.

Teillet, P. M., & Santer, R. (1991). Terrain Elevation and Sensor Altitude Dependence in a Semi-Analytical Atmospheric Code. *Canadian Journal of Remote Sensing*, *17*(1), 36–44.

Thenkabail, P. S., Lyon, J. G., & Huete, A. R. (2018). Advanced Applications in Remote Sensing of Agricultural Crops And Natural Vegetation, 2nd Edition, vol. 4. CRC Press - Taylor and Francis Group, 425 pages. DOI: 10.1201/9780429431166

Thome, K. J. (2001). Absolute radiometric calibration of Landsat-7 ETM+ using the reflectance-based method. *Remote Sensing of Environment*, *78*(1-2), 27–38. DOI: 10.1016/S0034-4257(01)00247-4

Thome, K. J., Helder, D. L., Aaron, D., & Dewald, J. D. (2004). Landsat-5 TM and Landsat-7 ETM+ Absolute Radiometric Calibration Using the Reflectance-Based Method. *IEEE Transactions on Geoscience and Remote Sensing*, *42*(12), 2777–2785. DOI: 10.1109/TGRS.2004.839085

Thomson, P. H. (1977). The carob in California. *Avocado Grower*, *1*(2), 36–39.

Thuiller, W., Lavorel, S., Araujo, M.-B., Sykes, M. T., & Prentice, I. C. (2005). Climate change threats to plant diversity in Europe. *Proceedings of the National Academy of Sciences of the United States of America*, *102*(23), 8245–8250. DOI: 10.1073/pnas.0409902102

Tilly, N., Reddig, F., Lussem, U., & Bareth, G. (2020). First Investigation of Mediterranean Oak Tree Vitality with High Resolution WorldView-3 Satellite data: Ccomparing Ten Vegetation Indices and Three Machine Learning Classifiers. *The Int. Archives of Photogrammetry, Remote Sensing and Spatial Information Sciences*, vol. XLIII-B3-2020, pp. 1069-1076.

Torresan, C., Berton, A., Carotenuto, F., Filippo, S., Gennaro, D., Gioli, B., Matese, A., Miglietta, F., Vagnoli, C., Zaldei, A., & Wallace, L. (2017). Forestry Applications of UAVs in Europe: A Review. *International Journal of Remote Sensing*, *38*(8-10), 2427–2447. DOI: 10.1080/01431161.2016.1252477

Tous, J., Romero, A., Plana, J., & Batlle, I. (1996). Current situation of carob plant material. *In Proceedings of the Third International Carob Symposium* (Tavira, Portugal) University of Lisbon, Portugal.

Ustin, S. L., Martens, S. N., Curtiss, B., & Vanderbilt, V. (1988). Use of high spectral resolution sensors to detect air pollution injury in conifer forests. In Fenstermaker, L.K. (Ed.), Remote Sensing Applications for Acid Deposition (pp. 72-85). Environmental Protection Agency (EPA), Las Vegas, NV, USA.

Van-Beek, J., Tits, L., Somers, B., Deckers, T., Janssens, P., & Coppin, P. (2016). Viewing Geometry Sensitivity of Commonly Used Vegetation Indices towards the Estimation of Biophysical Variables in Orchards. *Journal of Imaging*, *2*, 15. DOI: 10.1016/j.foreco.2021.119493

Vollenweider, P., & Günthardt-Goerg, M. S. (2005). Diagnosis of abiotic and biotic stress factors using the visible symptoms in foliage. *Environmental Pollution*, *137*(3), 455–465. DOI: 10.1016/j.envpol.2005.01.032

Wang, C., & Myint, S. W. (2015). A Simplified Empirical Line Method of Radiometric Calibration for Small UAS-Based Remote Sensing. *IEEE Journal of Selected Topics in Applied Earth Observations and Remote Sensing*, *8*(5), 1876–1885. DOI: 10.1109/JSTARS.2015.2422716

Wulder, M., White, J., Bentz, B., & Ebata, T. (2006). Augmenting the existing survey hierarchy for mountain pine beetle redattack damage with satellite remotely sensed data. *Forestry Chronicle*, *82*(2), 187–202. DOI: 10.5558/tfc82187-2

Yao, H., Qin, R., & Chen, X. (2019). Unmanned Aerial Vehicle for Remote Sensing Applications: A Review. *Remote Sensing (Basel)*, *11*(12), 1443. DOI: 10.3390/rs11121443

Ye, H., Huang, W., Huang, S., Cui, B., Dong, Y., Guo, A., Ren, Y., & Jin, Y. (2020). Recognition of Banana Fusarium Wilt Based on UAV Remote Sensing. *Remote Sensing (Basel)*, *2020*(12), 938. DOI: 10.3390/rs12060938

Yoder, B. J., & Waring, R. H. (1994). The Normalized Difference Vegetation Index of Small Douglas–Fir Canopies with Varying Chlorophyll Concentrations. *Remote Sensing of Environment*, *49*(1), 81–91. DOI: 10.1016/0034-4257(94)90061-2

Yu, R., Luo, Y., Zhou, Q., Zhang, X., Wu, D., & Ren, L. (2021). Early detection of pine wilt disease using deep learning algorithms and UAV-based multispectral imagery. *Forest Ecology and Management*, *497*, 119493.

Zhang, C., & Kovacs, J. M. (2012). The application of small UAS for precision agriculture: A review. *Precision Agriculture*, *13*, 693–712. DOI: 10.1007/s11119-012-9274-5

Zhang, Z., Liu, M., Liu, X., & Zhou, G. (2018). A New Vegetation Index Based on Multitemporal Sentinel-2 Images for Discriminating Heavy Metal Stress Levels in Rice. *Sensors (Basel)*, *18*(7), 2172. DOI: 10.3390/s18072172

Compilation of References

Aafi, A., Cuzin, F., & Tahiri, A. (2007). La faune sauvage et l'arganeraie: interactions et pressions. In: Actes du Premier Congrès International de l'Arganier. Essaouira.

Abdulridha, J., Batuman, O., & Ampatzidis, Y. (2019). UAV-Based Remote Sensing Technique to Detect Citrus Canker Disease Utilizing Hyperspectral Imaging and Machine Learning. *Remote Sensing (Basel)*, *11*(11), 1373. DOI: 10.3390/rs11111373

Acharya, R. P., Maraseni, T. N., & Cockfield, G. (2020). An ecosystem services valuation research framework for policy integration in developing countries: A case study from Nepal. *Sustainability (Basel)*, *12*(19), 8250. DOI: 10.3390/su12198250

Achour, A. (2014). *Contribution à l'étude de la phénologie de l'arganier et de la biodiversité au sein d'une parcelle clôturée. Evaluation d'un essai de régénération artificielle*. Thèse de Doctorat National. PP. 204. Université Ibn Zohr; Faculté des Sciences. Agadir.

Adous, F. E., Ghouat, M. A. E., & Ibannain, F. (2019). A Geo-Decisional Tool for the Management of the Agricultural Development Fund Under the Green Morocco Plan. In *Geospatial Technologies for Effective Land Governance* (pp. 160–173). IGI Global Scientific Publishing., DOI: 10.4018/978-1-5225-5939-9.ch010

AFI (Association Futaie Irrégulière) (2009). *Les traitements des Futaies Irrégulières. Valoriser les fonctions multiples de la forêt*. Convention France Bois Forêt/AFI 2009.

Afi, C., Telmoudi, M., Labbassi, S., Chabbi, N., Hallam, J., Msanda, F., & Ait Aabd, N. (2024). Assessing the Impact of Aridity on Argan Trees in Morocco: Implications for Conservation in a Changing Climate. *Resources*, *13*(10), 135. DOI: 10.3390/resources13100135

Agoumi, A., & Debbarh, A. (2025). Ressources en eau et bassins versants du Maroc: 50 ans de développement (1955–2005). 50 Ans de Développement Humains. *Perspectives*.

Ahtikoski, A., Rämö, J., Juutinen, A., Shanin, V., & Mäkipää, R. (2022). Continuous cover forestry and cost of carbon abatement on mineral soils and peatlands. *Frontiers in Environmental Science*, *10*, 837878. Advance online publication. DOI: 10.3389/fenvs.2022.837878

Aitken, S. N., & Whitlock, M. C. (2013). Assisted gene flow to facilitate local adaptation to climate change. *Annual Review of Ecology, Evolution, and Systematics*, *44*(1), 367–388. DOI: 10.1146/annurev-ecolsys-110512-135747

Alegria, C. (2023). Aboveground Biomass Mapping and Fire Potential Severity Assessment: A case study for eucalypts and shrubland areas in the central inland region of Portugal. *Forests*, *14*(9), 1795. DOI: 10.3390/f14091795

Allen, M. J., Moreno-Fernández, D., Ruiz-Benito, P., Grieve, S. W. D., & Lines, E. R. (2024). Low-cost tree crown dieback estimation using deep learning-based segmentation. *Environmental Data Science*, *3*(e18), 1–18. DOI: 10.1017/eds.2024.16

Allen, S. E., Grimshaw, H. M., & Rowland, A. P. (1986). Chemical Analysis. In Moore, P. D., & Chapman, S. B. (Eds.), *Methods of Plant Ecology* (pp. 285–344). Blackwell.

Altman, I., & Low, S. M. (Eds.). (1992). *Place Attachment*. Springer US., DOI: 10.1007/978-1-4684-8753-4

Alvites, C., Marchetti, M., Lasserre, B., & Santopuoli, G. (2022). LiDAR as a tool for assessing timber assortments: A systematic literature review. *Remote Sensing (Basel)*, *14*(18), 4466. DOI: 10.3390/rs14184466

Amami, B., Rhazi, L., Chaibi, M., Fauquette, S., Ougougdal, M. A., Charif, A., Ridaoui, M., Bouahim, S., Carré, M., & Daoud-Bouattour, A. (2013). Late Quaternary history of a Mediterranean temporary pool from western Morocco, based on sedimentological and palynological evidence. *Palaeogeography, Palaeoclimatology, Palaeoecology*, *392*, 281–292. DOI: 10.1016/j.palaeo.2013.09.024

Ameray, A., Bergeron, Y., Valeria, O., Montoro Girona, M., & Cavard, X. (2021). Forest carbon management: A review of Sylvicultural Practices and Management Strategies Across Boreal, Temperate and Tropical Forests. *Current Forestry Reports*, *7*(4), 245–266. DOI: 10.1007/s40725-021-00151-w

ANEF. (2024). ANEF: Recul de 70% des superficies forestières parcourues par le feu. *Le Desk*. https://ledesk.ma/encontinu/anef-recul-de-70-des-superficies-forestieres-parcourues-par-le-feu/

ANEF. (2024). Bilan des activités de l'Agence Nationale des Eaux et Forêts (ANEF) au titre de l'année 2024 (Rapport interne), 65 P.

ANEF. (2024). Forêts En Chiffres. http://www.eauxetforets.gov.ma/ForetsMarocaines/ForetsChiffres/Pages/Forets-En-Chiffres.aspx

ANEF. (2024). Forêts en chiffres. http://www.eauxetforets.gov.ma≡

ANEF. (2025). Répartition des superficies reboisées. Agence Nationale des Eaux et Forêts. Retrieved November 12, 2025, from http://www.eauxetforets.gov.ma/DeveloppementForestier/Reboisement/Pages/Reboisement.aspx

António, N., Tomé, M., Tomé, J., Soares, P., & Fontes, L. (2007). Effect of tree, stand, and site variables on the allometry of *Eucalyptus globulus* tree biomass. *Canadian Journal of Forest Research*, *37*(5), 895–906. DOI: 10.1139/X06-276

Arabi, M., Mechkirrou, L., El Malki, M., Alaoui, K., Chaieb, A., Maaroufi, F., Karmich, S. (2024). Overview of ecological dynamics in Morocco–Biodiversity, water scarcity, climate change, anthropogenic pressures, and energy resources–Navigating towards ecosolutions and sustainable development. *E3S Web of Conf.*, 527, 01001 (2024). DOI: 10.1051/e3sconf/202452701001

Arias, P. A., Bellouin, N., & Coppola, E. (2021). Technical Summary. In *Climate Change 2021: The Physical Science Basis Working Group I Contribution to the Sixth Assessment Report of the Intergovernmental Panel on Climate Change* (pp. 35–144). Cambridge University Press., DOI: 10.1017/9781009157896.002

Arion, I. D., Morar, I. M., Truta, A. M., Chereches, I. A., Isarie, V. I., & Arion, F. H. (2025). Internet of Things (IoT)-Based Applications in Smart Forestry: A Conceptual and Technological Analysis. *Forests*, *17*(1), 44. DOI: 10.3390/f17010044

ASD. (2015). *Analytical Spectral Devices. Technical Guide, 4th ed.; ASD Inc.: Boulder, CO, USA, 2015.* http://www.asdi.com/products-spectroradiometers.asp

Aspe, C., & Genin, D. (2014). À chacun sa biodiversité : Approches savantes et locales de la gestion des ressources et des milieux dans le Sud de la France et le Haut Atlas marocain. *Revue d'ethnoécologie*, (5).

Asrar, G., Fuchs, M., Kanemasu, E. T., & Hatfield, J. L. (1984). Estimating absorbed photosynthetic radiation and leaf area index from spectral reflectance in wheat. *Agronomy Journal*, *76*(2), 300–306. DOI: 10.2134/agronj1984.00021962007600020029x

Aubert, P.-M. (2014). Les évolutions de la politique forestière au Maroc : Entre réappropriation du modèle forestier français et idéalisation de la tribu. *Revue Forestière Française*, *305*(4), 305. DOI: 10.4267/2042/53627

Aubert, P.-M., & Sabir, M. (2014). Du technique au politique et retour : État des lieux et enjeux dans la gestion des forêts au Maroc. Introduction au dossier. *Revue Forestière Française*, (4), 295. DOI: 10.4267/2042/53626

Aurelle, D., Thomas, S., Albert, C., Bally, M., Bondeau, A., Boudouresque, C.-F., Cahill, A. E., Carlotti, F., Chenuil, A., Cramer, W., Davi, H., De Jode, A., Ereskovsky, A., Farnet, A.-M., Fernandez, C., Gauquelin, T., Mirleau, P., Monnet, A.-C., Prévosto, B., & Fady, B. (2022). Biodiversity, climate change, and adaptation in the Mediterranean. *Ecosphere*, *13*(4), e3915. DOI: 10.1002/ecs2.3915

Azedou, A., Amine, A., Kisekka, I., & Lahssini, S. (2025). Genetic algorithm optimization of ensemble learning approach for improved land cover and land use mapping: Application to Talassemtane National Park. *Ecological Indicators*, *177*, 113776. DOI: 10.1016/j.ecolind.2025.113776

Aziz, L., Flouchi, S., & Msiah, F.-E. (2014). *Acquis et défis de la gestion participative des ressources forestières : Cas du Parc National d'Ifrane*. DOI: 10.60569/1-A4

Aziz, D., Rafiq, S., Saini, P., Ahad, I., Gonal, B., Rehman, S. A., Rashid, S., Saini, P., Rohela, G. K., Aalum, K., Singh, G., Gnanesh, B. N., & Nabila Iliya, M. (2025). Remote sensing and artificial intelligence: Revolutionizing pest management in agriculture. *Frontiers in Sustainable Food Systems*, *9*, 1551460. Advance online publication. DOI: 10.3389/fsufs.2025.1551460

Badalamenti, E., Battipaglia, G., Gristina, L., Novara, A., Rühl, J., Sala, G., Sapienza, L., Valentini, R., & La Mantia, T. (2019). Carbon stock increases up to old growth forest along a secondary succession in Mediterranean island ecosystems. *PLoS One*, *14*(7), e0220194. DOI: 10.1371/journal.pone.0220194 PMID: 31339941

Bahri, M., Haboudane, D., Bannari, A., & Bonn, F. (2008). Mapping the dominant forest species of the Middle Atlas cedar ecosystem in Morocco: Comparison between methods based on spectral similarity. *Revue Télédétection*, vol. 7, no. 1-2-3-4, pp. 283-301.

Balkiz, O. (2016). *Assessment of the socio-economic values of goods and services provided by Mediterranean forest ecosystems: Düzlerçami Forest, Turkey*. Plan Bleu.

Bani-Aameur, F., & Ferradous, A. (2001). Fruit and seed variability in Moroccan argan tree (Argania spinosa). *Forest Genetics*, *8*(1), 39–45.

Bannari, A., Selouani, A., Nehari, W., Rhinane, H., El-Ghmari, A., & Oulbacha, S. (2024). Accuracy Evaluation of DEMs Derived from UAV data acquired over a Narrow-Deep Valley in the Middle-Atlas Mountain. *Proceedings of IEEE-IGARSS* (pp. 11449-11452). Athens, Greece, 17-12 July 2024. DOI: 10.1109/IGARSS53475.2024.10641039

Bannari, A., & Al-ali, Z. M. (2020). Ground Reflectance Factor Retrieval from Landsat (MSS, TM, ETM+, and OLI) Time Series Data based on Semi-empirical Line Approach and Pseudo-invariant Targets in Arid Landscape. *Proceedings of IEEE-IGARSS* (pp. 5990-5993). Waikoloa, Hawaii, USA, 26 September to 02 October 2020.

Bannari, A., Almansoori, T. A., Mohamed, A. M. A., El-Battay, A., & Hameid, N. (2018). Modeling Water Stress as an Indicator of Red Palm Weevil Infestation using Field Sampling, WorldView-3 Reflectance, and Laboratory Analysis. *Proceedings of IEEE-IGARSS* (pp. 9056-9059). Valencia, Spain, 23-27 July 2018. DOI: 10.1109/IGARSS.2018.8518687

Bannari, A., Asalhi, H., & Teillet, P. M. (2002). Transformed Difference Vegetation Index (TDVI) for Vegetation Cover Mapping. *Proceedings of IEEE-IGARSS*' (pp. 3053-3055), Toronto (Ontario), Canada, 9-13 July 2002. DOI: 10.1109/IGARSS.2002.1026867

Bannari, A., El-Ghmari, A., Rhinane, H., Selouani, A., & Loulad, S. (2024b). Vicarious Radiometric Calibration of UAV-Data in Mountainous area using Semi-Empirical Line Approach and *in-situ* Spectroradiometric Measurements. *Proceedings of IEEE-IGARSS* (pp. 11449-11452). Athens, Greece, 7-12 July 2024. DOI: 10.1109/IGARSS53475.2024.10642636

Bannari, A., Gibson, J. R., & Morin, D. (1997). The Necessity of Exterior Orientation Parameters for the Rigorous Geometric Correction of MEIS-II Airborne Digital Image. *Remote Sensing Reviews*, *16*(1-2), 135–156. DOI: 10.1080/02757259709532349

Bannari, A., Khurshid, K. S., Staenz, K., & Schwarz, J. (2007a). A Comparison of Hyperspectral Chlorophyll Indices for Wheat Crop Chlorophyll Content Estimation Using Laboratory Reflectance Measurements. *IEEE Transactions on Geoscience and Remote Sensing*, *45*(10), 3063–3073. DOI: 10.1109/TGRS.2007.897429

Bannari, A., Khurshid, S. K., Staenz, K., & Schwarz, J. (2008). Potential of Hyperion EO-1 Hyperspectral Data for Wheat Crop Chlorophyll Content Extraction in Precision Agriculture. *Canadian Journal of Remote Sensing*, *34*(1), 139–157. DOI: 10.5589/m08-001

Bannari, A., Mohamed, A. K., & Peddle, D. R. (2016). Bio-Physiological Parameters Retrieval and Statistical Analysis for Red Palm Weevils Stress-Attack Prediction Using WorldView-3 data. *Proceedings of IGARSS* (pp. 7129-7132). Beijing, China, 10-15 July 2016.

Bannari, A., Mohamed, A. M. A., & El-Battay, A. (2017). Water Stress Detection in Palm Threes Canopies as an Indicator for Red Palm Weevil Stress-Attack using WorldView-3 data. *Proceedings of IGARSS* (pp. 4000-4003). Texas, USA, 23-28 July 2017.

Bannari, A., Morin, D., Bénié, G. B., & Bonn, F. (1995b). A Theoretical Review of Different Mathematical Models of Geometric Corrections Applied to Remote Sensing Images. *Remote Sensing Reviews*, *13*(1-2), 27–47. DOI: 10.1080/02757259509532295

Bannari, A., Morin, D., Huete, A. R., & Bonn, F. (1995a). A Review of Vegetation indices. *Remote Sensing Reviews*, *13*(1-2), 95–120. DOI: 10.1080/02757259509532298

Bannari, A., Selouani, A., El-basri, M., Rhinane, H., El-Harti, A., & El-Ghmari, A. (2021). Multi-scales Analysis of DEMs Derived from Unmanned Aerial Vehicle in Precision Agriculture Context. *Proceedings of IEEE-IGARSS* (pp. 8285-8188). Brussels, Belgium, 12-14 July 2021. DOI: 10.1109/IGARSS47720.2021.9554020

Bannari, A., Teillet, P. M., & Richardson, G. (1999). The necessity of radiometric calibration and standardization of remote sensing data. *Canadian Journal of Remote Sensing*, *25*(1), 45–59. DOI: 10.1080/07038992.1999.10855262

Banque mondiale. (2010). *La génération des bénéfices environnementaux pour améliorer la gestion des bassins versants en Tunisie*. Auteurs: Croitoru, L., Daly-Hassen, H., Cherni, A., Sterk, G., Bird, N., Zanchi, G., Frieden, D., Oka, A., & Tran, L. Rapport no. 50192-TN.

Barakat, A., Rafai, M., Mosaid, H., Islam, M. S., & Saeed, S. (2023). Mapping of Water-Induced Soil Erosion Using Machine Learning Models: A Case Study of Oum Er Rbia Basin (Morocco). *Earth Systems and Environment*, *7*(1), 151–170. DOI: 10.1007/s41748-022-00317-x

Barbéro, M., Loisel, R., & Quézel, P. (1992). Biogeography, ecology and history of Mediterranean Pinus halepensis and Pinus brutia. *Plant Ecology*, *4*(1), 95–115.

Barbero, M., Loisel, R., Quézel, P., & Richardson, D. M. (1998). Pinewoods of the Mediterranean Basin. In *Ecology and Biogeography of Pinus*. Cambridge University Press.

Barbour, M. G., & Major, J. (1977). *Terrestrial vegetation of California*. University of California Davis.

Barré, P., Cécillon, L., Chenu, C., Martin, M., Vidal-Beaudet, L., & Eglin, T. (2020). La séquestration de carbone dans les sols agricoles, forestiers et urbains: État des lieux des méthodes d'évaluation et de quantification. *Etude et Gestion des Sols, 27*, 305–320.

Barrett, S. C. H. (2015). Clonality, sexual reproduction and evolution in plants. *Molecular Ecology, 24*(13), 2777–2796.

Barrière, O. (2011). Pluralisme juridique et patrimonialisation : Entre paradigmes de l'« appropriation » et du « patrimoine commun ». In *La question du patrimoine en Guyane* (Illustrated, p. 453). Ibis Rouge.

Barrière, O., & Barrière, C. (Eds.). (1997). *Le foncier-environnement : Fondements juridico-institutionnels pour une gestion viable des ressources naturelles renouvelables au Sahel*. Organisation des Nations unies pour l'Alimentation et l'Agriculture.

Basile-Doelsch, I., Balesdent, J., & Pellerin, S. (2020). Reviews and syntheses: The mechanisms underlying carbon storage in soil. *Biogeosciences, 17*(21), 5223–5242. DOI: 10.5194/bg-17-5223-2020

Basílio, J. J. N., Campoe, O. C., Queiroz, T. B., De Souza, C. R., Carneiro, R. L., Alvares, C. A., & Figura, M. A. (2024). Fine Root Density Dynamics and Carbon Stock of Eucalyptus spp.: Interplay of Age, Genotype, and Edaphoclimatic Conditions. *Plants, 13*(11), 1503. DOI: 10.3390/plants13111503 PMID: 38891311

Bates, E., Popović, M., Marsh, C., Clark, R., Kovac, M., & Bahadir-Kocer, B. (2025). Leaf Level Ash Dieback Disease Detection and Online Severity Estimation with UAVs. *In IEEE-Access*, 13, 55499-55511; DOI: 10.1109/ACCESS.2025.3541980

Batjes, N. H. (1996). Total carbon and nitrogen in the soils of the world. *European Journal of Soil Science, 47*(2), 151–163. DOI: 10.1111/j.1365-2389.1996.tb01386.x

Batlle, I., & Tous, J. (1997). Carob tree. *Ceratonia siliqua* L. Promoting the conservation and use of underutilized and neglected crops. 17 Institute of Plant Genetics and Crop Plant Research, Gatersleben. International Plant Genetic Resources Institute, Rome, Italy.

Batzer, D. P., & Wissinger, S. A. (1996). Ecology of insect communities in nontidal wetlands. *Annual Review of Entomology, 41*(1), 75–100. DOI: 10.1146/annurev.en.41.010196.000451 PMID: 15012325

Bauhus, J., Puettmann, K., & Messier, C. (2009). Silviculture for old-growth attributes. *Forest Ecology and Management, 258*(4), 525–537. DOI: 10.1016/j.foreco.2009.01.053

Baumgärtner, J. (2007). The insurance value of biodiversity in the provision of ecosystem services. *Natural Resource Modeling*, *20*(1), 87–127. DOI: 10.1111/j.1939-7445.2007.tb00202.x

Beau, R., & Larrère, C. (2018). *Penser l'anthropocène*. Sciences po, les presses.

Bedoui, C. (2020). Study of desertification sensitivity in Talh region (Central Tunisia) using remote sensing, GIS and the MEDALUS approach. *Geoenvironmental Disasters*, *7*(1), 16. DOI: 10.1186/s40677-020-00148-w

Beja, P., & Alcazar, R. (2003). Conservation of Mediterranean temporary ponds under agricultural intensification: An evaluation using amphibians. *Biological Conservation*, *114*(3), 317–643. DOI: 10.1016/S0006-3207(03)00051-X

Belghazi, T. (2013). Typologie de l'arganeraie et Estimation de La Biomasse Aérienne Des Taillis d'arganier (Plateau Des Haha-Essaouira, Maroc). Ph.D. Thesis, UCL-Université Catholique de Louvain, Ottignies-Louvain-la-Neuve, Belgium

Belghazi, T., Ponette, Q., Jonard, M., & Belghazi, B. (2016). Biomasse ligneuse et foliaire des taillis d'arganier dans le plateau des Haha, Maroc. *Forêt Méditerranéenne*, *XXXVII*(2), 157–172.

Ben Jamâa, M. L. & Mnara, S. (2000) — *Effect of the defoliations of Lymantria dispar on the cork growth.* Article, disponible via IOBC-WPRS.

Benabderrahmane, M. C., & Chenchouni, H. (2010). Assessing environmental sensitivity areas to desertification in Eastern Algeria using Mediterranean desertification and land use "MEDALUS" model. *Int J Sustain Water Environ Syst*, *1*(1), 5–10. DOI: 10.5383/swes.01.01.002

Benabid, A. (1985). Les écosystèmes forestiers du Maroc. *Forêt Méditerranéenne*, 53–64.

Benabid, A. (2000). Flore et écosystèmes du Maroc; *Evaluation et préservation de la biodiversité*. Edition Ibis Press, 2000. ISBN: 2-910728-13-7

Benabid, A. (1985). Les écosystèmes forestiers, préforestiers et presteppiques du Maroc: Diversité, répartition biogéographique et problèmes posés par leur aménagement. *Forêt Méditerranéenne*, *7*(1), 53–64.

Benabid, A. (2000). *Flore et écosystèmes du Maroc: Évaluation et préservation de la biodiversité*. Kalila Wa Dimna.

Benabou, A., Moukrim, S., Laaribya, S., Aafi, A., Chkhichekh, A., Maadidi, T. E., & Aboudi, A. E. (2022b). Mapping Ecosystem Services of Forest Stands : Case Study of Maamora, Morocco. *Geography, Environment, Sustainability*, *15*(1), 141–149. DOI: 10.24057/2071-9388-2021-047

Benabou, A., Moukrim, S., Lahssini, S., El Aboudi, A., Menzou, K., Elmalki, M., & Rhazi, L. (2022). Impact of climate change on potential distribution of *Quercus suber* in the conditions of North Africa. *Biosystems Diversity*, *30*(3), 289–294. DOI: 10.15421/012231

Benhizia, R., Kouba, Y., Szabó, G., Négyesi, G., & Ata, B. (2021). Monitoring the spatiotemporal evolution of the Green Dam in Djelfa Province, Algeria. *Sustainability (Basel)*, *13*(14), 7953. DOI: 10.3390/su13147953

Ben-Said, M., Chemchaoui, A., Etebaai, I., & Taher, M. (2025). Land use and land cover changes in Morocco: Trends, research gaps, and perspectives. *GeoJournal*, *90*(1), 44. DOI: 10.1007/s10708-024-11214-3

Benzyane, M. (2007). La gestion durable des ressources forestières au Maroc: Quelle stratégie? *Forêt Méditerranéenne*, *28*(1), 47–54.

Berie, H. T., & Burud, I. (2018). Application of unmanned aerial vehicles in earth resources monitoring: Focus on evaluating potentials for forest monitoring in Ethiopia. *European Journal of Remote Sensing*, *51*(1), 326–335. DOI: 10.1080/22797254.2018.1432993

Bernknopf, R., Kuwayama, Y., Gibson, R., Blakely, J., Mabee, B., Clifford, T. J., Quayle, B., Epting, J., Hardy, T., & Goodrich, D. (2019). The cost-effectiveness of satellite Earth observations to inform a post-wildfire response. *Resources for the Future*. https://media.rff.org/documents/Valuables_Wildfires.pdf

Bijaber, N., & Ahlafi, Z. (2005). *Études détaillées pour la cartographie des changements dans le cadre de l'Inventaire Forestier National*. Projet FORMA.

Bilintoh, T. (2019). *Remote Sensing for Mapping and Modeling Presence of Armyworm Infestation on Maize in Ejura, Ghana*. Master Thesis, Department of Geography, Michigan State University, United States, 50 pages.

Binkley, D., Campoe, O. C., Alvares, C., Carneiro, R. L., Cegatta, Í., & Stape, J. L. (2017). The interactions of climate, spacing and genetics on clonal Eucalyptus plantations across Brazil and Uruguay. *Forest Ecology and Management*, *405*, 271–283. DOI: 10.1016/j.foreco.2017.09.050

Bishop, J. T. (Ed.). (1999). *Valuing forests: A Review of Methods and Applications in Developing Countries*. IIED.

Bithas, P. S., Michailidis, E. T., Nomikos, N., Vouyioukas, D., & Kanatas, A. G. (2019). A Survey on Machine-Learning Techniques for UAV-Based Communications. *Sensors (Basel), 19*(23), 5170. DOI: 10.3390/s19235170

Blaustein, L., & Schwartz, S. S. (2001). Why study ecology in temporary pools? *Israel Journal of Zoology, 47*(4), 303–312. DOI: 10.1560/CKMU-Q2PM-HTGC-P9C8

Bliss, L. C., Marticorena, C., Matthei, O., & Quezada, M. (1998). Plant diversity in Chile: a comparative ecology of Mediterranean ecosystems. In Rundel, P. W. (Eds.), *Landscape disturbance and biodiversity in Mediterranean-type ecosystems*. Springer.

Boisvenue, C., Paradis, G., Eddy, I. M. S., McIntire, E. J. B., & Chubaty, A. M. (2022). Managing forest carbon and landscape capacities. *Environmental Research Letters, 17*(11), 114013. DOI: 10.1088/1748-9326/ac9919

Boix, D.. (2016). Pond research and management in Europe: “Small is beautiful”. *Hydrobiologia, 793*, 1–16.

Boix, D., Sabater, S., Gascón, S., Sala, J., Quintana, X. D., & Badosa, A. (2016). Ecology of Mediterranean temporary ponds. In Batzer, D., & Boix, D. (Eds.), *Invertebrates in Freshwater Wetlands* (pp. 229–255). Springer. DOI: 10.1007/978-3-319-24978-0_5

Bolívar, M. P. R., & Meijer, A. J. (2016). Smart Governance: Using a Literature Review and Empirical Analysis to Build a Research Model. *Social Science Computer Review, 34*(6), 673–692. DOI: 10.1177/0894439315611088

Bonis, A., Grillas, P., Van Wijck, C., & Lepart, J. (1993). The Effect Of Salinity On The Reproduction Of Coastal Submerged Macrophytes From Experimental Communities. *Journal of Vegetation Science, 4*(4), 461–468. DOI: 10.2307/3236073

Bonis, A., Lepart, J., & Grillas, P. (1995). Seed Bank Dynamics And Coexistence Of Annual MacrophytesIn A Temporary And Variable Habitat. *Oikos, 74*(1), 81–92. DOI: 10.2307/3545677

Bonsignore, C. P. (2012). Apate monachus (Fabricius, 1775), a bostrichid pest of pomegranate and carob trees in nurseries. *Plant Protection Science, 48*(2), 94–97. DOI: 10.17221/53/2011-PPS

Born, K., Christoph, M., & Fink, A. H. (2008). Moroccan Climate in the Present and Future: Combined View from Observational Data and Regional Climate Scenarios. In Zereini, F., & Hötzl, H. (Eds.), *Climatic Changes and Water Resources in the Middle East and North Africa. Springer Berlin Heidelberg, Berlin* (pp. 29–45). DOI: 10.1007/978-3-540-85047-2_4

Borrini, G., & Jaireth, H. (Eds.). (2007). *Sharing power : Learning-by-doing in co-management of natural resources throughout the world.* Earthscan.

Bouahim, S. (2010). Usage et conservation des mares temporaires méditerranéennes: cas des mares temporaires de la région de Benslimane (Maroc occidental). Thèse de doctorat, Université Hassan II, Casablanca & Université Montpellier II.

Bouahim, S., Rhazi, L., Amami, B., Waterkeyn, A., Rhazi, M., Saber, E. R., Zouahri, A., Van Den Broeck, M., Muller, S. D., Brendonck, L., & Grillas, P. (2014). Unravelling the impact of anthropogenic pressureon plant communities in Mediterranean temporary ponds. *Marine and Freshwater Research, 65*(10), 918–929. DOI: 10.1071/MF13194

Bouahim, S., Rhazi, L., Ernoul, L., Mathevet, R., Amami, B., Er-Riyahi, S., & Grillas, P. (2015). Combining vulnerability analysis and perceptions of ecosystem services in sensitive landscapes: A case from western Moroccan temporary wetlands. *Journal for Nature Conservation, 27*, 1–9. DOI: 10.1016/j.jnc.2015.05.003

Boubekraoui, H., Attar, Z., Maouni, Y., Ghallab, A., Saidi, R., & Maouni, A. (2024). Forest loss drivers and landscape pressures in a Northern Moroccan protected areas' network: Introducing a novel approach for conservation effectiveness assessment. *Conservation, 4*(3), 452–485. DOI: 10.3390/conservation4030029

Boubekraoui, H., Maouni, Y., Ghallab, A., Draoui, M., & Maouni, A. (2024). Deforestation drivers in northern Morocco: An exploratory spatial data analysis. *Environmental Research Communications, 6*(7), 071005. DOI: 10.1088/2515-7620/ad5ad6

Boudhar, A., Boudhar, S., & Ibourk, A. (2017). Input–output & water relations. *Journal of Economic Structures*, •••, 6.

Boudy, P. (1950). *Économie forestière nord-africaine. Tome II: Monographie et traitement des essences forestières.* Larose, Paris.

Boudy, P. (1952). *Guide du forestier en Afrique du Nord.* La Maison Rustique.

Bouguerra, S., Stiti, B., Khalfaoui, M., Jebari, S., Khaldi, A., & Berndtsson, R. (2024). Modeling Ecosystem Regulation Services and Performing Cost–Benefit Analysis for Climate Change Mitigation through Nature-Based Solutions Using InVEST Models. *Sustainability (Basel), 16*(16), 7201. DOI: 10.3390/su16167201

Boujraf, A., Dallahi, Y., Dahmani, J., Orlando, C. A., & Douira, A. (2023). Effect of plant Communities and Ecological Parameters on Soil Organic Carbon Stocks in the Mamora Forest, Morocco. *Journal of Ecological Engineering, 24*(5), 128–136. DOI: 10.12911/22998993/161616

Boukhris, I., Lahssini, S., Collalti, A., Moukrim, S., Santini, M., Chiti, T., & Valentini, R. (2023). Calibrating a Process-Based Model to Enhance Robustness in Carbon Sequestration Simulations: The Case of Cedrus atlantica (Endl.) Manetti ex Carrière. *Forests*, *14*(2), 401. DOI: 10.3390/f14020401

Boukrouh, S., Bouazzaoui, Y., El Aich, A., Mahyou, H., Chikhaoui, M., Ait Lafkih, M., N'Dorma, O., & Alados, C. (2024). Estimation of standing crop biomass in rangelands of the Middle Atlas mountains using remote sensing data. *African Journal of Range & Forage Science*, *41*(4), 244–259. DOI: 10.2989/10220119.2024.2360991

Boulmane, M., Oubrahim, H., Halim, M., & Bakker, M. R. (2017). The potential of Eucalyptus plantations to restore degraded soils in semi-arid Morocco (NWAfrica). *Annals of Forest Science*. 57-74pp. DOI: 10.1007/s13595-017-0652-z

Boulmane, M., Santa-Regina, I., Khia, A., & Oubrahim, H. (2014). Éstimation du stock de carbone organique dans l'écosystème des iliçaies du Moyen Atlas marocain. *Revue Nature et Technologie, 6*(2), 06-16. https://asjp.cerist.dz/en/article/43400

Boulmane, M., del Carmen Santa-Regina, M., Halim, M., Khia, A., Oubrahim, H., Abbassi, H., & Santa-Regina, I. (2015). Organic Carbon Storage in Evergreen Oak Forest Ecosystems of the Middle and High Moroccan Atlas Areas. *Open Journal of Forestry*, *5*(3), 260–273. DOI: 10.4236/ojf.2015.53023

Boulmane, M., Makhloufi, M., Bouillet, J. P., Saint-Andre, L., Satrani, B., Halim, M., & El Antry, S. (2010). Estimated Organic Carbon Stock in the Quercus ilex Moroccan Middle Atlas. *Acta Botanica Gallica, 157*, 451–467. DOI: 10.1080/12538078.2010.10516222

Boulton, A. J., & Brock, M. A. (1999). *Australian freshwater ecology: processes and management*. Gleneagles Publishing.

Bourbouze, A. (2003). Enjeux et débats sur la réorganisation des espaces pastoraux du sud méditerranéen. Actes du colloque international « Organisation spatiale et gestion des ressources et des territoires ruraux», 25-27 février 2003, Montpellier (France). 7 P.

Bourbouze, A., & Donadieu, R. (1987). L'Élevage sur parcours en régions méditerranéennes. Montpellier: CIHEAM/ IAM. *Options méditerranéennes*. 102 p.

Boutefeu, B. (2005). L'aménagement forestier en France : À la recherche d'une gestion durable à travers l'histoire. *Vertigo*, *6*(2), 6–2. DOI: 10.4000/vertigo.4446

Boutefeu, B. (2009). *La forêt mise en scène : Attentes des publics et scénarios de gestion de la forêt*. L'Harmattan.

Boutin, C., Jobin, B., & Bélanger, L. (1982). Écologie des mares temporaires méditerranéennes. *Ecologia Mediterranea*, *8*, 47–58.

Bozzini, A., Francini, S., Chirici, G., Battisti, A., & Faccoli, M. (2023). Spruce Bark Beetle Outbreak Prediction through Automatic Classification of Sentinel-2 Imagery. *Forests*, *14*(6), 1116. DOI: 10.3390/f14061116

Brahic, E., & Terreaux, J. P. (2009). *Evaluation économique de la biodiversité: Méthodes et exemples pour les forêts tempérées*. Editions Quae. DOI: 10.3917/quae.brahi.2009.01

Brendonck, L., & Williams, W. D. (2000). Biodiversity in wet-lands of dry regions (drylands). In Gopal, B., Junk, W. J., & Davis, J. A. (Eds.), *Biodiversity in Wetlands: assessment,Function and Conservation* (pp. 181–194). Backhuys.

Brennan, M., Fritsch, C., Cosgun, S., Dumarcay, S., Colin, F., & Gérardin, P. (2020). Quantitative and qualitative composition of bark polyphenols changes longitudinally with bark maturity in Abies alba Mill. *Annals of Forest Science*, *77*(1), 9. DOI: 10.1007/s13595-019-0916-x

Brock, M. A., Nielsen, D. L., Shiel, R. J., Green, J. D., & Langley, J. D. (2003). Drought and Aquatic Community Resilience: The Role Of Eggs And Seeds In Sediments Of Temporary Wetlands. *Freshwater Biology*, *48*(7), 1207–1218. DOI: 10.1046/j.1365-2427.2003.01083.x

Brovkina, O., Cienciala, E., Surový, P., & Janata, P. (2018). UAV for assessment of qualitative classification of Norway spruce in temperate forest stands. *Geo-Spatial Information Science*, *21*(1), 12–20. DOI: 10.1080/10095020.2017.1416994

Brown, C., Burns, A., & Arnell, A. (2018). A Conceptual Framework for Integrated Ecosystem Assessment. *One Ecosystem*, *3*, e25482. DOI: 10.3897/oneeco.3.e25482

Burkhard, B., & Maes, J. (Eds.). (2017). *Mapping Ecosystem Services*. Pensoft Publishers. (The primary ESMERALDA textbook), Sofia, 374 pp DOI: 10.3897/ab.e12837

Burkhard, B., Santos-Martin, F., Nedkov, S., & Maes, J. (2018). *An operational framework for integrated Mapping and Assessmentof Ecosystems and their Services (MAES). One Ecosystem*. Pensoft Publishers.

Busch, J., Engelmann, J., Cook-Patton, S. C., Griscom, B. W., Kroeger, T., & Possingham, H.. (2019). Low-cost CO_2 removal through tropical reforestation. *Nature Climate Change*, *9*, 463–466. DOI: 10.1038/s41558-019-0485-x

Cachinero-Vivar, A. M., Palacios-Rodríguez, G., Lara-Gómez, M. A., & Navarro-Cerrillo, R. M. (2021). Thinning Effect of C Sequestration along an Elevation Gradient of Mediterranean Pinus spp. Plantations. *Forests*, *2021*(12), 1583. DOI: 10.3390/f12111583

Camarero, J. J., Sánchez-Salguero, R., Sangüesa-Barreda, G., Lechuga, V., Viñegla, B., Seco, J. I., Taïqui, L., Carreira, J. A., & Linares, J. C. (2021). Drought, axe and goats. More variable and synchronized growth forecasts worsening dieback in Moroccan Atlas cedar forests. *The Science of the Total Environment*, *765*, 142752. DOI: 10.1016/j.scitotenv.2020.142752 PMID: 33082041

Candiago, S., Remondino, F., De Giglio, M., Dubbini, M., & Gattelli, M. (2015). Evaluating Multispectral Images and Vegetation Indices for Precision Farming Applications from UAV Images. *Remote Sensing (Basel)*, *7*(4), 4026–4047. DOI: 10.3390/rs70404026

Cardil, A., Vepakomma, U., & Brotons, L. (2017). Assessing Pine Processionary Moth Defoliation Using Unmanned Aerial Systems. *Forests*, *8*(10), 402. DOI: 10.3390/f8100402

Cardinale, B. J., Duffy, J. E., Gonzalez, A., Hooper, D. U., Perrings, C., Venail, P., & Naeem, S. (2012). Biodiversity loss and its impact on humanity. *Nature*, *486*(7401), 59–67. DOI: 10.1038/nature11148 PMID: 22678280

Carmona-Yáñez, M. D., Lucas-Borja, M. E., Zema, D. A., Jing, X., Kooch, Y., Garrido Gallego, P., Plaza-Alvarez, P. A., Zhou, G., & Delgado-Baquerizo, M. (2023). Influence of management and stand composition on ecosystem multifunctionality of Mediterranean tree forests. *Trees (Berlin)*, *37*(6), 1801–1816. DOI: 10.1007/s00468-023-02462-w

Catal, J., Navarro, A., & Calv.o, J. (2022). Mapping Cork Oak Mortality Using Multitemporal High-Resolution Satellite Imagery. *Remote Sens.*, 14, 2750.

Catarino, F. (1996). Carob - an old and sustainable Mediterranean crop: Its potential for the future. In: *Proc. Third International Carob Symposium* (Tavira, Portugal), University of Lisbon: Portugal.

Centre d'Analyse Stratégique (CAS) (2009). Approche économique de la biodiversité et des services liés aux écosystèmes. Contribution à la décision publique,

CESE. (2023). *Les écosystèmes forestiers du Maroc : Risques, défis et opportunités*. Conseil Economique, Social et Environnementa. [Auto-saisine]

Chakhchar, A., Haworth, M., El Modafar, C., Lauteri, M., Mattioni, C., Wahbi, S., & Centritto, M. (2017). An assessment of genetic diversity and drought tolerance in argan tree (Argania spinosa) populations: Potential for the development of improved drought tolerance. *Frontiers in Plant Science, 8*, 276. DOI: 10.3389/fpls.2017.00276 PMID: 28303146

Chaponniere, A., & Smakhtin, V. (2006). A review of climate change scenarios and preliminary rainfall trend analysis in the Oum er Rbia Basin, Morocco. *Working Paper 110 (Drought Series: Paper 8) Colombo, Sri Lanka: International Water Management Institute (IWMI)*, 15 pages.

Charles, L., & Kalaora, B. (2003). Sociologie et environnement en france. L'environnement introuvable ? *Écologie & politique, 27*(1), 31-57. Cairn.info. DOI: 10.3917/ecopo.027.0031

Charles, L., & Kalaora, B. (2001). La nature administrée. L'Europe, la France et la politique de la nature. [Cairn.info.]. *Le Debat, 116*(4), 47–63. DOI: 10.3917/deba.116.0047

Chavan, B. L., & Rasal, G. B. (2011). Sequestered Carbon Potential and Status of Eucalyptus Tree. *International Journal of Applied Engineering and Technology, 1*(1), 41–47.

Chave, J., Andalo, C., Brown, S., Cairns, M. A., Chambers, J. Q., Eamus, D., Fölster, H., Fromard, F., Higuchi, N., Kira, T., Lescure, J.-P., Nelson, B. W., Ogawa, H., Puig, H., Riéra, B., & Yamakura, T. (2005). Tree allometry and improved estimation of carbon stocks. *Oecologia, 145*(1), 87–99. Advance online publication. DOI: 10.1007/s00442-005-0100-x

Chávez-Durán, Á. A., Olvera-Vargas, M., Figueroa-Rangel, B., García, M., Aguado, I., & Ruiz-Corral, J. A. (2022). Mapping Homogeneous Response Areas for Forest Fuel Management Using Geospatial Data, K-Means, and Random Forest Classification. *Forests, 13*(12), 1970. DOI: 10.3390/f13121970

Chebli, Y., Chentouf, M., Ozer, P., Hornick, J.-L., & Cabaraux, J.-F. (2018). Forest and silvopastoral cover changes and its drivers in northern Morocco. *Applied Geography (Sevenoaks, England), 101*, 23–35. DOI: 10.1016/j.apgeog.2018.10.006

Chebli, Y., El Otmani, S., Elame, F., Moula, N., Chentouf, M., Hornick, J. L., & Cabaraux, J. F. (2021). Silvopastoral system in Morocco: Focus on their importance, strategic functions, and recent changes in the Mediterranean side. *Sustainability (Basel), 13*(19), 10744. DOI: 10.3390/su131910744

Cheddadi, R., Fady, B., François, L., Hajar, L., Suc, J.-P., Huang, K., Demarteau, M., Vendramin, G. G., & Ortu, E. (2009). Putative glacial refugia of Cedrus atlantica deduced from Quaternary pollen records and modern genetic diversity. *Journal of Biogeography*, *36*(7), 1361–1371. DOI: 10.1111/j.1365-2699.2008.02063.x

Cheng, F., Tian, J., He, J., He, H., Bao, G., Yang, Y., Liu, G., & Zhang, Z. (2024). China's future forest carbon sequestration potential under different management scenarios. *Trees, Forests and People*, *17*, 100621. DOI: 10.1016/j.tfp.2024.100621

Chiasson, G., Boucher, J. L., & Thibault, M. (2005). La forêt plurielle : Nouveau mode de gestion et d'utilisation de la forêt, le cas de la Forêt de l'Aigle. *VertigO*, (Volume 6 Numéro 2). DOI: 10.4000/vertigo.4298

Chkhichekh, A. (1993). Cartographie de la forêt de Bouhachem dans un objectif d'aménagement. Mémoire de 3ème cycle; ENFI. Salé. Maroc.

Chkhichekh, A. (2015). *Cartographie et étude de la diversité des groupements végétaux du Plateau Central marocain: cas du Site d'Intérêt Biologique et Ecologique (SIBE) d'El Harcha, Maroc.* Thèse de Doctorat National. PP. 103. Université Mohammed V; Faculté des Sciences. Rabat. Maroc.

Chkhichekh, A., El Aboudi, A., Aafi, A., Wahid, N., & Benabid, A. (2015, December). A contribution to the knowledge of the Moroccan forest ecosystems: Association of Quercetum rotundifolio-suberis in the Central Plateau of Morocco. *Plant Sociology*, *52*(2), 57–68.

Chowdhury, A., Naz, A., & Maiti, S. K. (2023). Variations in soil blue carbon sequestration between natural mangrove metapopulations and a mixed mangrove plantation: A case study from the world's largest contiguous mangrove forest. *Life (Basel, Switzerland)*, *13*(2), 271. DOI: 10.3390/life13020271

Collins, W. (1978). Remote sensing of crop type and maturity. *Photogrammetric Engineering and Remote Sensing*, *44*, 43–55.

Cook-Patton, S. C., Leavitt, S. M., Gibbs, D., Harris, N. L., Lister, K., Anderson-Teixeira, K. J., Briggs, R. D., Chazdon, R. L., Crowther, T. W., Ellis, P. W., Griscom, H. P., Herrmann, V., Holl, K. D., Houghton, R. A., Larrosa, C., Lomax, G., Lucas, R., Madsen, P., Malhi, Y., & Griscom, B. W. (2020). Mapping carbon accumulation potential of natural forest regrowth. *Nature*, *585*(7826), 545–550. DOI: 10.1038/s41586-020-2686-x

Coops, N. C., Achim, A., Arp, P., Bater, C. W., Caspersen, J. P., Côté, J.-F., Dech, J. P., Dick, A. R., van Ewijk, K., Fournier, R., Goodbody, T. R. H., Hennigar, C. R., Leboeuf, A., van Lier, O. R., Luther, J. E., MacLean, D. A., McCartney, G., Pelletier, G., Prieur, J.-F., & Woods, M. E. (2021). Advancing the application of remote sensing for forest information needs in Canada: Lessons learned from a national collaboration of university, industrial and government stakeholders. *Forestry Chronicle*, *97*(02), 109–126. DOI: 10.5558/tfc2021-014

Coops, N., Waring, R., Wulder, M., & White, J. (2009). Prediction and assessment of bark beetle-induced mortality of lodge pole pine using estimates of stand vigor derived from remotely sensed data. *Remote Sensing of Environment*, *113*(5), 1058–1066. DOI: 10.1016/j.rse.2009.01.013

Costa, D., Tavares, R. M., Baptista, P., & Lino-Neto, T. (2020). Cork Oak Endophytic Fungi as Potential Biocontrol Agents against Biscogniauxia mediterranea and Diplodia corticola. [— revue] [Open Access, PMC]. *Journal of Fungi (Basel, Switzerland)*, *6*(4), 287. DOI: 10.3390/jof6040287 PMID: 33202643

Costanza, R., d'Arge, R., de Groot, R., Farber, S., Grasso, M., Hannon, B., & van den Belt, M. (1997). The value of the world's ecosystem services and natural capital. *Nature*, *387*(6630), 253–260. DOI: 10.1038/387253a0

Croitoru L. (2007). Valuing the Non-Timber Forest Products in the Mediterranean Region. *Ecological Economics*, 63(4):768-775http://DOI: 10.1016/j.ecolecon.2007.01.014

Croitoru, L., & Daly-Hassen, H. (2010). Using Payments for Environmental Services to improve Conservation in a Tunisian Watershed. *Mountain Forum Bulletin*, (January), 89–91.

Cucherousset, J. (2006). Adaptations des organismes aux milieux temporaires. *Annales de Limnologie*, *42*, 1–12.

Curran, P. J. (1983). Multispectral remote sensing for the estimation of green leaf area index. *Philosophical Transactions of the Royal Society of London. Series A, Mathematical and Physical Sciences*, *309*(1508), 257–270. DOI: 10.1098/rsta.1983.0039

Dai, Q., Wang, T., Wei, P., & Fu, Y. (2023). Effects of Successive Planting of Eucalyptus Plantations on Tree Growth and Soil Quality. *Sustainability (Basel)*, *15*(8), 6746. DOI: 10.3390/su15086746

Dallahi, Y., Boujraf, A., Meliho, M., & Orlando, C. A. (2023). Assessment of forest dieback on the Moroccan Central Plateau using spectral vegetation indices. *Journal of Forestry Research*, *34*(3), 793–808. DOI: 10.1007/s11676-022-01525-x

Dallahi, Y., Malaainine, M. E., Hbiak, I., Boujraf, A., Ould Abidine, M. M., Orlando, C. A., Meliho, M., El Mderssa, M., & Minoubi, A. (2023). Contribution to the Modeling of the Organic Matter of Moroccan Forest Soils within the Context of Global Change – Case Study of the Central Plateau. *Ecological Engineering and Environmental Technology*, *24*(8), 261–271. DOI: 10.12912/27197050/172004

Dalponte, M., Marzini, S., Solano-Correa, Y. T., Tonon, G., Vescovo, L., & Gianelle, D. (2020). Mapping forest windthrows using high spatial resolution multispectral satellite images. *International Journal of Applied Earth Observation and Geoinformation*, *93*, 102206. DOI: 10.1016/j.jag.2020.102206

Daly-Hassen, H., Annabi, M., & King-Okumu, C. (2019). Social and private profitability of tree-based adaptation options to climate change in a dryland area of Tunisia. *New Medit*, n°2/2019. pp. 89-104.

Daly-Hassen, H., Croitoru, L., Tounsi, K., Aloui, A., & Jebari, S. (2012). *Evaluation économique des biens et services des forêts tunisiennes*. DGF/FAO.

Daly-Hassen, H., Pettenella, D., & Ahmed, T. J. (2010). Economic instruments for the sustainable management of Mediterranean watersheds. *Forest Systems*, *19*(2), 141–155. DOI: 10.5424/fs/2010192-01310

Danish, A., Wazir Aitizaz Ahsan, A. H., Begum, F., Ashiq, H., Chitsan, L., & Hussain, F. (2025). *Assessing the impact of land use and land cover changes on soil properties and carbon sequestration in the upper Himalayan Region of Gilgit*. Sustainable Chemistry One World., DOI: 10.1016/j.scowo.2024.100038

Dash, J. P., Pearse, G. D., & Wat, M. S. (2018). UAV Multispectral Imagery Can Complement Satellite Data for Monitoring Forest Health. *Remote Sensing (Basel)*, *10*(8), 1216. DOI: 10.3390/rs10081216

Dash, J. P., Watt, M. S., Pearse, G. D., Heaphy, M., & Dungey, H. S. (2017). Assessing Very High Resolution UAV Imagery for Monitoring Forest Health during a Simulated Disease Outbreak. *ISPRS Journal of Photogrammetry and Remote Sensing*, *131*, 1–14. DOI: 10.1016/j.isprsjprs.2017.07.007

Datt, B. (1999). A New Reflectance Index for Remote Sensing of Chlorophyll Content in Higher Plants: Tests Using Eucalyptus Leaves. *Journal of Plant Physiology*, *154*(1), 30–36. DOI: 10.1016/S0176-1617(99)80314-9

de Arano, I. M., Garavaglia, V., & Farcy, C. (2016). Chapitre 5—Les forêts face au changement global. In *Mediterra 2016 : Zéro gaspillage en Méditerranée* (p. 117-138). Presses de Sciences Po. DOI: 10.3917/scpo.cihea.2016.01.0117

De Oliveira, T. W. G., Rubilar, R., Sanquetta, C. R., Corte, A. P. D., & Behling, A. (2021). Simultaneous estimation as an alternative to young *eucalyptus* aboveground biomass modeling in ecophysiological experiments. *Acta Scientiarum. Agronomy, 43*, e52126. DOI: 10.4025/actasciagron.v43i1.52126

Defarges, P. M. (2022). *La gouvernance* (6e éd. mise à jour). Que sais-je ? Deneault, A. (2013). *Gouvernance : Le management totalitaire*. Lux éd. [diff. Harmonia Mundi].

Deil, U. (2005). A review on habitats, plant traits and vegetation of ephemeral wetlands – a global perspective. *Phytocoenologia, 35*(2-3), 533–705. DOI: 10.1127/0340-269X/2005/0035-0533

Dennison, P. E., Brunelle, A. R., & Carter, V. A. (2010). Assessing canopy mortality during a mountain pine beetle outbreak using GeoEye-1 high spatial resolution satellite data. *Remote Sensing of Environment, 114*(11), 2431–2435. DOI: 10.1016/j.rse.2010.05.018

Descola, P. (2015). *Par-delà nature et culture*. Gallimard.

DGF. 2010. *Résultats du deuxième inventaire forestier et pastoral*. Ministère de la Défense Nationale, Ministère de l'Agriculture, des Ressources Hydrauliques et de la Pêche, et Ministère de l'Enseignement Supérieur et de la Recherche Scientifique.

Di Matteo, G., Luzzi, G., Basile, A., Sposato, A., Bertini, G., Neri, U., Pennelli, B., Napoli, R., & Nardi, P. (2023). Carbon concentrations and carbon storage capacity of three old-growth forests in the Sila National Park, Southern Italy. *Journal of Forestry Research, 34*(1), 233–242. DOI: 10.1007/s11676-022-01549-3

Diaci, J., Kerr, G., & O'hara, K. (2011). Twenty-first century forestry: Integrating ecologically based, uneven-aged silviculture with increased demands on forests. *Forestry, 84*(5), 463–465. DOI: 10.1093/forestry/cpr053

Diamantoglou, S., & Mitrakos, K. (1981). Leaf longevity in Mediterranean evergreen sclerophylls. Chapter. In Margaris, N. S., & Mooney, H. A. (Eds.), *Components of Productivity of Mediterranean-climate Regions - Basic and Applied Aspects. Tasks for Vegetation Science* (Vol. 4, pp. 17–19). Springer. DOI: 10.1007/978-94-009-8683-1_4

Díaz-Pines, E., Rubio, A., Miegroet, H. V., Montes, F., & Benito, M. (2011). Does tree species composition control soil organic carbon pools in Mediterranean mountain forests? *Forest Ecology and Management, 262*(10), 1895–1904. DOI: 10.1016/j.foreco.2011.02.004

DJI. (2020). *P4 Multi-spectral, User Manual*. 55 pages. https://www.dji.com/p4.multispectral

Dobignard & Chatelain. (2011). Index synonymique de la flore d'Afrique du Nord. *Éditions des conservatoires et jardins botaniques. Genève.* 430 p.

Du, H., Zeng, F., Peng, W., Wang, K., Zhang, H., Liu, L., & Song, T. (2015). Carbon storage in a eucalyptus plantation chronosequence in southern China. *Forests*, *6*(6), 1763–1778. DOI: 10.3390/f6061763

Eitel, J. U., Vierling, L. A., Litvak, M. E., Long, D. S., Schulthess, U., Ager, A. A., Krofcheck, D. J., & Stoscheck, L. (2011). Broadband, red-edge information from satellites improves early stress detection in a New Mexico conifer woodland. *Remote Sensing of Environment*, *115*(12), 3640–3646. DOI: 10.1016/j.rse.2011.09.002

El Jazouli, A., Barakat, A., Khellouk, R., Rais, J., & El Baghdadi, M. (2019). Remote sensing and GIS techniques for prediction of land use land cover change effects on soil erosion in the high basin of the Oum Er Rbia River (Morocco). *Remote Sensing Applications: Society and Environment*, *13*, 361–374. DOI: 10.1016/j.rsase.2018.12.004

El Madihi, M. (2020). Étude de la biodiversité végétale et des traits fonctionnels des plantes des zones humides temporaires du Maroc le long d'un gradient de latitude et de pression anthropique. Thèse de doctorat, Université Mohammed V, Rabat.

El Madihi, M., Rhazi, M., Bouahim, S., Amami, B., Waterkeyn, A., & Grillas, P. (2017). Functional traits of plant communities in Mediterranean temporary pools under different disturbance regimes. *Plant Ecology & Diversity*, *10*(4), 291–303. DOI: 10.1080/17550874.2017.1346716

El Mderssa, M. E., Belghazi, B., Benjelloun, H., Zennouhi, O., Nassiri, L., & Ibijbijen, J. (2019). Estimation of Carbon Sequestration; Using Allometric Equations; in Azrou Cedar Forests (Cedrus atlantica Manetti) in the Central Middle Atlas of Morocco under Climate Change. *Open Journal of Forestry*, *09*(03), 214–225. DOI: 10.4236/ojf.2019.93011

El Mderssa, M., Benjelloun, H., Zaher, H., Zennouhi, O., Nassiri, L., & Ibijbijen, J. (2019). Estimating the sequestration potential of organic carbon in forest soils in the Central Middle Atlas: A tool to fight climate change. *Atlas Journal of Biology*, •••, 603–610. DOI: 10.5147/ajb.v0i0.198

El Mderssa, M., Elmalki, M., Whalen, J. K., Ikraoun, H., Aliyat, F. Z., Dallahi, Y., Abbas, Y., Nassiri, L., & Ibijbijen, J. (2024). Forest stand and soil types determine soil organic carbon storage in the Middle Atlas region of Morocco using machine learning models. *All Earth*, *36*(1), 1–10. DOI: 10.1080/27669645.2024.2400432

El Mderssa, M., Malki, F., Ikraoun, H., Nassiri, L., & Ibijbijen, J. (2022). Determination des parametres d'evaluation du stock de carbone dans les ecosystemes forestiers (Cedrus atlantica Manetti, cedre de l'Atlas au Maroc): Methodes specifiques et generiques. *Bois et Forêts des Tropiques*, *351*, 67–77. DOI: 10.19182/bft2022.351.a36330

El Moussaoui, E. H., Moumni, A., Khabba, S., Er-Raki, S., Ait Hssaine, B., Chehbouni, A., & Lahrouni, A. (2025). Bibliometric and review analysis of argan trees studies: Global research trends and challenges. *Agroforestry Systems*, *99*(5), 132. DOI: 10.1007/s10457-025-01228-2

El Yousfi, S. M. (1988). *La dégradation forestière dans le Sud marocain. Exemple de l'arganeraie d'Admine (Sous) entre 1969 et 1986*. Mémoire de 3ème cycle [P; Rabat.]. *IAV Hassan*, *II*, 117.

Elbahi, A., Dugon, M., Oubrou, W., El Bekkay, M., Hermas, J., & Lawton, C. (2023). A geospatial approach to assess habitat diversity and inform conservation efforts using a case study from North Africa. *African Journal of Ecology*, *61*(2), 422–450. DOI: 10.1111/aje.13130

El-Bouhali, A., Amyay, M., & Ech-Chahdi, K. E. O. (2024). Combined impact of drought and land use changes on water resources in the Tabular Middle Atlas, Morocco. *Revista de Estudios Andaluces*, *48*(48), 202–220. DOI: 10.12795/rea.2024.i48.10

Ellatifi, M. (2012). L'économie de la forêt et des produits forestiers au Maroc: Bilan et perspectives. *Pessac: Thèse de Doctorat En Sciences Économiques, Université Montesquieu-Bordeaux IV*. https://www.audentia-gestion.fr/economie/PDF/Thse_-_Ellatifi.pdf

Ellert, B. H., Vanden Bygaart, A. J., & Bremer, E. (2007). Measuring Change in Soil Organic Carbon Storage. In Carter, M. R., & Gregorich, E. G. (Eds.), *Soil Sampling and Methods of Analysis* (pp. 49–62). CRC Press.

Elmoussaoui, E., Moumni, A., & Lahrouni, A. (2021). Cartography of moroccan argan tree using combined optical and sar imagery integrated with digital elevation model. *The International Archives of the Photogrammetry, Remote Sensing and Spatial Information Sciences, XLVI-4-W5-2021*, 211–217. ISPRS TC IV (WG IV-1)
The 6th International Conference on Smart City Applications - 27–29 October 2021, Virtual Safranbolu, Turkey. DOI: 10.5194/isprs-archives-XLVI-4-W5-2021-211-2021

Emberger, L. (1939). Aperçu général sur la végétation du Maroc: commentaire de la carte phytogéographique du Maroc 1:1.500.000 *Mémoire hors-série de la Société des Sciences Naturelles du Maroc*

Enríquez-de-Salamanca, Á. (2024). Environmental and social impacts of carbon sequestration. *Integrated Environmental Assessment and Management*, *20*(6), 1812–1838. DOI: 10.1002/ieam.4925

Erkan, N., Güner, Ş. T., & Aydın, A. C. (2023). Thinning effects on carbon stocks in *Pinus brutia* plantations. *Carbon Balance and Management*, *18*(1). Advance online publication. DOI: 10.1186/s13021-023-00226-0

Estrada, J. S., Fuentes, A., Reszka, P., & Auat-Cheein, F. (2023). Machine learning assisted remote forestry health assessment: A comprehensive state of the art review. *Frontiers in Plant Science*, *14*(1139232), 1–25. DOI: 10.3389/fpls.2023.1139232

Etienne, L. R. (1997). Patrimonialité plutôt que propriété : Approches de politiques foncières en Afrique et dans l'océan indien dans une perspective de gestion patrimoniale et environnementale. In *Droits de propriété et environnement* (pp. 321–334). Dalloz.

Fadil, S., Sebari, I., Ajerame, M. M., Ajeddour, R., El Maghraoui, I., Ait El kadi, K., Zefri, Y., & Jabrane, M. (2024). An Integrating Framework for Biomass and Carbon Stock Spatialization and Dynamics Assessment Using Unmanned Aerial Vehicle LiDAR (LiDAR UAV) Data, Landsat Imagery, and Forest Survey Data in the Mediterranean Cork Oak Forest of Maamora. *Land (Basel)*, *13*(5), 688. DOI: 10.3390/land13050688

Fadil, S., Sebari, I., Bouhaloua, M., & Kadi, K. A. E. (2020). Opportunités d'utilisation de la technologie drone au niveau des écosystèmes forestiers. *Revue Marocaine des Sciences Agronomiques et Vétérinaires*, *8*(4). https://www.agrimaroc.org/index.php/Actes_IAVH2/article/view/888

Fadil, S., Sebari, I., Bouhaloua, M., Mohamed, B., Ait El Kadi, K., Kellouch, S., Lfalah, H., & Meghraoui, K. (2022). Biomass and Carbon Stock Quantification in Cork Oak Forest of Maamora Using a New Approach Based on the Combination of Aerial Laser Scanning Carried by Unmanned Aerial Vehicle and Terrestrial Laser Scanning Data. *Forests*, *13*(8), 1211. DOI: 10.3390/f13081211

Fadil, S., Sebari, I., Moulay, M. A., & El kadi, K. A. (2024). Modeling and spatialization of biomass and carbon stock using unmanned Aerial Vehicle Lidar (Lidar-UAV) metrics and forest inventory in cork oak forest of Maamora. *Regional Science Policy & Practice*, *16*(11), 100127. Advance online publication. DOI: 10.1016/j.rspp.2024.100127

Fady, B., Esposito, E., Abulaila, K., Aleksic, J. M., Alia, R., Alizoti, P., Apostol, E.-N., Aravanopoulos, P., Ballian, D., Bou Dagher Kharrat, M., ... Vendramin, G. G. (2022). Forest genetics research in the Mediterranean Basin: Bibliometric analysis, knowledge gaps, and perspectives. Current Forestry Reports, 8, 277–298. DOI: , B., Lefèvre, F., Reynaud, M., Bou Dagher-Kharrat, M., Vendramin, G. G. & Bariteau, M. (2003). Gene flow among different taxonomic units: evidence from nuclear and cytoplasmic markers in Cedrus plantation forests. Theoretical and Applied Genetics, 107, 1132–1138.DOI: 10.1007/s40725-022-00169-Fady

FAO (2012). *Les chênaies méditerranéennes: écologie, dynamique et gestion.*

FAO, & IUSS. (2006). *World reference base for soil resources: a framework for international classification, correlation and communication*. FAO, Rome, Italia

FAO. (1989). *Bilan de trente ans de développement pastoral dans le bassin méditerranéen. Organisation des Nations Unies pour l'Alimentation et l'Agriculture*. FAO.

FAO. (2001). *Forestry Curriculum Development and Revision*. https://www.fao.org/4/y1842e/y1842e00.htm#TopOfPage

FAO. (2010). *Les forêts et le changement climatique, Document de travail 9. Les forêts et le changement climatique dans la région du Proche-Orient. Organisation des Nations Unies pour l'Alimentation et l'Agriculture*. FAO.

FAO. (2013). *Etat des forêts méditerranéennes 2013*. l'Organisation des Nations Unies pour l'alimentation et l'agriculture.

FAO. (2013). *State of Mediterranean forests (2013)*. FFEM, Silva Mediterranea, Plan Bleu. https://www.fao.org/docrep/017/i3226e/i3226e.pdf [15 February 2016]

FAO. (2015c). *Évaluation des ressources forestière mondiales 2015. Répertoire de données de FRA 2015*. FAO.

FAO. (2018). *SOFO 2018—La situation des forêts du monde 2018*. www.fao.org. https://www.fao.org/state-of-forests/fr/

FAO. (2020). *Détermination des valeurs de productions des essences forestières et des prélèvements de bois de feu et pastoraux au Maroc dans le cadre du Projet IKI (réalisé par le consultant national Naggar mustapha). Organisation des Nations Unies pour l'Alimentation et l'Agriculture*. FAO.

FAO. (2020). *Gestion intégrée des incendies de forêts au Maroc. Stratégie nationale intersectorielle & plan d'action 2020-2030.*

FAO. (2020). Global Forest Resources Assessment 2020.

FAO. (2021). *Gestion intégrée des incendies de forêts au Maroc Stratégie nationale intersectorielle & plan d'action 2020-2030—Google Search.* https://www.google.com/search?client=safari&rls=en&q=Gestion+int%C3%A9gr%C3%A9e+des+incendies+de+for%C3%AAts+au+Maroc+Strat%C3%A9gie+nationale+intersectorielle+%26+plan+d%E2%80%99action+2020-2030&ie=UTF-8&oe=UTF-8

FAO. (2022). *Grazing with trees: A silvopastoral approach to managing and restoring drylands* (Haddad, F. F., Herrera, P. M., & Besbes, B., Eds.). Vol. 187). Food & Agriculture Organisation.

FAO. (2022). *La situation des forêts du monde (SOFO) 2022 : Des solutions forestieres pour une relance verte et... Des economies inclusives, resilientes et durables.* FOOD & AGRICULTURE ORG.

FAO. (2025). *The status of Mediterranean forests.* FAO., DOI: 10.4060/cd6921en

FAO-FOSA. (2000). *Etude prospective du secteur forestier en Afrique (FOSA)-Algérie Organisation des Nations Unies pour l'Alimentation et l'Agriculture.* FAO.

FAO-FOSA. (2001). *Etude prospective du secteur forestier en Afrique (FOSA)-Tunisie Organisation des Nations Unies pour l'Alimentation et l'Agriculture.* FAO.

FAO-NENA. (2022). *A turning point for drought management: Upscaling the silvopastoral approach.* Food and Agriculture Organization, Regional Office for the Near East and North Africa.

Farcy, C., & Huybens, N. (Eds.). (2016). *Forêts, savoirs et motivations.* L'Harmattan.

Farineau, J., & Morot-Gaudry, J. F. (2018). La photosynthèse: Processus physiques, moléculaires et physiologiques. *Editions Quae. Hors Collection,* 460.

Favero, A., Daigneault, A., & Sohngen, B. (2020). Forests: Carbon sequestration, biomass energy, or both? *Science Advances, 6*(13), eaay6792. Advance online publication. DOI: 10.1126/sciadv.aay6792

Fay, G. (1986). Désagrégation des collectivités et dégradation des milieux dans le Haut Atlas marocain. *Revue de l'Occident Musulman et de la Mediterranee, 41*(1), 234–248. DOI: 10.3406/remmm.1986.2121

Fennane, M., Ibn Tattou, M., & El Oualidi, J. (2014). Flore pratique du Maroc.: Vol. 3. *Série Botanique 40.* Travaux de l'Institut Scientifique.

Ferreira, L., De Andrade Sandim, A. S., Lopes, D. A., Sousa, J. J., Lopes, D. M. M., Silva, M. E. C. M., & Pádua, L. (2025). Estimating Biomass in Eucalyptus globulus and Pinus pinaster Forests Using UAV-Based LiDAR in Central and Northern Portugal. *Land (Basel), 14*(7), 1460. DOI: 10.3390/land14071460

Feyerabend, P. (1975). *Against method: Outline of an anarchistic theory of knowledge*. Verso.

Filahi, S., Tramblay, Y., Mouhir, L., & Diaconescu, E. P. (2017). Projected changes in temperature and precipitation indices in Morocco from high-resolution regional climate models. *International Journal of Climatology, 37*(14), 4846–4863. DOI: 10.1002/joc.5127

Foody, G. M. (2002). Status of land cover classification accuracy assessment. *Remote Sensing of Environment, 80*(1), 185–201. DOI: 10.1016/S0034-4257(01)00295-4

Foster, A., Rahimzadeh-Bajgiran, P., Daigneault, A., & Weiskittel, A. (2024). Cost-effectiveness of remote sensing technology for spruce budworm monitoring in Maine, USA. *Forests Monitor, 1*(1), 66–98. DOI: 10.62320/fm.v1.i1.14

Foucault, M. (2004). *Sécurité, territoire, population : Cours au Collège de France, 1977-1978.* Seuil : Gallimard.

Fraval, A., Bourret, J., Guy, P., Arbez, M., & Brenot, C. (1997). *Forêts.* Editions Quae. https://books.google.com/books?hl=fr&lr=&id=rz7lmDAefo0C&oi=fnd&pg=PA6&dq=Fraval+A,+Villemant+C.+Les+actions+anthropiques.+La+Ma%C3%A2mora+et+ses+ennemies.+Les+dossiers+de+l%E2%80%99environnement+de+l%E2%80%99INRA.+For%C3%AAts+1997&ots=Qz6sBwXVsg&sig=H5ZMJjf2jWcT8M0O-hx1LAmjQ5U

Fry, A. S., Weaver, M. J., Libala, N., Ralekhetla, M. M., Gusha, B., Mtati, N., & Palmer, C. G. (2024). Stepping stones to transformative participatory governance capabilities in a rural landscape restoration project. *Ecosystems and People (Abingdon, England), 20*(1), 2425378. DOI: 10.1080/26395916.2024.2425378

Gabira, M. M., Girona, M. M., DesRochers, A., Kratz, D., Da Silva, R. B. G., Duarte, M. M., De Aguiar, N. S., & Wendling, I. (2023). The impact of planting density on forest monospecific plantations: An overview. *Forest Ecology and Management, 534*, 120882. DOI: 10.1016/j.foreco.2023.120882

Gallego-Fernandez, J. B., Garcia-Mora, M. R., & Garcia-Novo, F. (1999). Small wetlands lost: A biological conservation hazard in Mediterranean landscapes. *Environmental Conservation, 26*(3), 190–199. DOI: 10.1017/S0376892999000272

Ganatsas, P., Tsakaldimi, M., Karydopoulos, T., Papaemannuil, A., & Papadopoulos, S. (2024). Long-term effect of different forest thinning intensity on carbon sequestration rates and potential uses in climate change mitigation actions. *Mitigation and Adaptation Strategies for Global Change*, *29*(1), 3. DOI: 10.1007/s11027-023-10102-4

Gauquelin T., Michon G., Joffre R., Duponnois R., Genin D., 2018. Mediterranean forests, land use and climate change: a social-ecological perspective. *Regional Environmental Change*, 2018, 18 (3), pp.623-636. ff10.1007/s10113-016-0994-3ff. ffhal-01594954f Hanley N. & Barbier E.B. (2009). *Pricing nature: Cost-benefit analysis and environmental policy*, Edward Elgar Publishing, USA.

Geels, F. W. (2019). Socio-technical transitions to sustainability: A review of criticisms and elaborations of the Multi-Level Perspective. *Current Opinion in Environmental Sustainability*, *39*, 187–201. DOI: 10.1016/j.cosust.2019.06.009

Ghaffariyan, M. R. (2025). Silvopastoral systems as a strategy for drought resilience: A short international review. *Silva Balcanica*, *26*(1), 29–37. DOI: 10.3897/silvabalcanica.26.e127074

Giannakopoulos, C., Bindi, M., Moriondo, M., LeSager, P., & Tin, T. (2005). Climate change impacts in the Mediterranean resulting from a _oC global temperature rise. A report for WWF, Published in July 2005 by WWF, the global conservation organization, Gland, Switzerland.

Gil, L., & Varela, M. C. (2008). *EUFORGEN Technical Guidelines for genetic conservation and use for cork oak (Quercus suber)*. Bioversity International.

Girardin, C. A. J., Jenkins, S., Seddon, N., Allen, M., Lewis, S. L., Wheeler, C. E., Griscom, B. W., & Malhi, Y. (2021). Nature-based solutions can cool the planet. *Nature*, *593*(7858), 191–196. DOI: 10.1038/d41586-021-01241-2

Gitelson, A.A., Keydan, G.P., & Merzlyak, M.N. (2006). Three-band model for non-invasive estimation of chlorophyll, carotenoids, and anthocyanin contents in higher plant leaves. *Geophys. Res. Lett.,* 33(11), L11402, 1-5.

Gitelson, A. A., Kaufman, Y. J., & Merzlyak, M. N. (1996). Use of a green channel in remote sensing of global vegetation from EOS-MODIS. *Remote Sensing of Environment*, *58*(3), 289–298. DOI: 10.1016/S0034-4257(96)00072-7

Gitelson, A. A., Stark, R., Grits, U., Rundquist, D., Kaufman, Y., & Derry, D. (2002b). Vegetation and soil lines in visible spectral space: A concept and technique for remote estimation of vegetation fraction. *International Journal of Remote Sensing*, *23*(13), 2537–2562. DOI: 10.1080/01431160110107806

Gitelson, A., Kaufman, Y. J., Stark, R., & Rundquist, D. (2002a). Novel algorithms for remote estimation of vegetation fraction. *Remote Sensing of Environment, 80*(1), 76–87. DOI: 10.1016/S0034-4257(01)00289-9

Godbout, J. (2014a). *La participation contre la démocratie : Suivi de L'engagement: une fidélité* (Deuxième édition). Liber.

Godbout, J. (2014b). *La participation contre la démocratie : Suivi de L'engagement: une fidélité* (Deuxième édition). Liber.

Goffman, E., & Joseph, I. (2009). *Les cadres de l'expérience*. de Minuit.

Gomat, H., Bengo, D. M., Ifo, S. A., Mankessi, F., Ossendza, J., Mabiala, P., Zeller, B., & Saint-Andre, L. (2024). What Is the Impact of Ploughing on Eucalyptus Plantations in Bambou Mingali (Plateaux Batéké, Republic of Congo)? *Agricultural Sciences, 15*(10), 1132–1146. DOI: 10.4236/as.2024.1510061

Gomez, D. F., Ritger, H. M. W., Pearce, C., Eickwort, J., & Hulcr, J. (2020). Ability of Remote Sensing Systems to Detect Bark Beetle Spots in the Southeastern US. *Forests, 11*(11), 1167. DOI: 10.3390/f11111167

González-García, M., Hevia, A., Majada, J., Rubiera, F., & Barrio-Anta, M. (2015). Nutritional, carbon and energy evaluation of Eucalyptus nitens short rotation bioenergy plantations in northwestern Spain. *iForest - Biogeosciences and Forestry, 9*(2), 303–310. DOI: 10.3832/ifor1505-008

Goodbody, T. R. H., Coops, N. C., Marshall, P. L., Tompalski, P., & Crawford, P. (2017). Unmanned aerial systems for precision forest inventory purposes: A review and case study. *Forestry Chronicle, 93*(01), 71–81. DOI: 10.5558/tfc2017-012

Goxe, A. (2010). Chapitre 5. L'évaluation des politiques territoriales au regard du développement durable. In B. Zuindeau (Éd.), *Développement durable et territoire : Nouvelle édition originale* (p. 71-81). Presses universitaires du Septentrion. DOI: 10.4000/books.septentrion.15389

Grillas, P., Gauthier, P., Yavercovski, N. & Perennou, C. (2004). *Mediterranean Temporary Pools. Volume 1: Issues relating to conservation, functioning and management*. Tour du Valat.

Grillas, P., Rhazi, L., Lefebvre, G., El Madihi, M., & Poulin, B. (2021). Foreseen impact of climate change on temporary ponds located along a latitudinal gradient in Morocco. *Inland Waters, 11*(4), 492–507. Advance online publication. DOI: 10.1080/20442041.2021.1962688

Griscom, B. W., Adams, J., Ellis, P. W., Houghton, R. A., Lomax, G., Miteva, D. A., Schlesinger, W. H., Shoch, D., Siikamäki, J. V., Smith, P., Woodbury, P., Zganjar, C., Blackman, A., Campari, J., Conant, R. T., Delgado, C., Elias, P., Gopalakrishna, T., Hamsik, M. R., & Fargione, J. (2017). Natural climate solutions. *Proceedings of the National Academy of Sciences of the United States of America, 114*(44), 11645–11650. DOI: 10.1073/pnas.1710465114

Gugliuzzo, A., Mazzeo, G., Mansour, R., & Tropea-Garzia, G. (2019). Carob pests in the Mediterranean region: Bio-ecology, natural enemies and management options. *Phytoparasitica, 47*(5), 605–628. DOI: 10.1007/s12600-019-00766-7

Gumus, V., El Moçayd, N., Seker, M., & Seaid, M. (2024). Future drought projections in Morocco and potential impact on agriculture. *Journal of Environmental Management, 367*, 122019. DOI: 10.1016/j.jenvman.2024.122019

Hadjadj, K., & Letreuch Belarouci, A. (2017). *Synthèse bibliographique sur le thuya de berbérie (Tetraclinis articulata* (Vahl) Mast.). *Geo-Eco-Trop, 41*(1), 13–27.

Hadri, A., Saidi, M. E. M., Saouabe, T., & El Fels, A. E. A. (2021). Temporal trends in extreme temperature and precipitation events in an arid area: Case of chichaoua mejjate region (morocco). *Journal of Water and Climate Change, 12*(3), 895–915. DOI: 10.2166/wcc.2020.234

Haines-Young, R., Potschin, M., & Kienast, F. (2012). Indicators of ecosystem service potential at European scales: Mapping marginal changes and trade offs. *Ecological Indicators, 21*, 39–53. DOI: 10.1016/j.ecolind.2011.09.004

Hakamada, R. E., Hubbard, R. M., Moreira, G. G., Stape, J. L., Campoe, O., & De Barros Ferraz, S. F. (2020). Influence of stand density on growth and water use efficiency in Eucalyptus clones. *Forest Ecology and Management, 466*, 118125. DOI: 10.1016/j.foreco.2020.118125

Hakam, O., Ongoma, V., Beniaich, A., Meskour, B., El Kadi, M. A., & Brouziyne, Y.. (2024). Climate change impact on argan distribution. *Modeling Earth Systems and Environment, 10*(4), 5529–5652. DOI: 10.1007/s40808-024-02077-1

Hammi, S., Simonneaux, V., Cordier, J. B., Genin, D., Alifriqui, M., Montes, N., & Auclair, L. (2010). Can traditional forest management buffer forest depletion? Dynamics of Moroccan High Atlas Mountain forests using remote sensing and vegetation analysis. *Forest Ecology and Management, 260*(10), 1861–1872. DOI: 10.1016/j.foreco.2010.08.033

Hamouch, C., Chaaouan, J., & Bouiss, C. (2025). Spatial-temporal assessment of soil erosion using the RUSLE model in the upstream Inaouène watershed, Northern Morocco. *Natural Hazards Review*, *5*(1), 148–156. DOI: 10.1016/j.nhres.2024.08.002

Hansen, M. C., Potapov, P. V., Moore, R., Hancher, M., Turubanova, S. A., Tyukavina, A., Thau, D., Stehman, S. V., Goetz, S. J., Loveland, T. R., Kommareddy, A., Egorov, A., Chini, L., Justice, C. O., & Townshend, J. R. G. (2013). High-Resolution Global Maps of 21st-Century Forest Cover Change. *Science*, *342*(6160), 850–853. DOI: 10.1126/science.1244693 PMID: 24233722

Hao, X. B. C., Ball, J. L. B., Culley, M. R., Carter, & Parkin, G. W. (2008). "Chapter 57 Soil Density and Porosity." In M. R. Carter & E. G. Gregorich (2nd Eds., *Soil Sampling and Methods of Analysis* (pp: 743–759). Boca Raton, FL: Taylor & Francis Group

Harroucha, R., & Chaouni, A.-A. (2023). A GIS-based decision support system for the management of the Moroccan habous property. *GeoJournal*, *88*(4), 3905–3922. DOI: 10.1007/s10708-023-10847-0

Hawryło, P., Bednarz, B., Wężyk, P., & Szostak, M. (2018). Estimating defoliation of Scots pine stands using machine learning methods and vegetation indices of Sentinel-2. *European Journal of Remote Sensing*, *51*(1), 194–204. DOI: 10.1080/22797254.2017.1417745

HCEFLCD. (2015). *Stratégie nationale des zones humides du Maroc 2015–2024.* Haut-Commissariat aux Eaux et Forêts et à la Lutte Contre la Désertification, Rabat.

HCEFLCD. (2018). *Rapport des activités 2018, Haut-Commis-sariat aux Eaux et Forêts et à la Lutte contre la Désertification, Rabat Chellah, Maroc.*

Heckenhahn, J., & Drupp, M. A. (2024). Relative Price Changes of Ecosystem Services: Evidence from Germany, *Environmental & Resource Economics,* Springer. *European Association of Environmental and Resource Economists*, *87*(3), 833–880.

Heiden, U., d'Angelo, P., Schwind, P., Karlshöfer, P., Müller, R., Zepp, S., Wiesmeier, M., & Reinartz, P. (2022). Soil Reflectance Composites: Improved Thresholding and Performance Evaluation. *Remote Sensing (Basel)*, *14*(18), 4526. DOI: 10.3390/rs14184526

Hermet, G. (2004). Un régime à pluralisme limité? À propos de la gouvernance démocratique. *Revue Française de Science Politique*, *54*(1), 159–178. DOI: 10.3917/rfsp.541.0159

Hernández-Morcillo, M., Burgess, P., Mirck, J., Pantera, A., & Plieninger, T. (2018). Scanning agroforestry-based solutions for climate change mitigation and adaptation in Europe. *Environmental Science & Policy*, *80*, 44–52. DOI: 10.1016/j.envsci.2017.11.013

Herzenni, A. (2008). *Aires forestières communautaires conservées dans le Haut Atlas central*. IUCN.

Hildebrandt, G. (1987). TOY or TOOL—Fernerkundung aus dem Weltraum: Spiel- oder Werkzeug für die Forstwirtschaft? *Forstwissenschaftliches Centralblatt*, *106*(1), 141–168. DOI: 10.1007/BF02741143

Hirigoyen, A., Acuna, M., Rachid-Casnati, C., Franco, J., & Navarro-Cerrillo, R. (2021). Use of optimization modeling to assess the effect of timber and carbon pricing on harvest scheduling, carbon sequestration, and net present value of eucalyptus plantations. *Forests*, *12*(6), 651. DOI: 10.3390/f12060651

Hirschmugl, M., Gallaun, H., Dees, M., Datta, P., Deutscher, J., Koutsias, N., & Schardt, M. (2017). Methods for Mapping Forest Disturbance and Degradation from Optical Earth Observation Data: A Review. *Current Forestry Reports*, *3*(1), 32–45. DOI: 10.1007/s40725-017-0047-2

Hlal, M., El Monhim, B., Chenal, J., Munyaka, J.-C. B., Azmi, R., Sbai, A., Cwick, G., & Hichou, B. B. (2025). Application of Deep Learning and Geospatial Analysis in Soil Loss Risk in the Moulouya Watershed, Morocco. *Water (Basel)*, *17*(9), 1351. DOI: 10.3390/w17091351

Hoegh-Guldberg, O., Hughes, L., McIntyre, S., Lindenmayer, D. B., Parmesan, C., Possingham, H. P., & Thomas, C. D. (2008). Assisted colonization and rapid climate change. *Science*, *321*(5887), 345–346. DOI: 10.1126/science.1157897 PMID: 18635780

Holopainen, M., Vastaranta, M., & Hyyppä, J. (2014). Outlook for the next generation's precision forestry in Finland. *Forests*, *5*(7), 1682–1694. DOI: 10.3390/f5071682

Holzinger, A., Schweier, J., Gollob, C., Nothdurft, A., Hasenauer, H., Kirisits, T., Häggström, C., Visser, R., Cavalli, R., Spinelli, R., & Stampfer, K. (2024). From Industry 5.0 to Forestry 5.0: Bridging the gap with Human-Centered Artificial Intelligence. *Current Forestry Reports*, *10*(6), 442–455. DOI: 10.1007/s40725-024-00231-7 PMID: 39464642

Honnay, O., & Bossuyt, B. (2005). Prolonged clonal growth: Escape route or route to extinction? *Oikos*, *108*(2), 427–432. DOI: 10.1111/j.0030-1299.2005.13569.x

Hornero, A., Zarco-Tejada, P. J., Quero, J. L., North, P. R. J., Ruiz-Gómez, F. J., Sánchez-Cuesta, R., & Hernandez-Clemente, R. (2021). Modelling hyperspectral- and thermal-based plant traits for the early detection of Phytophthora-induced symptoms in Oak decline. *Remote Sensing of Environment*, *263*, 112570. DOI: 10.1016/j.rse.2021.112570

Houghton, R. A., & Nassikas, A. A. (2018). Negative emissions from stopping deforestation. *Global Change Biology*, *24*(1), 350–359. DOI: 10.1111/gcb.13876

Hrdina, A., & Romportl, D. (2017). Evaluating Global Biodiversity Hotspots – Very Rich and Even More Endangered. *Journal of Landscape Ecology*, *10*(1), 108–115. DOI: 10.1515/jlecol-2017-0013

Huete, A. R. (1988). A soil-adjusted vegetation index (SAVI). *Remote Sensing of Environment*, *25*(3), 295–309. DOI: 10.1016/0034-4257(88)90106-X

Huete, A. R., Didan, K., Miura, T., Rodriguez, E. P., Gao, X., & Ferreira, L. G. (2002). Overview of the radiometric and biophysical performance of the MODIS vegetation indices. *Remote Sensing of Environment*, *83*(1), 195–213. DOI: 10.1016/S0034-4257(02)00096-2

Hufty, M. (2005). La gouvernance internationale de la biodiversité. *Etudes Internationales*, *32*(1), 5–29. DOI: 10.7202/704254ar

Hutchinson, G. E. (1957). Concluding remarks. *Cold Spring Harbor Symposia on Quantitative Biology*, *22*(0), 415–427. DOI: 10.1101/SQB.1957.022.01.039

IF. (2018). *L'Institut de France Le changement climatique*. Académie des Sciences.

Ikraoun, H., El Mderssa, M., Aliyat, F.Z., Nassiri, L., Ibijbijen I. (2024). Assessing Carbon Storage and Biomass in Moroccan Cork Oak Forests. *Ecological Engineering & Environmental Technology*. 25(6), 43–53 pp. https://doi.org/DOI: 10.12912/27197050/186550

Ikraoun, H., El Mderssa, M., El Mouridi, Z., Nassiri, L., & Ibijbijen, J. (2022). Carbon stock assessment and anlysis of physiochemeical characteristics under seven woody forest formations in Oulmes Central Plateau, Morocco. *Soil & Environment*, *41*(1), 55–62. DOI: 10.25252/SE/2021/222684

Innes, J. (1993). *Forest health: its assessment and status*. Published by CAB International.

IPCC (Intergovernmental Panel on Climate Change). (2023). AR6 Synthesis Report. *Climatic Change*, •••, 2023.

IPCC. (2006). *Guidelines for National Greenhouse Gas Inventories* (Vol. 4). Agriculture, Forestry and Other Land Use.

IPCC. (2019). *2019 Refinement to the 2006 IPCC Guidelines for National Greenhouse Gas Inventories. IPCC.* https://www.ipcc.ch/report/2019-refinement-to-the-2006-ipcc-guidelines-for-national-greenhouse-gas-inventories

IPCC. (2021). *Climate Change 2021 – The Physical Science Basis.* Cambridge University Press.

IPCC. (2021). *Climate Change 2021: The Physical Science Basis. Contribution of Working Group I to the Sixth Assessment Report of the Intergovernmental Panel on Climate Change* (Masson-Delmotte, V., Zhai, P., Pirani, A., Connors, S. L., Péan, C., Berger, S., Caud, N., Chen, Y., Goldfarb, L., Gomis, M. I., Huang, M., Leitzell, K., Lonnoy, E., Matthews, J. B. R., Maycock, T. K., Waterfield, T., Yelekçi, O., Yu, R., & Zhou, B., Eds.). Cambridge University Press., DOI: 10.1017/9781009157896

Jacobs, S. W. L., & Brock, M. A. (1993). Wetlands of Australia: southern (temperate) Australia. In Whigham, D. F., Dykyjov, D., & Hejny, S. (Eds.), *Wetlands of the World: Inventory, Ecology and Management* (Vol. 1, pp. 244–304). Kluwer Academic Publishers.

Jacobs, S. W. L., & Wilson, K. L. (1996). A biogeographical analysis of the freshwater plants of Australasia. *Australian Systematic Botany*, *9*(2), 169–183. DOI: 10.1071/SB9960169

Jana, B., Biswas, S., Majumder, M., Roy, P., & Mazumdar, A. (2009). Carbon Sequestration Rate and Aboveground Biomass Carbon Potential of Four Young Species. *Journal of Ecology and the Natural Environment*, *1*(2), 15–24.

Jebari, S., Bouguerra, S., Chebil, A., Lamia, B., & Touzi, S. (2025). Quantification de la dégradation du paysage agricole et évaluation économique d'un aménagement approprié: Cas du bassin versant Douimis (Tunisie). *New Medit*, *5*(2025). Advance online publication. DOI: 10.30682/nm2505g

Jenkins, D. G., & Buikema, A. L.Jr. (1998). Do similar communities develop in similar sites? *Ecological Monographs*, *68*(3), 421–443. DOI: 10.1890/0012-9615(1998)068[0421:DSCDIS]2.0.CO;2

Jessop, B. (1998). The rise of governance and the risks of failure : The case of economic development. *International Social Science Journal*, *50*(155), 29–45. DOI: 10.1111/1468-2451.00107

Jin, S., & Sader, S. A. (2005). MODIS time-series imagery for forest disturbance detection and quantification of patch size effects. *Remote Sensing of Environment*, *99*(4), 462–470. DOI: 10.1016/j.rse.2005.09.017

Jones, M. W., Abatzoglou, J. T., Veraverbeke, S., Andela, N., Lasslop, G., Forkel, M., Smith, A. J. P., Burton, C., Betts, R. A., van der Werf, G. R., Sitch, S., Canadell, J. G., Santín, C., Kolden, C. A., Doerr, S. H., & Le Quéré, C. (2022). Global and regional trends and drivers of fire under climate change. *Reviews of Geophysics*, 60(3), e2020RG000726. DOI: 10.1029/2020RG000726

Jopia, A., Zambrano, F., Pérez-Martínez, W., Vidal-Páez, P., Molina, J., & Mardones, F. (2020). Time-Series of Vegetation Indices (VNIR/SWIR) Derived from Sentinel-2 (A/B) to Assess Turgor Pressure in Kiwifruit. *ISPRS International Journal of Geo-Information*, *9*(11), 641. DOI: 10.3390/ijgi9110641

Joyce, C. (2012). Preface: Wetland services and management. *Hydrobiologia*, *692*(1), 1–3. DOI: 10.1007/s10750-012-1179-9

Jumiyati, S. (2025). Economic Valuation of Forest Ecosystem Services: Methods, Applications, and Policy Implications. *Journal of Forestry*, *1*(1).

Kalaora, B., Poupardin, D., & Larrère, R. (2023). *Le corps forestier dans tous ses états : De la Restauration à la Belle Époque* (Reproduction en fac-similé). l'Harmattan.

Kalaora, B. (1993). *Le musée vert : Radiographie du loisir en forêt*. L'Harmattan.

Kampen, M., Lederbauer, S., Mund, J.P., & Immitzer, M. (2019). UAV-Based Multispectral Data for Tree Species Classification and Tree Vitality Analysis. Dreiländertagung der DGPF, der OVG und der SGPF in Wien, Österreich - Publikationen der DGPF, Band 28, pp. 623-639.

Karam, B. (2016). *Estimation de la valeur économique et sociale des services rendus par les écosystèmes forestiers méditerranéens, Biosphère de Jabal Moussa, Liban*. Plan Bleu.

Karimi, H., Staudhammer, C. L., Therrell, M. D., Kleindl, W. J., Mungai, L. M., Amanambu, A. C., & Jones, C. N. (2025). Mapping Tradeoffs and Synergies in Ecosystem Services as a Function of Forest Management. *Land. MDPI*, *14*(8), 1–26.

Karpouzli, E., & Malthus, T. (2003). The empirical line method for the atmospheric correction of IKONOS imagery. *International Journal of Remote Sensing*, *24*(5), 1143–1150. DOI: 10.1080/0143116021000026779

Karrou, M. (2006). *Evaluation des dispositifs d'alerte précoce à la sécheresse à l'échelle nationale (cas du Maroc).* [Rapport d'expertise nationale effectuée dans le cadre du projet SMAS au Maroc (LIFE05TCY/TN/000150).].

Kasekete, D. K., Ligot, G., Mweru, J. M., Drouet, T., Rousseau, M., Moango, A., & Bourland, N. (2022). Growth, Productivity, Biomass and Carbon Stock in Eucalyptus saligna and Grevillea robusta Plantations in North Kivu, Democratic Republic of the Congo. *Forests*, *13*(9), 1508. DOI: 10.3390/f13091508

Keeley, J. E., & Zedler, P. H. (1998). Characterization and global distribution of vernal pools. In Witham, C. C. (Ed.), *Vernal pool ecosystems* (pp. 1–14). California Native Plant Society.

Kerry, R. G., Montalbo, F. J. P., Das, R., Patra, S., Mahapatra, G. P., Maurya, G. K., Nayak, V., Jena, A. B., Ukhurebor, K. E., Jena, R. C., Gouda, S., Majhi, S., & Rout, J. R. (2022). An overview of remote monitoring methods in biodiversity conservation. *Environmental Science and Pollution Research International*, *29*(53), 80179–80221. DOI: 10.1007/s11356-022-23242-y PMID: 36197618

Khalfaoui, M., Daly-Hassen, H., Stiti, B., & Jebari, S. (2020). Toward Decision-Making Support: Valuation and Mapping of New Management Scenarios for Tunisian Cork Oak Forests. *Forests*, *11*(2), 197. DOI: 10.3390/f11020197

Kim, M. S., Daughtry, C. S. T., Chappelle, E. W., McMurtrey, J. E., III, & Walthall, C. L. (1994). The use of high spectral resolution bands for estimating absorbed photosynthetically active radiation (APAR). *Proceedings of the 6th Symposium on Physical Measurements and Signatures in Remote Sensing* (pp. 299-306). Val D'Isère, France.

King, J. L. (1998). Loss of diversity as a consequence of habitat destruction in California vernal pools. In *Ecology, conservation, and management of vernal pool ecosystems—Proceedings from a 1996 conference*. California Native Plant Society.

Kirchhoff, M., Engelmann, L., Zimmermann, L. L., Seeger, M., Marzolff, I., Aït Hssaine, A., & Ries, J. B. (2019). Geomorphodynamics in Argan Woodlands, South Morocco. *Water (Basel)*, *11*(10), 2193. DOI: 10.3390/w11102193

Klapwijk, M. J., Boberg, J., Bergh, J., Bishop, K., Björkman, C., & Ellison, D.. (2018). Capturing complexity: Forests & mitigation. *Global Environmental Change*, *52*, 238–247. DOI: 10.1016/j.gloenvcha.2018.07.012

Komarolya, D. V., & Maleknia, R. (2025). Identifying indicators of sustainable management of urban forests : A qualitative approach. *Trees, Forests and People*, *22*, 101084. DOI: 10.1016/j.tfp.2025.101084

Kuenzer, C., Ottinger, M., Wegmann, M., Guo, H., Wang, C., Zhang, J., Dech, S., & Wikelski, M. (2014). Earth observation satellite sensors for biodiversity monitoring: Potentials and bottlenecks. *International Journal of Remote Sensing*, *35*(18), 6599–6647. DOI: 10.1080/01431161.2014.964349

Kurbanov, E., Vorobev, O., Lezhnin, S., Sha, J., Wang, J., Li, X., Cole, J., Dergunov, D., & Wang, Y. (2022). Remote sensing of forest burnt area, burn severity, and post-fire recovery: A review. *Remote Sensing (Basel)*, *14*(19), 4714. DOI: 10.3390/rs14194714

Laaribya, S. (2023). A review analysis of the degradation of cork oak forests in North Atlantic, Morocco. *Journal of Biometry Studies*, *3*(2), 54–59. DOI: 10.61326/jofbs.v3i2.05

Laaribya, S., & Belghazi, B. (2016). Radial growth of Atlas cedar. *Nature & Technology*, *14*, 19–32.

Lahssini, S., & Vanuxem, S. (2016). Les Paiements pour Services Écologiques ou la reconquête de la Mâamora par ses habitants? *Développement durable et territoires. Économie, géographie, politique, droit, sociologie*, (Vol. 7, n°1). *Développement Durable & Territoires*, *7*(1). Advance online publication. DOI: 10.4000/developpementdurable.11165

Le Clec'h, S., Oszwald, J., Jégou, N., Dufour, S., Cornillon, P. A., Miranda, I. de S., Gonzaga, L., Grimaldi, M., Gond, V., & Arnauld de Sartre, X. (2013). Cartographier le carbone stocké dans la végétation: Perspectives pour la spatialisation d'un service écosystémique. *Bois et Forêts des Tropiques*, *316*(316), 35–47. DOI: 10.19182/bft2013.316.a20529

Leal Filho, W., Matandirotya, N. R., Lütz, J. M., Alemu, E. A., Brearley, F. Q., Baidoo, A. A., Kateka, A., Ogendi, G. M., Adane, G. B., Emiru, N., & Mbih, R. A. (2021). Impacts of climate change to African indigenous communities and examples of adaptation responses. *Nature Communications*, *12*(1), 6224. DOI: 10.1038/s41467-021-26540-0 PMID: 34711857

Lefebvre, H. (2000). *La production de l'espace* (4e éd). Anthropos.

Legendre, P., & Gauthier, O. (2014). Statistical methods for temporal and space–time analysis of community composition data. *Proceedings. Biological Sciences*, *281*(1778), 20132728. DOI: 10.1098/rspb.2013.2728 PMID: 24430848

Lehmann, J. R. K., Nieberding, F., Prinz, T., & Knoth, C. (2015). Analysis of Unmanned Aerial System–Based CIR Images in Forestry–a New Perspective to Monitor Pest Infestation Levels. *Forests*, *6*(3), 594–612. DOI: 10.3390/f6030594

Lemes, A. P., Garcia, A. R., Pezzopane, J. R. M., Brandão, F. Z., Watanabe, Y. F., Cooke, R. F., Sponchiado, M., de Paz, C. C. P., Camplesi, A. C., Binelli, M., & Gimenes, L. U. (2021). Silvopastoral system is an alternative to improve animal welfare and productive performance in meat production systems. *Scientific Reports, 11*(1), 14092. DOI: 10.1038/s41598-021-93609-7 PMID: 34238990

Leng, Y., Li, W., Ciais, P., Sun, M., Zhu, L., Yue, C., Chang, J., Yao, Y., Zhang, Y., Zhou, J., Li, Z., Wang, X., Xi, Y., & Peng, S. (2024). Forest aging limits future carbon sink in China. *One Earth, 7*(5), 822–834. DOI: 10.1016/j.oneear.2024.04.011

Lepoutre, B. (1965). Régénération artificielle du chêne-liège et équilibre climacique de la subéraie en forêt de la Maâmora. *Annales de la Recherche Forestière du Maroc, 9*, 1–188.

Liang, J., & Gamarra, J. G. P. (2020). The importance of sharing global forest data in a world of crises. *Scientific Data, 7*(1), 424. DOI: 10.1038/s41597-020-00766-x

Lieutier, F., Day, K. R., Battisti, A., Grégoire, J. C., & Evans, H. F. (2004). *Bark and wood boring insects in living trees in Europe, a synthesis*. Springer. DOI: 10.1007/978-1-4020-2241-8

Li, H., Wang, S., Bai, X., Luo, W., Tang, H., Cao, Y., Wu, L., Chen, F., Li, Q., Zeng, C., & Wang, M. (2018). Global carbonate carbon sink. *The Science of the Total Environment, 643*, 157–170. DOI: 10.1016/j.scitotenv.2018.06.196

Linares, J. C., Taïqui, L., & Camarero, J. J. (2011). Increasing drought sensitivity and decline of Atlas cedar (Cedrus atlantica) in the Moroccan Middle Atlas forests. *Forests, 2*(3), 777–796. DOI: 10.3390/f2030777

Liu, M., Zheng, S., Pendall, E., Smith, P., Liu, J., Li, J., Fang, C., Li, B., & Nie, M. (2025). Unprotected carbon dominates decadal soil carbon increase. *Nature Communications, 16*(1), 2008. DOI: 10.1038/s41467-025-57354-z PMID: 40011482

Louhaichi, M., Hassan, S., Jamel, K., & Ouled Belgacem, A. (2023). Challenges and opportunities for promoting use of native species for the rehabilitation of degraded silvopastoral systems in Tunisia. International Center for Agricultural Research in the Dry Areas (ICARDA). Beirut, Lebanon. https://hdl.handle.net/20.500.11766/68814

Luhmann, N. (2001). *La légitimation par la procédure*. Presses de l'Université Laval. DOI: 10.1515/9782763712598

Lumbreras, A., Marques, J., Belo, A., Cristo, M., Fernandes, M., Galioto, D., Machado, M., Mira, A., Sá-Sousa, P., Silva, R., Sousa, L. G., & Pinto-Cruz, C. (2016). Assessing the conservation status of Mediterranean temporary ponds using biodiversity: A new tool for practitioners. *Hydrobiologia*, *782*(1), 187–199. DOI: 10.1007/s10750-016-2697-7

M'Hirit, O. (1994). *Le pin d'Alep (Pinus halepensis Mill.): Ecologie et sylviculture en zones arides et semi-arides*. FAO.

M'Hirit, O. (1999). *La forêt Marocaine: Enjeux et perspective d'une gestion durable*. Ministère Chargé des Eaux et Forêts.

M'Hirit, O., Benzyane, M., Benchekroun, F., El Yousfi, S. M., & Bendaanoun, M. (1998). *L'arganier, une espèce fruitière-forestière à usages multiples*. Éditions Mardaga.

Mace, G. M., Norris, K., & Fitter, A. H. (2012). Biodiversity and ecosystem services: A multilayered relationship. *Trends in Ecology & Evolution*, *27*(1), 19–26. DOI: 10.1016/j.tree.2011.08.006 PMID: 21943703

Machouri. N., Laouina, A., & Sfa, M. (2020). *Les Suberaies Atlantiques et leurs espaces periforestiers: Problematique de degradation et Possibilite de l'elaboration d'une Strategie Participative De Developpement Durable*. Retrieved August 5, 2025, from https://lerma-flsh.uca.ma/wp-content/uploads/2025/06/Les-Suberaies-Atlantiques-et-leurs-espaces.pdf

Mackay-Smith, T. H., Spiekermann, R. I., Richards, D. R., Harcourt, N., & Burkitt, L. L. (2025). An integrative approach to silvopastoral system design: Perspectives, potentials and principles. *New Zealand Journal of Agricultural Research*, *68*(2), 218–258. DOI: 10.1080/00288233.2023.2298922

Magnussen, S., Nord-Larsen, T., & Riis-Nielsen, T. (2018). Lidar supported estimators of wood volume and aboveground biomass from the Danish national forest inventory (2012–2016). *Remote Sensing of Environment*, *211*, 146–153. DOI: 10.1016/j.rse.2018.04.015

Mahdad, M. Y. (2013). *Situation et perspectives d'amélioration du caroubier (Ceratonia siliqua L.) dans la wilaya de Tissemsilt*. Thèse de Maîtrise, Faculté des Sciences de la Technologie, Département des Sciences et de la Nature et de la Vie, Université de Tissemsilt, Algérie, 71 pages.

Mahdi, M. (2002). *Mutations sociales et réorganisation des espaces steppiques*. Imp. Najah El Jadida.

Malki, F., Al Karkouri, J., Sabir, M., & El Mderssa, M. (2022a). Spatio-temporal evolution of Maamora forest. *E3S Web of Conferences 337 0100, 10.*

Malki, F., Al Karkouri, J., Sabir, M., Ibijbijen, J., El Mderssa, M., Ikraoun, H., & Dallahi, Y. (2022b). Assessment of the storage potential of organic carbon in Maamora forest soils: A strategic guidance tool for reforestation. *Soil & Environment*, *41*(1). Advance online publication. DOI: 10.25252/SE/2022/232782

Manfreda, S., McCabe, M. F., Miller, P. E., Lucas, R., Madrigal, V. P., Mallinis, G., Dor, E. B., Helman, D., Estes, L., & Ciraolo, G.. (2018). On the Use of Unmanned Aerial Systems for Environmental Monitoring. *Remote Sensing (Basel)*, *10*(4), 641. DOI: 10.3390/rs10040641

Manzoni, S., Trofymow, J. A., Jackson, R. B., & Porporato, A. (2010). Stoichiometric controls on carbon, nitrogen, and phosphorus dynamics in decomposing litter. *Ecological Monographs*, *80*(1), 89–106. DOI: 10.1890/09-0179.1

Marchane, A., Tramblay, Y., Hanich, L., Ruelland, D., & Jarlan, L. (2017). Climate change impacts on surface water resources in the Rheraya catchment (High Atlas, Morocco). *Hydrological Sciences Journal*, *62*(6), 979–995. DOI: 10.1080/02626667.2017.1283042

Marghadi, S. (2009). Intégration des méthodes d'aide à la décision dans l'aménagement multifonctionnel des forêts au Maroc (Doctoral dissertation, UCL-Université Catholique de Louvain).

Marien, J.-N., & Idrissi Azami, A. (2011). Le développement des plantations clonales d'eucalyptus au Maroc: Une dynamique toujours plus actuelle. *Forêt Méditerranéenne*, *32*(1), 31–38.

Mastrorilli, M., Rana, G., Verdiani, G., Tedeschi, G., Fumai, A., & Russo, G. (2018). Economic Evaluation of Hydrological Ecosystem Services in Mediterranean River Basins Applied to a Case Study in Southern Italy. *Water (Basel)*, *10*(3), 241. DOI: 10.3390/w10030241

Ma, T., Zhang, C., Ji, L., Zuo, Z., Beckline, M., Hu, H., Li, X., & Xiao, X. (2024). Development of forest aboveground biomass estimation, its problems and future solutions: A review. *Ecological Indicators*, *159*, 111653. Advance online publication. DOI: 10.1016/j.ecolind.2024.111653

Matesanz, S., & Valladares, F. (2014). El papel de la plasticidad fenotípica en la respuesta de la vegetación mediterránea al cambio global. In Herrero, A., & Zavala, M. A. (Eds.), *Impactos, Vulnerabilidades y Adaptación de los Bosques y la Biodiversidad de España frente al cambio climático (chapitre, pages N/D)*. MAGRAMA.

Mavsar, R., Herreros, F., Varela, E., Gouriveau, F., & Duclerq, M. (2014). *Méthodes et outils d'évaluation socio-économiques des biens et services rendus par les écosystèmes boisés méditerranéens*. Projet FFEM/Plan Bleu, Composante 2. CTFC et EFIMED, 113 p.

McLachlan, J. S., Hellmann, J. J., & Schwartz, M. W. (2007). A framework for debate of assisted migration in an era of climate change. *Conservation Biology*, *21*(2), 297–302. DOI: 10.1111/j.1523-1739.2007.00676.x PMID: 17391179

Mechergui, T. (2019). Influence of mulching and tree shelters on 4-year survival and growth of zeen oak (Quercus canariensis) seedlings. Journal of Forestry Research.

Médail, F., Michaud, H., Molina, J., Paradis, G., & Loisel, R. (1998). Conservation de la flore et de la végétation des mares temporaires dulçaquicoles et oligotrophes de France méditerranéenne. *Ecologia Mediterranea*, *24*(2), 119–134. DOI: 10.3406/ecmed.1998.1856

MedWet. (2018). *MedWet Framework for Action 2019–2030: Wetlands for sustainable development in the Mediterranean region*. Tour du Valat.

Meliho, M., Boulmane, M., Khattabi, A., Dansou, C. E., Orlando, C. A., Mhammdi, N., & Noumonvi, K. D. (2023). Spatial Prediction of Soil Organic Carbon Stock in the Moroccan High Atlas Using Machine Learning. *Remote Sensing (Basel)*, *15*(10), 2494. DOI: 10.3390/rs15102494

Menzou, K. (2020). Forêts marocaines et usages. La formation en économie sociale et solidaire des forestiers. In J.-L. Laville (Éd.), *Comment former à l'économie sociale et solidaire?* Presses universitaires de Rennes.

Menzou, K. (2024). *Dégradation des espaces forestiers au Maroc : Risques et action publique (Cas de l'espace forestier de la Mamora* [Thèse en Aménagement et Développement durable.]. l'Institut National d'Aménagement et d'Urbanisme (INAU) de Rabat.

Merlo, M., & Croitoru, L. (Eds.). (2005). *Valuing Mediterranean forests: Towards total economic value*. CABI Publishing. DOI: 10.1079/9780851999975.0000

Merzlyak, M. N., Gitelson, A. A., Chivkunova, O. B., & Rakitin, V. Y. (1999). Non-destructive optical detection of pigment changes during leaf senescence and fruit ripening. *Physiologia Plantarum*, *106*(1), 135–141. DOI: 10.1034/j.1399-3054.1999.106119.x

Messier, C., Bauhus, J., Doyon, F., Maure, F., Sousa-Silva, R., Nolet, P., Mina, M., Aquilué, N., Fortin, M.-J., & Puettmann, K. (2019). The functional complex network approach to foster forest resilience to global changes. *Forest Ecosystems*, *6*(1), 1–16. DOI: 10.1186/s40663-019-0166-2

Messier, C., Puettmann, K. J., & Coates, K. D. (Eds.). (2013). *Managing forests as complex adaptive systems: building resilience to the challenge of global change.* Routledge. DOI: 10.4324/9780203122808

Mhirit, O., & Benchekroun, F. (2006). *Les écosystèmes forestiers et périforestiers : Situation, enjeux et perspectives pour 2025* (Le maroc possible : une offre de débat pour une ambition collective, p. 393-483). http://www.rdh50.ma

Michez, A., Piégay, H., Lisein, J., Claessens, H., & Lejeune, P. (2016). Classification of Riparian Forest Species and Health Condition Using Multi–Temporal and Hyperspatial Imagery from Unmanned Aerial System. *Environmental Monitoring and Assessment*, *188*(3), 1–19. DOI: 10.1007/s10661-015-4996-2

Mihi, A., Ghazela, R., & Wissal, D. (2022). Mapping potential desertification-prone areas in North-Eastern Algeria using logistic regression model, GIS, and remote sensing techniques. *Environmental Earth Sciences*, *81*(15), 385. DOI: 10.1007/s12665-022-10513-7 PMID: 35891927

Milian, J. (2007). Le dilemme entre développement et protection dans les montagnes du Maroc-le cas des parcs du Moyen Atlas. *Géocarrefour*, *82*(4), 177–186. DOI: 10.4000/geocarrefour.3002

Millennium Ecosystem Assessment. (2005). *Ecosystems and human well-being: Synthesis*. Island Press.

Minarik, R., & Langhammer, J. (2016). Use of a Multispectral UAV Photogrammetry for Detection and Tracking of Forest Disturbance Dynamics. Proceedings of *International Archives of Photogrammetry Remote Sens. and Spatial Informational Science* (Vol. XLI-B8, pp. 1-8), XXIII ISPRS Congress, Prague, 23 June 2026.

Modeste, M., Abdellatif, K., Nadia, M., & Zhang, H. (2016). *Impact of land use and vegetation cover on risks of erosion in the Ourika watershed (Morocco)*. https://idl-bnc-idrc.dspacedirect.org/bitstream/10625/55962/1/IDL-55962.pdf

Mohajane, M., Essahlaoui, A., Oudija, F., El Hafyani, M., & Cláudia Teodoro, A. (2017). Mapping Forest Species in the Central Middle Atlas of Morocco (Azrou Forest) through Remote Sensing Techniques. *ISPRS International Journal of Geo-Information*, *6*(9), 275. DOI: 10.3390/ijgi6090275

Mohammed, M. (2008). Les tendances en matière de propriété forestière, de mode de faire-valoir des ressources forestières et d'arrangements institutionnels. *Cbneih*, 175.

Montouroy, Y. (2014). Le changement dans la gouvernance globale des forêts:La recentralisation de l'action européenne contre le bois illégal. *Gouvernement et action publique, 3*(1), 107-128. DOI: 10.3917/gap.141.0107

Moreno-Fernández, D., Viana-Soto, A., Camarero, J. J., Zavala, M. A., Tijerín, J., & García, M. (2021). Using spectral indices as early warning signals of forest dieback: The case of drought-prone Pinus pinaster forests. *The Science of the Total Environment, 793*, 148578. DOI: 10.1016/j.scitotenv.2021.148578

Morgan, R. B., Herrmann, V., Kunert, N., Bond-Lamberty, B., Muller-Landau, H. C., & Anderson-Teixeira, K. J. (2021). Global patterns of forest autotrophic carbon fluxes. *Global Change Biology*, *27*(12), 2840–2855. DOI: 10.1111/gcb.15574 PMID: 33651480

Mori, A. S., Lertzman, K. P., & Gustafsson, L. (2017). Biodiversity and ecosystem services in forest ecosystems: A research agenda for applied forest ecology. *Journal of Applied Ecology*, *54*(1), 12–27. DOI: 10.1111/1365-2664.12669

Moroccan Forest Ecosystems: Risks, Challenges and Opportunities. (n.d.). *CESE*. Retrieved December 30, 2025, from https://www.cese.ma/docs/ecosystemes-forestiers-du-maroc-risques-defis-et-opportunites/

Motohka, T., Nasahara, K. N., Oguma, H., & Tsuchida, S. (2010). Applicability of Green-Red Vegetation Index for Remote Sensing of Vegetation Phenology. *Remote Sensing (Basel)*, *2*(10), 2369–2387. DOI: 10.3390/rs2102369

Mouafik, M., Chakhchar, A., Fouad, M., & El Aboudi, A. (2024). Remote Sensing Technologies for Monitoring Argane Forest Stands: A Comprehensive Review. *Geographies*, *4*(3), 441–461. DOI: 10.3390/geographies4030024

Moukrim, S., Benabou, A., Lahssini, S., Aafi, A., Chkhichekh, A., Moudden, F., Ben Bammou, M., El Aboudi, A., & Laaribya, S. (2022b). Spatio-temporal analysis of North African forest cover dynamics using time series of vegetation indices–case of the Maamora forest (Morocco). *Biosystems Diversity*, *30*(4), 372–379. DOI: 10.15421/012236

Moukrim, S., Lahssini, S., Alaoui, H. M., Rifai, N., Arahou, M., & Rhazi, L. (2018). Modélisation de la distribution spatiale des espèces endémiques pour leur conservation: Cas de l'Argania spinosa (L.) Skeels. *Revue d'Écologie*, *73*(2), 153–166. DOI: 10.3406/revec.2018.1923

Moukrim, S., Lahssini, S., Naggar, M., Lahlaoi, H., Rifai, N., Arahou, M., & Rhazi, L. (2019). Local community involvement in forest rangeland management : Case study of compensation on forest area closed to grazing in Morocco. *The Rangeland Journal*, *41*(1), 43–53. DOI: 10.1071/RJ17119

Moukrim, S., Lahssini, S., Rhazi, M., Alaoui, H. M., Benabou, A., Wahby, I., El Madihi, M., Arahou, M., & Rhazi, L. (2019). Climate change impacts on potential distribution of multipurpose agro-forestry species: Argania spinosa (L.) Skeels as case study. *Agroforestry Systems*, *93*(4), 1209–1219. DOI: 10.1007/s10457-018-0232-8

Moukrim, S., Lahssini, S., Rhazi, M., Menzou, K., El Madihi, M., Rifai, N., Bouziani, Y., Azedou, A., Boukhris, I., & Rhazi, L. (2022). Climate Change Impact on Potential Distribution of an Endemic Species *Abies marocana* Trabut. *Ekologia (Bratislava)*, *41*(4), 329–339. DOI: 10.2478/eko-2022-0034

Moukrim, S., Lahssini, S., Rifai, N., Menzou, K., Mharzi-Alaoui, H., Labbaci, A., Rhazi, M., Wahby, I. W., El Madihi, M., & Rhazi, L. (2020). Modélisation de la distribution potentielle de *Cedrus atlantica* Manetti au Maroc et impacts du changement climatique. *Bois et Forêts des Tropiques*, *344*, 3–16. DOI: 10.19182/bft2020.344.a31888

Moukrim, S., Tellal, J., Lahssini, S., Boudjaj, A., Rhazi, M., El Madihi, M., Ben Bammou, M., Moukrim, M., Saber, E.-R., Wahby, I., Rhazi, L., & Grillas, P. (2025). Long-time dynamics in a North African coastal Ramsar site, the Sidi Boughaba Lake. *Ekologia (Bratislava)*, *44*(2), 147–159. DOI: 10.2478/eko-2025-0017

Mourjane, M., Tabyaoui, H., & Hammichi, F. E. (2024). Integration of DInSAR and SBAS radar techniques to delineate the main areas of river recharge by sediments using ALOS-PALSAR-2 data: Oued Ghis case in Morocco. *BIO Web of Conferences, 109*, 01028. DOI: 10.1051/bioconf/202410901028

Muller, S. D., Daoud-Bouattour, A., Fauquette, S., Bottollier-Curtet, M., Rifai, N., Robles, M., Saber, E.-R., El Madihi, M., Moukrim, S., & Rhazi, L. (2022). Holocene history of peatland communities of central Rif (Northern Morocco). *Geobios*, *70*, 35–53. DOI: 10.1016/j.geobios.2021.12.001

Murugesan, K., & Nanjappan, R. (2025). Novel hybrid RNN K-Means algorithm for the forest cover type prediction. *Proceedings of AIP Conference* (vol. 3258, 1: 020015). Madurai, India, 6-17 April 2025. DOI: 10.1063/5.0271542

Nadir, B. (2008). *Domanialité et environnement : Cas des eaux et forêts* (1ère éd). Editions IDGL.

Naggar M., (2000). Éléments de base d'une stratégie de sylvopastoralisme en Afrique du Nord. *Options méditerranéennes*, série A, n° 39, pp. 191-202.

Naggar M., Lahssni S., (2017). L'aménagement forestier et les enjeux de restauration des écosystèmes forestiers au Maroc. *Revue Forêt Méditerranéenne* t. XXXVIII, n° 1. PP: 77-82.

Naggar, M. ((2021). Renouveau des pratiques pastorales en forêt au Maroc: De la tradition à l'aménagement. *Éditions universitaires européennes*. 87 p.

Naggar, M., & Lahssini, S. (2015). La filière caroubier au service du développement socioéconomique des territoires forestiers de montagne. *Actes du XIVe Congrès forestier mondial* (8 pages). Durban, Afrique du Sud, 7-11 Septembre 2025.

Naggar, M. (2018). Stratégie sylvopastorale et renouveau des pratiques pastorales en forêt au Maroc. *Revue Forestière Française*, *70*(5), 487–501. DOI: 10.4267/2042/70133

Naggar, M., & Lahssini, S. (2017). L'aménagement forestier et les enjeux de restauration des écosystèmes forestiers au Maroc. *Forêt Méditerranéenne*, *38*(1), 49–62.

Nancy, J.-L. (2007). L'être-avec de l'être-là. *Cahiers philosophiques, 111*(3), 66-78. Cairn.info. DOI: 10.3917/caph.111.0066

Nancy, J.-L. (2009). Démocratie finie et infinie. In *Démocratie, dans quel état?* (p. 77-94). La Fabrique Éditions. Cairn.info. DOI: 10.3917/lafab.colle.2009.01.0077

Nandal, A., Yadav, S. S., Rao, A. S., Meena, R. S., & Lal, R. (2023). Advance methodological approaches for carbon stock estimation in forest ecosystems. *Environmental Monitoring and Assessment*, *195*(2), 315. DOI: 10.1007/s10661-022-10898-9

Naseri, M. H., Jouibary, S. S., & Habashi, H. (2023). Analysis of forest tree dieback using Ultra-Cam and UAV-imagery. *Scandinavian Journal of Forest Research*, *38*(6), 392–400. DOI: 10.1080/02827581.2023.2231349

Näsi, R., Honkavaara, E., Lyytikäinen-Saarenmaa, P., Blomqvist, M., Litkey, P., Hakala, T., Viljanen, N., Kantola, T., Tanhuanpää, T., & Holopainen, M. (2015). Using UAV–Based Photogrammetry and Hyperspectral Imaging for Mapping Bark Beetle Damage at Tree–Level. *Remote Sensing (Basel)*, *7*(11), 15467–15493. DOI: 10.3390/rs71115467

Navarro, A., Catalao, J., & Calvao, J. (2019). Assessing the Use of Sentinel-2 Time Series Data for Monitoring Cork Oak Decline in Portugal. *Remote Sensing (Basel)*, *11*(21), 2515. DOI: 10.3390/rs11212515

Navarro-Cerrillo, R. M., Ruiz-Gómez, F. J., Camarero, J. J., Castillo, V., Barberá, G. G., Palacios-Rodríguez, G., Navarro, F. B., Blanco, J. A., Imbert, J. B., Cachinero-Vivar, A. M., Molina, A. J., & del Campo, A. D. (2022). Long-Term Carbon Sequestration in Pine Forests under Different Silvicultural and Climatic Regimes in Spain. *Forests*, *13*(3), 450. DOI: 10.3390/f13030450

Navarro-Rosales, F., Fernández-Habas, J., Reyna-Bowen, L., Gómez, J. A., & Fernández-Rebollo, P. (2023). Subsoiling for planting trees in dehesa system: Long-term effects on soil organic carbon. *Agroforestry Systems*, *97*(4), 699–710. DOI: 10.1007/s10457-023-00820-8

NDC. (2025). Contribution Déterminée au niveau National CDN 3.0 du Maroc 2026-2025. Royaume du Maroc, 107. https://unfccc.int/sites/default/files/2025 10/ MOROCCO%20NDC%203.0%20_30.9.25.pdf

Nelson, R. (2013). How did we get here? An early history of forestry lidar[1]. *Canadian Journal of Remote Sensing, 39*(sup1), S6–S17. DOI: 10.5589/m13-011

Nicolet, P., Biggs, J., Fox, G., Hodson, M. J., Reynolds, C., Whitfield, M., & Williams, P. (2004). The wetland plant and macroinvertebrate assemblages of temporary ponds in England and Wales. *Biological Conservation*, *120*(2), 261–278. DOI: 10.1016/j.biocon.2004.03.010

Ni, M., Wu, Q., Li, G., & Li, D. (2025). Remote Sensing Technology for Observing Tree Mortality and Its Influences on Carbon–Water Dynamics. *Forests*, *16*(2), 194. DOI: 10.3390/f16020194

Odum Eugene, P. (1969). The Strategy of Ecosystem Development. *Science*, *164*(3877), 262–270. DOI: 10.1126/science.164.3877.262

Oertli, B., Céréghino, R., Hull, A., & Miracle, R. (2010). Pond conservation in Europe. *Hydrobiologia*, *634*(1), 1–9. DOI: 10.1007/s10750-009-9891-9

Okacha, A., Salhi, A., Arari, K., El Badaoui, K., & Lahrichi, K. (2023). Soil erosion assessment using the RUSLE model for better planning: A case study from Morocco. *Modeling Earth Systems and Environment*, *9*(3), 3721–3729. DOI: 10.1007/ s40808-023-01731-4

Ortaçeşme, V., Özkan, B., & Karagüzel, O. (2002). An Estimation of the Recreational Use Value of Kursunlu Waterfall Nature Park by the Individual Travel Cost Method. *Turkish Journal of Agriculture and Forestry*, *26*, 57–62.

Orwa, C., Mutua, A., Kindt, R., Jamnadass, R., & Simons, A. (2009) — Agroforestry Database — *surtout sur les rendements, la longévité, la capacité de rejets.*

Ost, F. (2003). *La nature hors la loi. L'écologie à l'épreuve du droit*. La Découverte. Cairn.info. https://www.cairn.info/la-nature-hors-la-loi--9782707139368.htm

Ost, F. (1995). *La nature hors la loi : L'écologie à l'épreuve du droit*. Découverte.

Ostrom, E., & Baechler, L. (2010). *Gouvernance des biens communs : Pour une nouvelle approche des ressources naturelles*. De Boeck.

Otsu, K., Pla, M., Duane, A., Cardil, A., & Brotons, L. (2019). Estimating the Threshold of Detection on Tree Crown Defoliation Using Vegetation Indices from UAS Multispectral Imagery. *Drones (Basel)*, *3*(4), 80. DOI: 10.3390/drones3040080

Otsu, K., Pla, M., Vayreda, J., & Brotons, L. (2018). Calibrating the Severity of Forest Defoliation by Pine Processionary Moth with Landsat and UAV Imagery. *Sensors (Basel)*, *18*(10), 3278. DOI: 10.3390/s18103278

Otto, S. P., & Gerstein, A. C. (2006). Why have sex? The population genetics of sex and recombination. *Biological Reviews of the Cambridge Philosophical Society*, *81*(4), 647–679. PMID: 16856849

Ouadah, N. (2016). *Estimation de la valeur économique et sociale des services rendus par les écosystèmes forestiers méditerranéens, Parc National de Chréa, Algérie*. Plan Bleu.

Oubrahim, H., Boulmane, M., Bakker, M. R., Augusto, L., & Halim, M. (2015). Carbon storage in degraded cork oak (Quercus suber) forests on flat lowlands in Morocco. [early view]. *IForest (Viterbo)*. Advance online publication. DOI: 10.3832/ifor1364-008

Oumasst, A., Tiouidji, F. E., Tabi, S., Zahidi, A., El Mousadik, A., El Finti, A., Aitlhaj, A., & Hallam, J. (2024). Development of Allometric Equations to Determine the Biomass of Plant Components and the Total Storage of Carbon Dioxide in Young Mediterranean Argan Trees. *Sustainability (Basel)*, *16*(11), 4592. DOI: 10.3390/su16114592

Palacios-Rodríguez, G., Quinto, L., Lara-Gómez, M. A., Pérez-Romero, J., Recio, J. M., Álvarez-Romero, M., Cachinero-Vivar, A. M., Hernández-Navarro, S., & Navarro-Cerrillo, R. M. (2022). Carbon Sequestration in Carob (Ceratonia siliqua L.) Plantations under the EU Afforestation Program in Southern Spain Using Low-Density Aerial Laser Scanning (ALS) Data. *Forests*, *13*(2), 285. DOI: 10.3390/f13020285

Palahi, M., Mavsar, R., & Birot, Y. (2008). Mediterranean forests under focus. *International Forestry Review*, *10*(4), 676–688. DOI: 10.1505/ifor.10.4.676

Pan, Y., Birdsey, R. A., Fang, J., Houghton, R., Kauppi, P. E., Kurz, W. A., Phillips, O. L., Shvidenko, A., Lewis, S. L., Canadell, J. G., Ciais, P., Jackson, R. B., Pacala, S. W., McGuire, A. D., Piao, S., Rautiainen, A., Sitch, S., & Hayes, D. (2011). A large and persistent carbon sink in the world's forests. *Science*, *333*(6045), 988–993. DOI: 10.1126/science.1201609

Paris Corona, P., Quatrini, V., Schirru, M., Dettori, S., & Puletti, N. (2018). Towards the economic valuation of ecosystem production from cork oak forests in Sardinia (Italy). iForest - Biogeosciences and Forestry, 11(5), 660-667. https://doi.org/doi: 10.3832/ifor2558-011

Parsapour, M. K., Sohrabi, H., Soltani, A., & Iranmanesh, Y. (2013). Allometric equations for estimating biomass in four poplar species at Charmahal and Bakhtiari province. *Tahqiqat-i Jangal va Sinubar-i Iran*, *21*(3), 517–528. DOI: 10.22092/ijfpr.2014.4730

PCI-Geomatics. (2018). *Using PCI Software*. PCI-Geomatics.

Pedlar, J. H., McKenney, D. W., Beaulieu, J., Colombo, S. J., McLachlan, J. S., & O'Neill, G. A. (2012). Placing forestry in the assisted migration debate. *Bioscience*, *62*(9), 835–842. DOI: 10.1525/bio.2012.62.9.10

Peluso, N. L., & Vandergeest, P. (2001). Genealogies of the Political Forest and Customary Rights in Indonesia, Malaysia, and Thailand. *The Journal of Asian Studies*, *60*(3), 761–812. DOI: 10.2307/2700109

Penuelas, J., Baret, F., & Filella, I. (1995). Semi-empirical indices to assess carotenoids/chlorophyll a ratio from leaf spectral reflectance. *Photosynthetica*, *31*, 221–230.

Penuelas, J., Filella, I., Biel, C., Serrano, L., & Save, R. (1993). The reflectance at the 950-970 nm region as an indicator of plant water status. *International Journal of Remote Sensing*, *14*(10), 1887–1905. DOI: 10.1080/01431169308954010

Penuelas, J., Gamon, J., Freeden, A., Merino, J., & Field, C. (1994). Reflectance indices associated with physiological changes in nitrogen and water limited sunflower leaves. *Remote Sensing of Environment*, *48*(2), 135–146. DOI: 10.1016/0034-4257(94)90136-8

Pereira, A. P. A., Cherubin, M. R., de Araujo, A. S. F., Santana, M. C., de Medeiros, E. V., & da Costa, D. P. (2024). Soil quality in mono and mixed eucalyptus. *Sustainability*, *16*(6), 2534. DOI: 10.3390/su16062534

Pérez-Cutillas, P., & Salhi, A. (2024). Long-term hydroclimatic projections and climate change scenarios at regional scale in Morocco. *Journal of Environmental Management, 371, 123254.* Piao S, Yue C, Ding J, Guo Z. (2022). Perspectives on the role of terrestrial ecosystems in the 'carbon neutrality' strategy. Science China Earth Sciences65:1178–1186. https://doi.org/DOI: 10.1016/j.jenvman.2024.123254

Peri, P. L., Chará, J., Viñoles, C., Dube, F., Sotomayor, A., Kremer, K., Bahamonde, H., Varella, A., Paniagua, P., Enriquez, A., Teglia, C., Bussoni, A., & Lencinas, M. V. (2024). Current trends in silvopastoral systems. *Agroforestry Systems*, *98*(7), 1945–1953. DOI: 10.1007/s10457-024-01093-5

Picard, N., Boyemba-Bosela, F., & Rossi, V. (2015). Reducing the error in biomass estimates strongly depends on model selection. *Annals of Forest Science*, *72*(6), 6. Advance online publication. DOI: 10.1007/s13595-014-0434-9

Picard, N., Laurent, S. A., & Henry, M. (2012). *Manuel de construction d'équations allométriques pour l'estimation du volume et la biomasse des arbres: de la mesure de terrain à la prédiction*. CIRAD-FAO.

Pingoud, K., Ekholm, T., Sievänen, R., Huuskonen, S., & Hynynen, J. (2018). Trade-offs between carbon and harvest. *Journal of Environmental Management*, *210*, 96–103. DOI: 10.1016/j.jenvman.2017.12.076

Pinto-Cruz, C., Molina, J. A., Loidi, J., & Costa, J. C. (2009). Plant communities as a tool in temporary pond conservation. *Hydrobiologia*, *634*(1), 11–24. DOI: 10.1007/s10750-009-9885-7

Pix4D-Mapper-V4.1. (2017). *Manual of Pix-4D Mapper*. https://support.pix4d.com/hc/en-us/articles/204272989-Offline-Getting-Started-and-Manual-pdf. Manual-Pix4D-Mapper.pdf

Plan Bleu (2014). *Méthodes et outils d'évaluation socio-économique des biens et services rendus par les écosystèmes boisés méditerranéens*, Rapport technique préparé par le CTFC et EFIMED, Projet FFEM.

Puettmann, K. J., Wilson, S. M., Baker, S. C., Donoso, P. J., Drössler, L., Amente, G., Harvey, B. D., Knoke, T., Lu, Y., Nocentini, S., Putz, F. E., Yoshida, T., & Bauhus, J. (2015). Silvicultural alternatives to conventional even-aged forest management-what limits global adoption? *Forest Ecosystems*, *2*(1), 8. DOI: 10.1186/s40663-015-0031-x

Pyke, C. P. (2005). Assessing climate change impacts on vernal pool ecosystems and endemic branchiopods. *Ecosystems (New York, N.Y.)*, *8*(1), 95–105. DOI: 10.1007/s10021-004-0086-y

Qarro M., Berkat O., M'Hamedi M., Yessef M., (2000). L'aménagement sylvopastoral: contraintes et développement durable, 2000. Rupture: Revue Options Méditerranéennes: Série A. *Séminaires Méditerranéens*; n. 39. PP: 171-174.

Qarro, M. (2006). Le sylvopastoralisme et la gestion durable des écosystèmes naturels arides. 14th International Soil Conservation Organisation - Conférence Water Management and Soil Conservation in Semi-Arid Environments, Marrakech, Morocco, May 14-19, 2006, 12 p.

Quézel, P. (1998). La végétation des mares transitoires à isoetes en région méditerranéenne, intérêt patrimonial et conservation. *Ecologia Mediterranea*, *24*(2), 111–117. DOI: 10.3406/ecmed.1998.1855

Quézel, P., & Médail, F. (2003). *Écologie et biogéographie des forêts du bassin méditerranéen*. Elsevier.

Quézel, P., & Médail, F. (2003). *Ecologie et biogéographie des forets du bassin méditerranéen*. Elsevier., http://www.lavoisier.fr/livre/environnement/ecologie-et-biogeographie-des-forets-du-bassin-mediterraneen/quezel/descriptif-9782842994518

Quintana, X. D., Boix, D., Gascón, S., & Sala, J. (2015). Monitoring temporary ponds. *Limnetica*, *34*, 1–20.

Ramdani, M. (1986). Contribution à l'étude écologique des mares temporaires du Maroc. Thèse de doctorat, Université de Rabat.

Ramírez-Valiente, J. A., Santos-del-Blanco, L., Alía, R., Robledo-Arnuncio, J. J., & Climent, J. (2022). Adaptation of Mediterranean forest species to climate: Lessons from common garden experiments. *Journal of Ecology*, *110*(5), 1022–1042. DOI: 10.1111/1365-2745.13730

Ram, J. C., Dagar, K., Lal, G., Singh, V., Toky, V. S., Tanwar, S. R., & Dar, M.K., & Chauhan. (2011). Biodrainage to combat waterlogging, increase farm productivity and sequester carbon in canal command areas of northwest India. *Current Science*, *100*(11), 1673–1680.

Reusch, T. B. H., & Wood, T. E. (2007). *Molecular ecology of global ch*ange. *Molecular Ecology*, *16*(19), 3973–3992. DOI: 10.1111/j.1365-294X.2007.03454.x PMID: 17894755

Rhazi, L., Grillas, P., Saber, E., Rhazi, M., Brendonck, L., & Waterkeyn, A. (2012). Vegetation of Mediterranean temporary pools: A fading jewel? *Hydrobiologia*, *689*(1), 23–36. DOI: 10.1007/s10750-011-0679-3

Rhazi, L., Grillas, P., Tan Ham, L., & El Khyari, D. (2001). The seed bank and the between-years dynamics of the vegetation of a Mediterranean temporary pool (NW Morocco). *Ecologia Mediterranea*, *27*(1), 69–88. DOI: 10.3406/ecmed.2001.1907

Rhazi, L., Rhazi, M., Grillas, P., & Khyari, D. (2006). Richness and structure of plant communities in temporary pools from western Morocco: Influence of human activities. *Hydrobiologia*, *570*(1), 197–203. DOI: 10.1007/s10750-006-0180-6

Ribeiro, S. C., Soares, C. P. B., Fehrmann, L., Jacovine, L. A. G., & Von Gadow, K. (2015). Aboveground and belowground biomass and carbon estimates for clonal *eucalyptus* trees in southeast brazil. *Revista Árvore*, *39*(2), 353–363. DOI: 10.1590/0100-67622015000200015

Ribot, J. C., & Peluso, N. L. (2003). A Theory of Access*. *Rural Sociology*, *68*(2), 153–181. DOI: 10.1111/j.1549-0831.2003.tb00133.x

Ricciardi, A., & Simberloff, D. (2009). Assisted colonization is not a viable conservation strategy. *Trends in Ecology & Evolution*, *24*(5), 248–253. DOI: 10.1016/j.tree.2008.12.006 PMID: 19324453

Rigolot, E. (2004). Fire, fuel management and landscape dynamics in Mediterranean forests. *Annals of Forest Science*, *61*(6), 599–610.

Rivas Martinez, S. (1982). Etages bioclimatiques, secteurs chorologiques et série de végétation de l'Espagne méditerranéenne. *Ecologia Mediterranea*, *8*(1), 275–288. DOI: 10.3406/ecmed.1982.1954

Robertson, G., Linser, S., & Köhl, M. (2024). Forest disturbance and damage: Perspectives for forest monitoring and reporting. *Forests Monitor*, *1*(1), 39–65. DOI: 10.62320/fm.v1.i1.10

Rocha, A. A., Borges, W. R., Von Huelsen, M. G., De Oliveira, E., Sousa, F. R., Maciel, S. T. R., De Almeida Rocha, J., & Jacobson, T. K. B. (2024). Imaging tree root systems using ground penetrating radar (GPR) data in Brazil. *Frontiers in Earth Science (Lausanne)*, *12*, 1353572. Advance online publication. DOI: 10.3389/feart.2024.1353572

Rodríguez, M., & Carlos, J. (2001). Organic carbon content under different types of land use and soil in peninsular Spain. *Biology and Fertility of Soils*, *33*(1), 53–61. DOI: 10.1007/s003740000289

Romagny, B., Aderghal, M., Auclair, L., Ilbert, H., & Lemeilleur, S. (2018). Communs en crise. Agdals, terres collectives, forêts et terroirs au Maroc. *Revue internationale des études du développement, 233*(1), 53-73. Cairn.info. DOI: 10.3917/ried.233.0053

Romagny, B., Auclair, L., & Elgueroua, A. (2008). La gestion des ressources naturelles dans la vallée des Aït Bouguemez (Haut Atlas) : La montagne marocaine à la recherche d'innovations institutionnelles. *Mondes en développement, 141*(1), 63-80. Cairn.info. DOI: 10.3917/med.141.0063

Romero, B., Scotti, I., Fady, B., & Ganteaume, A. (2023). Fire frequency, as well as stress response and developmental gene control serotiny level variation in a widespread pioneer Mediterranean conifer, *Pinus halepensis. Ecology and Evolution, 13*(3), e9919. DOI: 10.1002/ece3.9919 PMID: 36960240

Rosanvallon, P. (2014). *La contre-démocratie : La politique à l'âge de la défiance*. Éd. Points.

Rosset, V., Lehmann, A., & Oertli, B. (2010). Warmer and richer? Predicting the impact of climate warming on species richness in small temperate water bodies. *Global Change Biology, 16*(8), 2376–2387. DOI: 10.1111/j.1365-2486.2010.02206.x

Roumane, A., Bhukuth, A., Bazin, D., Roumane, A., Bhukuth, A., & Bazin, D. J. (2010). *La gestion des ressources forestières au Maroc: Une approche par le capital social.* https://www.academia.edu/download/97757843/6570607.pdf

Rouse, J. W., Haas, R. W., Schell, J. A., Deering, D. W., & Harlan, J. C. (1974). *Monitoring the vernal advancement and retro-gradation (Green wave effect) of natural vegetation.* NASA/GSFCT Type III Final Report Number: NASA-CR-139243, Greenbelt, MD, USA, 250 pages.

Roy, D. P., Zhang, H. K., Ju, J., Gomez-Dans, J. L., Lewis, P. E., Schaaf, C. B., Sun, Q., Li, J., Huang, H., & Kovalskyy, V. (2016). A general method to normalize Landsat reflectance data to nadir BRDF adjusted reflectance. *Remote Sensing of Environment, 176*, 255–271. DOI: 10.1016/j.rse.2016.01.023

Ruffault, J., Curt, T., Moron, V., Trigo, R. M., Mouillot, F., Koutsias, N., Pimont, F., Martin-StPaul, N., Barbero, R., Dupuy, J.-L., Russo, A., & Belhadj-Khedher, C. (2020). Increased likelihood of heat-induced large wildfires in the Mediterranean Basin. *Scientific Reports, 10*(1), 13790. DOI: 10.1038/s41598-020-70069-z PMID: 32796945

Ruiz-Peinado, R.. (2013). Thinning impacts on carbon. *Forest Ecology and Management.*

Ruiz-Peinado, R., Bravo-Oviedo, A., López-Senespleda, E., Montero, G., & Río, M. (2013). Do thinnings influence biomass and soil carbon stocks in Mediterranean maritime pinewoods? *European Journal of Forest Research, 132*(2), 253–262. DOI: 10.1007/s10342-012-0672-z

Ruiz-Peinado, R., Del Rio, M., & Montero, G. (2011). New models for estimating the carbon sink capacity of Spanish softwood species. *Forest Systems*, *20*(1), 1. Advance online publication. DOI: 10.5424/fs/2011201-11643

Rullan-Silva, C. D., Oltho, A. E., Delgado de la Mata, J. A., & Pajares-Alonso, J. A. (2013). Remote Monitoring of Forest Insect Defoliation - A Review. *Forest Systems*, *22*(3), 377–391. DOI: 10.5424/fs/2013223-04417

Russo, F., Maselli, G., & Nestico, A. (2023). Forest ecosystem services: Economic evaluation of carbon sequestration on a large scale. *Valori e Valutazioni*, *33*, 17–30. DOI: 10.48264/VVSIEV-20233303

Ryan, C. M., Berry, N. J., & Joshi, N. (2014). Quantifying the causes of deforestation and degradation and creating transparent REDD+ baselines: A method and case study from central Mozambique. *Applied Geography (Sevenoaks, England)*, *53*, 45–54. DOI: 10.1016/j.apgeog.2014.05.014

Saadani, Y. (2015). La stratégie de développement durable des forêts et parcours en Tunisie 2015-2024 - Pour une transition vers l'économie verte. Revue Foret Méditerranéenne, T. XXXVI, n°2, 2015. PP: 235-240.SGG, 2002. Arrêté du ministre chargé des eaux et forêts fixant les limites, conditions et modalités de demande et d'octroi de la compensation pour mise en défens du domaine forestier à exploiter ou à régénérer. Bulletin officiel n°5000 publié par le Secrétariat Général du Gouvernement (SGG), PP:456-457.

Saber, E. (2006). *Approche de la dynamique spatio-temporelle des géosystèmes forêt et mares temporaires de la province de Benslimane (Maroc occidental)* (Doctoral dissertation, Université de Provence, Aix-Marseille I).

Sabir, M., Sagno, R., Tchintchin, Q., Zaher, H., & Benjelloun, H. (2020). SOC stocks in Morocco. In *Carbone des sols en Afrique* (pp. 71–90). IRD éditions.

Saddik, A., Hssaisoune, M., Belaqziz, S., Labbaci, A., Tairi, A., Meskour, B., & Bouchaou, L. (2024). Assessing the health and yield of argan trees in Morocco's unique ecosystem: A multispectral and machine learning approach. *International Journal of Remote Sensing*, *45*(24), 9192–9220. DOI: 10.1080/01431161.2024.2407560

Saint -André. L., Vallet, P., Pignard, G., Dupouey, J.L., Colin, A., Loustau, D., Le Bas, C., Meredieu, C., Caraglio, Y., Porté, A., Hamza, N., Cazin, A., Nouvellon, Y., & Dhôte, J.F.(2010). Estimating carbon stocks in forest stands: 1. Methodological developments. In D. Loustau (ed.), *Forests, carbon cycle and climate change* (pp. 79-100). Versailles: Ed. Quae

Salamí, E., Barrado, C., & Pastor, E. (2014). UAV flight experiments applied to the remote sensing of vegetated areas. *Remote Sensing (Basel)*, *6*(11), 11051–11081. DOI: 10.3390/rs61111051

Sallak, B. (2019). *Les risques hydrologiques d'inondations et la problématique d'aménagement des territoires de piémont: Cas de Taghzirt*. Thèse de Doctorat, Université Sultan Moulay Slimane, Béni-Mellal, Maroc, 289 pages.

Şandric, I., Irimia, R., Petropoulos, G. P., Anand, A., Srivastava, P. K., Pleşoianu, A., Faraslis, I., Stateras, D., & Kalivas, D. (2022). Tree's Detection and Health's assessment from ultra-high resolution UAV imagery and deep learning. *Geocarto International*, *37*(25), 10459–10479. DOI: 10.1080/10106049.2022.2036824

Sarraf, M., & Jorio, A. (2010). Land Degradation: The Case of Morocco. *The Cost of Environmental Degradation*, 75.

Sattout, E. J., Talhouk, S. N., & Caligaria, P. D. S. (2007). Economic value of cedar relics in Lebanon: An application of contingent valuation method for conservation. *Ecological Economics*, *61*(2-3), 315–322. DOI: 10.1016/j.ecolecon.2006.03.001

Sauvage, Ch. 1961: *Recherches géobotaniques sur les subéraies marocaines*. Trav. Inst. Sci. Cherif. Serv. Bot., 21: 462p.

Sbay, H., & Zine El Abidine, A. (1994). *Croissance et production du cèdre de l'Atlas au Maroc*. Revue Annales de la Recherche Forestière au Maroc.

Scarascia-Mugnozza, G., Oswald, H., Piussi, P., & Radoglou, K. (2000). Forests of the Mediterranean region: Gaps in knowledge and research needs. *Forest Ecology and Management*, *132*(1), 97–109. DOI: 10.1016/S0378-1127(00)00383-2

Schläpfer, D., Popp, C., & Richter, R. (2020). Drone data Atmospheric Correction Concept for Multi – and Hyperspectral Imagery - The DROACOR Model. *The Int. Arch. of Photogram. Remote Sens. and Spatial Inf. Sci.*, XLIII-B3-2020. *Commission III, WG*, *III*(4), 473–478.

Schmutz, J. (2001). La Doctrine mediévale des causes et la théologie de la nature pure (XIIIe-XVIIe siècles). *Revue Thomiste*, *101*(1-2), 217–264.

Schwartz, S. S., & Jenkins, D. G. (2000). Temporary aquatic habitats: Constraints and opportunities. *Aquatic Ecology*, *34*(1), 3–8. DOI: 10.1023/A:1009944918152

Schwerz, F., Neto, D. D., Caron, B. O., Tibolla, L. B., Sgarbossa, J., Eloy, E., Elli, E. F., & Carvalho, L. G. (2020). Carbon stocks, partitioning, and wood composition in short-rotation forestry system under reduced planting spacing. *Annals of Forest Science*, *77*(3), 67. DOI: 10.1007/s13595-020-00974-w

Schwörer, C., Morales-Molino, C., Gobet, E., Henne, P. D., Pasta, S., Pedrotta, T., Van Leeuwen, J. F. N., Vannière, B., & Tinner, W. (2024). Simulating past and future fire impacts on Mediterranean ecosystems. *Journal of Ecology*, *112*(5), 954–970. DOI: 10.1111/1365-2745.14293

Sebbar, B., Moumni, A., Lahrouni, A., Chehbouni, A., Belghazi, T., & Maksoudi, B. (2022). Remotely Sensed Phenology Monitoring and Land-cover Classification for the Localization of the Endemic Argan Tree in the Southern-west of Morocco. *Journal of Sustainable Forestry*, *41*(10), 1014–1028. DOI: 10.1080/10549811.2021.1897018

SEIFN. (2014). Service des Etudes et de l'Inventaire Forestier National, Maroc forestier in http://www.eauxetforets.gov.ma/ForetsMarocaines/ForetsChiffres/Pages/Forets-En-Chiffres.aspx

Semlitsch, R. D., & Bodie, J. R. (2003). Biological criteria for buffer zones around wetlands and riparian habitats. *Conservation Biology*, *17*(5), 1219–1228. DOI: 10.1046/j.1523-1739.2003.02177.x

Serres, M. (1990). *Le contrat naturel*. Editions F. Bourin.

SGG. (2021). Arrêté du ministre de l'agriculture, de la pêche maritime, du développement rural et des eaux et forêts, modifiant et complétant l'arrêté fixant les limites, conditions et modalités de demande et d'octroi de la compensation pour mise en défens du domaine forestier à exploiter ou à régénérer. Bulletin officiel n°7080 publié par le Secrétariat Général du Gouvernement (SGG), PP:564-565.

Shah, I. H., Manzoor, M. A., Ji'nhui, W., Li, X., Hameed, M. K., Rehaman, A., Li, P., Zhang, Y., Niu, Q., & Chang, L. (2024). Comprehensive review: Effects of climate change and greenhouse gases emission relevance to environmental stress on horticultural crops and management. *Journal of Environmental Management*, *351*, 119978. DOI: 10.1016/j.jenvman.2023.119978

Shahi, T. B., Balasubramaniam, T., Sabir, K., & Nayak, R. (2025). Pasture monitoring using remote sensing and machine learning: A review of methods and applications. *Remote Sensing Applications: Society and Environment*, *37*, 101459. DOI: 10.1016/j.rsase.2025.101459

Shang, R., Chen, J. M., Xu, M., Lin, X., Li, P., Yu, G., He, N., Xu, L., Gong, P., Liu, L., Liu, H., & Jiao, W. (2023). China's current forest age structure will lead to weakened carbon sinks in the near future. *Innovation (Cambridge (Mass.))*, *4*(6), 100515. Advance online publication. DOI: 10.1016/j.xinn.2023.100515

Shirkey, G., Anctil, A., John, R., Kolluru, V., Mungai, L., & Kashongwe, H.. (2024). Opportunities for NbS. *Environmental Research Letters*.

Shi, Y. (2020). Remotely sensing the species of individual trees. *International J. of Applied Earth Observation and Geo-information, 73*, 207–219. DOI: 10.1016/j.jag.2018.06.018

Silvertown, J. (2008). The evolutionary maintenance of sexual reproduction: Evidence from the ecological distribution of asexual reproduction in clonal plants. *International Journal of Plant Sciences, 169*(1), 157–168. DOI: 10.1086/523357

Simenel, R., Mohammed, A., Sabir, M., & Auclair, L. (2016). Cairn, borne ou belvédère? Quand le naturalisme et l'analogisme négocient la limite entre espace cultivé et forêt au Maroc. *Anthropologica, 58*(1), 60–76. DOI: 10.3138/anth.581.A04

Simonneaux, V., Cheggour, A., Deschamps, C., Mouillot, F., Cerdan, O., & Le Bissonnais, Y. (2015). Land use and climate change effects on soil erosion in a semi-arid mountainous watershed (High Atlas, Morocco). *Journal of Arid Environments, 122*, 64–75. DOI: 10.1016/j.jaridenv.2015.06.002

Simou, M. R., Houran, N., Benayad, M., Maanan, M., Loulad, S., & Rhinane, H. (2024). Monitoring forest degradation over four decades using remote sensing and machine learning classification algorithms in bouskoura, morocco. *The International Archives of the Photogrammetry, Remote Sensing and Spatial Information Sciences, XLVIII-4-W9-2024*, 337–342. GeoAdvances 2024 – 8th International Conference on GeoInformation Advances - 11–12 January 2024, Istanbul, Türkiye. DOI: 10.5194/isprs-archives-XLVIII-4-W9-2024-337-2024

Sims, D. A., & Gamon, J. A. (2002). Relationships between Leaf Pigment Content and Spectral Reflectance Across a Wide Range of Species, Leaf Structures and Developmental Stages. *Remote Sensing of Environment, 81*(2-3), 337–354. DOI: 10.1016/S0034-4257(02)00010-X

Sinaga, K. P., & Yang, M. S. (2020). Unsupervised K-Means Clustering Algorithm. *IEEE Access : Practical Innovations, Open Solutions, 8*, 80716–80727. DOI: 10.1109/ACCESS.2020.2988796

Slater, P. N., Biggar, S. F., Holm, R. G., Jackson, R. D., Mao, Y., Moran, M. S., Palmer, J. M., & Yuan, B. (1987). Reflectance- and radiance-based methods for the in-flight absolute calibration of multispectral sensors. *Remote Sensing of Environment, 22*(1), 11–37. DOI: 10.1016/0034-4257(87)90026-5

Smith, P., Adams, J., Beerling, D. J., Beringer, T., Calvin, K. V., & Fuss, S.. (2019). Land management options. *Annual Review of Environment and Resources, 7*, 40.

Solano, F., Mansi, C., Baliva, M., Canestrelli, D., Chiarucci, A., Manicone, R., Maugeri, M., Modica, G., Palli, J., Quilghini, G., & Piovesan, G. (2025). Mediterranean strictly protected forests are cooler. *Agricultural and Forest Meteorology, 375*, 110858. DOI: 10.1016/j.agrformet.2025.110858

Sombroek, W. G., Nachtergaele, F. O., & Hebel, A. (1993). Amounts, Dynamics, Sequestering of Carbon in Tropical ans Subtropical Soils. AMBIO. V(22). N°7. 417-426pp

Song, C., Woodcock, C. E., Seto, K. C., Lenney, M. P., & Macomber, S. A. (2001). Classification and change detection using Landsat TM data: When and how to correct atmospheric effects. *Remote Sensing of Environment, 75*(2), 230–244. DOI: 10.1016/S0034-4257(00)00169-3

Spruce, J., Sader, S., Ryan, R., Smoot, J., Kuper, P., Ross, K., Prados, D., Russell, J., Gasser, G., McKellip, R., & Hargrove, W. (2011). Assessment of MODIS NDVI time series data products for detecting forest defoliation by gypsy moth outbreaks. *Remote Sensing of Environment, 115*(2), 427–437. DOI: 10.1016/j.rse.2010.09.013

Stone, C. D. (1974). *Should trees have standing? Toward legal rights for natural objects*. Kaufmann. DOI: 10.2307/1228012

Su, L., Heydari, M., Mousavi, S. R., & Prévosto, B. (2024). Role of environmental and stand factors on forest dieback: An approach using structural equation modelling and machine learning. *Forest ecology and Management*, 562, 121927, ISSN 0378-1127.

Supiot, A. (2015). *La gouvernance par les nombres*. Inst. d'Études Avancées de Nantes. [u.a.]

TEEB. (2010). *The Economics of Ecosystems and Biodiversity: Ecological and Economic Foundations* (Kumar, P., Ed.). Earthscan.

Teillet, P. M., & Santer, R. (1991). Terrain Elevation and Sensor Altitude Dependence in a Semi-Analytical Atmospheric Code. *Canadian Journal of Remote Sensing, 17*(1), 36–44.

Terrab, A., Talavera, S., Arista, M., Paun, O., Stuessy, T. F., & Tremetsberger, K. (2006). *Cedrus atlantica* (Endl.) Manetti ex Carrière: Biogeography and genetic differentiation across the Moroccan Atlas Mountains. *Plant Ecology, 187*, 133–147.

Thaiki, M., Bounoua, L., & Dekkaki, H. C. (2023). Using Satellite Data to Characterize Land Surface Processes in Morocco. *Remote Sensing (Basel), 15*(22), 5389. https://ntrs.nasa.gov/citations/20230017899. DOI: 10.3390/rs15225389

The IUCN Red List of Threatened Species. (n.d.). IUCN Red List of Threatened Species. Retrieved October 31, 2025, from https://www.iucnredlist.org/en

Thenkabail, P. S., Lyon, J. G., & Huete, A. R. (2018). Advanced Applications in Remote Sensing of Agricultural Crops And Natural Vegetation, 2nd Edition, vol. 4. CRC Press - Taylor and Francis Group, 425 pages. DOI: 10.1201/9780429431166

Thiéry, A. (1987). Les mares temporaires: Écologie et biogéographie. *Annales de Limnologie*, *23*, 91–105.

Thiéry, A. (1997). Temporary pools: Ecological and biogeographical aspects. *Hydrobiologia*, *345*, 1–9.

Thomas, S. C., & Martin, A. R. (2012). Carbon Content of Tree Tissues: A Synthesis. *Forests*, *3*(2), 332–352. DOI: 10.3390/f3020332

Thome, K. J. (2001). Absolute radiometric calibration of Landsat-7 ETM+ using the reflectance-based method. *Remote Sensing of Environment*, *78*(1-2), 27–38. DOI: 10.1016/S0034-4257(01)00247-4

Thome, K. J., Helder, D. L., Aaron, D., & Dewald, J. D. (2004). Landsat-5 TM and Landsat-7 ETM+ Absolute Radiometric Calibration Using the Reflectance-Based Method. *IEEE Transactions on Geoscience and Remote Sensing*, *42*(12), 2777–2785. DOI: 10.1109/TGRS.2004.839085

Thomson, P. H. (1977). The carob in California. *Avocado Grower*, *1*(2), 36–39.

Thuiller, W., Lavorel, S., Araujo, M.-B., Sykes, M. T., & Prentice, I. C. (2005). Climate change threats to plant diversity in Europe. *Proceedings of the National Academy of Sciences of the United States of America*, *102*(23), 8245–8250. DOI: 10.1073/pnas.0409902102

Tian, L., Wu, X., Tao, Y., Li, M., Qian, C., Liao, L., & Fu, W. (2023). Review of Remote Sensing-Based Methods for Forest Aboveground Biomass Estimation: Progress, Challenges, and Prospects. *Forests*, *14*(6), 1086. DOI: 10.3390/f14061086

Tilly, N., Reddig, F., Lussem, U., & Bareth, G. (2020). First Investigation of Mediterranean Oak Tree Vitality with High Resolution WorldView-3 Satellite data: Ccomparing Ten Vegetation Indices and Three Machine Learning Classifiers. *The Int. Archives of Photogrammetry, Remote Sensing and Spatial Information Sciences*, vol. XLIII-B3-2020, pp. 1069-1076.

Torresan, C., Berton, A., Carotenuto, F., Filippo, S., Gennaro, D., Gioli, B., Matese, A., Miglietta, F., Vagnoli, C., Zaldei, A., & Wallace, L. (2017). Forestry Applications of UAVs in Europe: A Review. *International Journal of Remote Sensing*, *38*(8-10), 2427–2447. DOI: 10.1080/01431161.2016.1252477

Tous, J., Romero, A., Plana, J., & Batlle, I. (1996). Current situation of carob plant material. *In Proceedings of the Third International Carob Symposium* (Tavira, Portugal) University of Lisbon, Portugal.

Tuan, Y.-F. (2011). *Space and place : The perspective of experience* (7. print). Univ. of Minnesota Press.

UN. (2017). *Adoption du cs forets*. https://press.un.org/fr/201/ag11977.doc.htm

UNCCD. (2022). Global Land Outlook: Second edition. United Nations Convention to Combat Desertification. Bonn, Germany.

United Nations. (1998). Kyoto Protocol to the United Nations Framework Convention on Climate Change.

United Nations. (2015). Paris Agreement, 27

Ustin, S. L., Martens, S. N., Curtiss, B., & Vanderbilt, V. (1988). Use of high spectral resolution sensors to detect air pollution injury in conifer forests. In Fenstermaker, L.K. (Ed.), Remote Sensing Applications for Acid Deposition (pp. 72-85). Environmental Protection Agency (EPA), Las Vegas, NV, USA.

Valavanidis, A., & Vlachogianni, T. (2011). Ecosystems and biodiversity hotspots in the Mediterranean basin threats and conservation efforts. *Sci Adv Environ Toxicol Ecotoxicol*, *10*, 1–24.

Van den Broeck, M. (2016). *Monitoring and conservation of Mediterranean temporary ponds*. PhD thesis, Catholic University of Leuven, Belgium, 181 p.

Van Engelen, V. W. P., Batjes, N. H., Dijkshoorn, J. A., & Huting, J. R. M. (2005). *Harmonized Global Soil Resources Database (Final Report)*. (ISRIC Report 2005/06). ISRIC. https://edepot.wur.nl/46477

Van Kooten, G. C., Binkley, C. S., & Delcourt, G. (1995). Carbon taxes & forest rotations. *American Journal of Agricultural Economics*, *77*(2), 365–374. DOI: 10.2307/1243546

Van-Beek, J., Tits, L., Somers, B., Deckers, T., Janssens, P., & Coppin, P. (2016). Viewing Geometry Sensitivity of Commonly Used Vegetation Indices towards the Estimation of Biophysical Variables in Orchards. *Journal of Imaging*, *2*, 15. DOI: 10.1016/j.foreco.2021.119493

Vidalou, J.-B. (2017). *Être forêts : Habiter des territoires en lutte*. Zones.

Vieira, J., Nabais, C., & Campelo, F. (2022). Dry and hot years drive growth decline of *Pinus halepensis* at its southern range limit in the Moroccan High Atlas Mountains. *Trees (Berlin)*, *36*(5), 1585–1595. DOI: 10.1007/s00468-022-02314-z

Viera, M., & Rodríguez-Soalleiro, R. (2019). A Complete Assessment of Carbon Stocks in Above and Belowground Biomass Components of a Hybrid *Eucalyptus* Plantation in Southern Brazil. *Forests*, *10*(7), 536. DOI: 10.3390/f10070536

Vollenweider, P., & Günthardt-Goerg, M. S. (2005). Diagnosis of abiotic and biotic stress factors using the visible symptoms in foliage. *Environmental Pollution*, *137*(3), 455–465. DOI: 10.1016/j.envpol.2005.01.032

Walkley, A., & Black, I. A. (1934). An examination of the Degtjareff method for determining soil organic matter, and a proposed modification of the chromic acid titration method. *Soil Science*, *37*(1), 29–38. DOI: 10.1097/00010694-193401000-00003

Wang, C., & Myint, S. W. (2015). A Simplified Empirical Line Method of Radiometric Calibration for Small UAS-Based Remote Sensing. *IEEE Journal of Selected Topics in Applied Earth Observations and Remote Sensing*, *8*(5), 1876–1885. DOI: 10.1109/JSTARS.2015.2422716

Weiskopf, S. R., Isbell, F., Arce-Plata, M. I., Di Marco, M., Harfoot, M., Johnson, J., Lerman, S. B., Miller, B. W., Morelli, T. L., Mori, A. S., Weng, E., & Ferrier, S. (2024). Biodiversity loss reduces global terrestrial carbon storage. *Nature Communications*, *15*(1), 4354. DOI: 10.1038/s41467-024-47872-7

White, J. C., Wulder, M. A., Hermosilla, T., Coops, N. C., & Hobart, G. W. (2017). A nationwide annual characterization of 25 years of forest disturbance and recovery for Canada using Landsat time series. *Remote Sensing of Environment*, *194*, 303–321. DOI: 10.1016/j.rse.2017.03.035

Williams, A. P., Abatzoglou, J. T., Gershunov, A., Guzman-Morales, J., Bishop, D. A., & Balch, J. K.. (2019). Anthropogenic climate change & wildfires in California. *Earth's Future*, *7*(8), 892–910. DOI: 10.1029/2019EF001210

Williams, D. D. (2006). *The Biology of temporary waters*. Oxford Biology.

World Bank. (1995). Une stratégie de développement des parcours en zones arides et semi-arides. Rapport technique Maroc. 82 p.

World Bank. (2023). Notes sur les forêts algériennes- Gestion durable des forêts pour lutter contre les incendies de forêts. 112 P.

Wulder, M. A., Loveland, T. R., Roy, D. P., Crawford, C. J., Masek, J. G., Woodcock, C. E., Allen, R. G., Anderson, M. C., Belward, A. S., Cohen, W. B., Dwyer, J., Erb, A., Gao, F., Griffiths, P., Helder, D., Hermosilla, T., Hipple, J. D., Hostert, P., Hughes, M. J., & Zhu, Z. (2019). Current status of Landsat program, science, and applications. *Remote Sensing of Environment*, *225*, 127–147. DOI: 10.1016/j.rse.2019.02.015

Wulder, M., White, J., Bentz, B., & Ebata, T. (2006). Augmenting the existing survey hierarchy for mountain pine beetle redattack damage with satellite remotely sensed data. *Forestry Chronicle*, *82*(2), 187–202. DOI: 10.5558/tfc82187-2

Wunder, S. (2005). *Payments for environmental services: Some nuts and bolts*. CIFOR.

Yao, H., Qin, R., & Chen, X. (2019). Unmanned Aerial Vehicle for Remote Sensing Applications: A Review. *Remote Sensing (Basel)*, *11*(12), 1443. DOI: 10.3390/rs11121443

Ye, H., Huang, W., Huang, S., Cui, B., Dong, Y., Guo, A., Ren, Y., & Jin, Y. (2020). Recognition of Banana Fusarium Wilt Based on UAV Remote Sensing. *Remote Sensing (Basel)*, *2020*(12), 938. DOI: 10.3390/rs12060938

Yoder, B. J., & Waring, R. H. (1994). The Normalized Difference Vegetation Index of Small Douglas–Fir Canopies with Varying Chlorophyll Concentrations. *Remote Sensing of Environment*, *49*(1), 81–91. DOI: 10.1016/0034-4257(94)90061-2

Yu, R., Luo, Y., Zhou, Q., Zhang, X., Wu, D., & Ren, L. (2021). Early detection of pine wilt disease using deep learning algorithms and UAV-based multispectral imagery. *Forest Ecology and Management*, *497*, 119493.

Zacharias, I., & Zamparas, M. (2010). Mediterranean temporary ponds: A disappearing ecosystem. *Biodiversity and Conservation*, *19*(14), 3827–3834. DOI: 10.1007/s10531-010-9933-7

Zadbagher, E., Marangoz, A. M., & Becek, K. (2023). Characterizing and estimating forest structure using active remote sensing: An overview. *Advances in Remote Sensing*, *3*(1), 38–46.

Zaher, H., Rhoufacha, M., Dahmani, H. (2025). Carbon stock dynamics along a cork oak stand chronosequence: implications for climate-smart forest management. *Journal of the Saudi Society of Agricultural Sciences*. 24-45 pp. DOI: 10.1007/s44447-025-00054-3

Zaher, H., Sabir, M., Benjelloun, H., & Hounzandji, P. I. (2020). Effect of forest land use change on carbohydrates, physical soil quality and carbon stocks in Moroccan cedar area. *Journal of Environmental Management, 254*, 109544. Advance online publication. DOI: 10.1016/j.jenvman.2019.109544

Zedler, P. H. (1987). The ecology of Southern California vernal ponds: A community profile. *US Fish and Wildlife Service Biological Report., 85*, 7–11.

Zhang, C.. (2019). Effects of thinning on carbon. *Journal of Forestry Research.*

Zhang, C., & Kovacs, J. M. (2012). The application of small UAS for precision agriculture: A review. *Precision Agriculture, 13*, 693–712. DOI: 10.1007/s11119-012-9274-5

Zhang, H., Song, T., Wang, K., Du, H., Yue, Y., Wang, G., & Zeng, F. (2014). Biomass and Carbon Storage in an Age-Sequence of Cyclobalanopsis Glauca Plantations in Southwest China. *Ecological Engineering, 73*, 184–191. DOI: 10.1016/j.ecoleng.2014.09.008

Zhang, L., Zhang, Z., Guo, L., Cui, X., Butnor, J. R., Li, S., Cao, X., & Chen, X. (2024). A simple method for estimating the coarse lateral root biomass of shrubs using ground-penetrating radar: Validation by Caragana microphylla Lam. in Inner Mongolia. *The Science of the Total Environment, 919*, 170897. DOI: 10.1016/j.scitotenv.2024.170897 PMID: 38346659

Zhang, P., Cui, Z., Liu, X., & Xu, D. (2022). Above-Ground biomass and nutrient accumulation in ten *eucalyptus* clones in Leizhou peninsula, southern China. *Forests, 13*(4), 530. DOI: 10.3390/f13040530

Zhang, Z., Liu, M., Liu, X., & Zhou, G. (2018). A New Vegetation Index Based on Multitemporal Sentinel-2 Images for Discriminating Heavy Metal Stress Levels in Rice. *Sensors (Basel), 18*(7), 2172. DOI: 10.3390/s18072172

Zribi, L., Chaar, H., Khaldi, A., Hanchi, B., Mouillot, F., & Gharbi, F. (2016). Estimate of biomass and carbon pools in disturbed and undisturbed oak forests in Tunisia. *Forest Systems, 25*(2), e060. DOI: 10.5424/fs/2016252-08062

About the Contributors

Said Moukrim is an agricultural engineer and researcher specializing in ecology and natural resource management. He holds a PhD in Biological Sciences from Mohammed V University in Rabat, focusing on the impact of climate change on Moroccan forest ecosystems. His research applies species distribution modeling to assess ecological responses of key forest species. Currently a professor at Mohammed V University in Rabat, Dr. Moukrim previously served as Head of the Forest Rangeland and Silvopastoral Service at the Moroccan Department of Water and Forests. He has contributed to national and international studies on forest resilience. His work spans climate change adaptation, forest restoration, and land degradation. He has published in peer-reviewed journals and presented at national and international conferences. He also trains researchers and officials in forest management and climate resilience.

Saïd Lahssini is a professor of higher education specializing in forest management and inventory. He holds a PhD in Forest Management from the Université catholique de Louvain, Belgium, and has nearly 24 years of experience in forestry research, digitalization, and training engineering. Currently, he teaches at the National School of Forestry Engineers (ENFI) in Morocco, where he leads courses on forest inventory, afforestation, growth modeling, and forest productivity. He has also contributed as a consultant for organizations such as FAO, GIZ, and UNDP, focusing on ecosystem restoration, biodiversity mapping, and climate change adaptation. His expertise includes geospatial technologies, data science applications in forestry, and the integration of digital tools for environmental monitoring. Dr. Lahssini has authored numerous scientific publications and actively participates in international conferences on forestry and environmental sustainability.

***

Younes Abbas is a Professor at the Faculty of Polydisciplinary Studies of Beni Mellal, part of Sultan Moulay Slimane University. He currently serves as Head of the Department of Biology–Geology and previously acted as coordinator of academic programs within the department. Dr. ABBAS holds a PhD in Plant Ecophysiology and has more than 17 years of research experience, particularly through his work in collaboration with the Forest Research Center. His scientific interests focus on plant sciences and ecology, with a strong emphasis on forest ecosystems. His research is centered on plant ecophysiology and plant–microorganism interactions, with recent integration of advanced approaches such as metabolomics and genomics to better understand plant responses to environmental constraints. His work contributes to elucidating adaptive mechanisms in plants, particularly in Mediterranean and semi-arid forest ecosystems. Professor ABBAS has been involved in several national and international collaborative research projects, including FRQ–CNRST, PRAD, and ANPMA programs. He maintains active scientific collaborations with leading institutions such as Laval University (Canada), the Texas Biotechnology Center (USA), Valorhiz (France), and King Abdullah University of Science and Technology (Saudi Arabia). He is the author of more than 100 peer-reviewed scientific publications in international journals. In addition to his research activities, he actively contributes to academic leadership, graduate supervision, and the development of interdisciplinary research programs addressing key environmental and socio-economic challenges.

Mohamed Ankri is a Water and Forest Engineer at the National Agency for Water and Forests (ANEF) in Morocco. He obtained his engineering degree from the National School of Forest Engineers in Salé, with a specialization in forest management.

Ali Azedou, PhD is a researcher specializing in geospatial technologies and forest ecosystem management at National School of Forestry Engineering (ENFI). His work focuses on integrating remote sensing, artificial intelligence, and participatory governance frameworks to address climate change impacts on Mediterranean forest ecosystems, with particular emphasis on Morocco's digital forestry transition.

Abderrazak Bannari received a B.Sc. in surveying engineering (Hassan-II Agronomic and Veterinary Institute, Rabat, Morocco) and a M.Sc. in Geomatic and remote sensing sciences (Laval University, Québec, Canada), and a Ph.D. in remote sensing form Sherbrooke University (Quebec, Canada). He joined SPAR Aerospace Limited in Montreal (Quebec, Canada) as a researcher where he worked on the conceptualization of small satellite sensors. He was a visiting scientist at

JPL-NASA (Pasadena, U.S.A.), the University of Arizona (Tucson, U.S.A.), and the Canada Center for Remote Sensing (Ottawa, Canada). After his retirement as a remote sensing and GIS professor from the University of Ottawa (Ontario, Canada), he was acting as a chair of the Department of Geo-informatics at the Arabian Gulf University, Kingdom of Bahrain. He is the author of 2 books, 10 book chapters, and more than 265 articles, and he supervised more than 64 graduate students. Currently, he is running a private company "Space Pix-Map International Inc." for international scientific consultations in remote sensing, Geomatics, and environmental applications (Ottawa, Canada).

Aissa Chkhichekh. A trained forestry engineer, he grew up in an isolated forestry post surrounded by thuya stands, an environment that profoundly shaped his early connection to nature and forest ecosystems. A graduate of the National Forestry Engineering School of Salé, he began his career in forest management, then moved into forest and watershed management studies, before transitioning to technical and academic teaching in forestry. In 2015, he earned a doctorate in ecology. He has contributed to numerous management studies covering all of Morocco's forest ecosystems and is the author and co-author of many scientific articles. He continues his commitment to the development of forest ecosystems as an independent expert and part-time lecturer at the National Forestry Engineering School (ENFI) at the Faculty of Sciences in Rabat and at the Royal school for Forest (IRTSEF), and provides continuing education for researchers and officials in forest management.

Hamed Daly Hassen is a Full Professor of Higher Education and Senior Researcher with more than three decades of experience in forest and natural resource economics. He earned his Ph.D. in Forest Economics from AgroParisTech–ENGREF in 1995. His career has been dedicated to linking scientific research with public policy, with the aim of fostering sustainable resource management in Tunisia and across the Mediterranean region. In addition to his academic work, he has actively contributed to international initiatives such as IPBES, COST Actions, UNEP, and various cooperation projects, further promoting sustainability at both regional and global levels. He also brings significant leadership experience, having served as Director-General of the National Observatory of Agriculture (ONAGRI) and as CEO of the National Oil Office (ONH), where he oversaw key strategic developments in Tunisia's agricultural sector.

Saliha Diarra is a forest engineer and a second-year PhD candidate in Ecology and Natural Ecosystems Management. His research focuses on ecological modelling and biogeography, particularly on understanding the impacts of climate change on biodiversity. He is particularly interested in integrating advanced remote sensing and modelling approaches to support the conservation and sustainable management of ecosystems. He is also engaged in exploring nature-based solutions for climate change mitigation and adaptation, including carbon sequestration and accounting in forests and agroecosystems. Through his research, he aims to address pressing environmental challenges and inform evidence-based decision-making in biodiversity conservation and climate resilience.

Abderrahmene El-Ghmari obtained a M.Sc. and Ph.D. degrees, respectively, from the Institute Polytechnic of Lorraine (France) and the University of Strasbourg (France). As well as, he obtained the scientific accreditation (habilitation) recognizing his aptitude to manage and supervise scientific research activities from the University of Avignon and the Pays de Vaucluse (France). His research interests are covering the management of forest cover dieback due to acid rain, as well as the risk assessment of carob forests. He joined the Sultan Moulay Slimane University (Béni-Mellal, Morocco) since 1996 where he developed new programs on the applications of geomatics to environmental impact assessment, hydrology and water resource management, and climate change. In addition to his administrative responsibilities as Vice-Dean of research, he supervised more than 25 M.Sc. and Ph.D. students and published around 50 scientific articles. Prof. EL-Ghmari is involved in several regional, national and international associations for the development and protection of Carob forests.

Mohamed El Mderssa is an Associate Professor at the Polydisciplinary Faculty of Beni Mellal, Sultan Moulay Slimane University, Morocco. He holds a PhD in ecology and environmental sciences and is a trained forest engineer. He is actively involved in higher education, research supervision, and participation in academic juries and scientific committees. His research focuses on forest ecology, soil science, carbon sequestration, and sustainable natural resource management. His work emphasizes the integration of field data, remote sensing, and artificial intelligence approaches to assess ecosystem services and support climate change adaptation and mitigation strategies in Mediterranean forest ecosystems. He maintains scientific collaborations with various national institutions as well as international academic partners within the framework of research projects focused on the sustainable management of forest resources.

Abdellatif Khattabi is an agronomic engineer from IAV Hassan II and Ecole Supérieure du Bois, Paris (1981), he holds an MS in Economics (1988), a PhD in Forestry (1992) from the University of Idaho, and an MS in ICT from Louis Pasteur University (2003). Full Professor at ENFI (1994–2024), he lectures at five Moroccan universities and is an Affiliated Researcher at the Royal Institute for Strategic Studies.

Before 1994, he spent 13 years as a scientist at the National Center for Forestry Research. Specializing in integrated natural resources management, climate change adaptation, and rural development, he has authored numerous scientific publications and reports. A Lead Author for the IPCC Fifth Assessment Report, Fulbright Alumni, and President of the Regional Science Association of Morocco, he has consulted for international organizations including the World Bank, UNESCO, UNDP, FAO, GIZ, ...He is a member of several professional associations and an IZA Research Fellow.

Najib Magri currently works at the Centre National de la Recherche Forestière (Département des Eaux et Forêts). Najib does research in Forestry, Geoinformatics (GIS), Space Remote Sensing and Information Systems (Business Informatics). His current projects are Fauna and Flora Biodiversity, Bryophytes, Watershed Management, Carbon sequestration...

Kamal Menzou is a teacher and researcher specializing in natural resource governance and environmental public policies. He holds a doctorate in Spatial Planning and Territorial Development from the National Institute of Spatial Planning and Urbanism in Rabat and a doctorate in Political Science from Mohammed V University in Rabat. Dr. Menzou has accumulated over 25 years of experience in the territorial management of forest areas. He also contributed to the management of international forestry programs with the EU, GIZ, AFD, FAO, etc., when he was head of the international cooperation department within the Moroccan forestry administration. He has contributed to studies on the management of forest areas in several regions of Morocco. He also has over 15 years of experience in teaching and supervising scientific research and served as President of the Third International Colloquium of Political Science Researchers in Rabat in 2014. Currently, he teaches at the National School of Forestry Engineers (ENFI) and is the author and co-author of numerous scientific articles in national and international journals. He is also an active member of CEMED (Center of Experts 'Mastery' for Studies and Development), contributing to research and development initiatives.

Mohammed El Madihi is an Associate Professor at the Faculty of Sciences, Chouaib Doukkali University (El Jadida, Morocco). He teaches general ecology and plant ecology at undergraduate and graduate levels. His research focuses on climate change impacts on ecosystems, with particular emphasis on Mediterranean wetlands, plant biodiversity, forest ecosystems, and temporary ponds. His work integrates ecological approaches to understand the structure, functioning, and resilience of plant communities under environmental stress. He has contributed to several national and international research projects related to biodiversity conservation and sustainable ecosystem management in Mediterranean environments.

Mustapha Naggar, a forestry engineer who worked at the Moroccan Ministry of Forestry from 1985 to 2018, an expert specializing in the collaborative planning and management of forest and silvopastoral resources, and former Head of the Forest Planning Division. Author and co-author of several scientific articles and books on collaborative and sustainable management.

Laila Rhazi is a professor at Mohammed V University of Rabat (Morocco), specializing in plant ecology and wetland biodiversity. Her research focuses on the structure, diversity, and dynamics of plant communities in Mediterranean temporary ponds under global change, including climate and anthropogenic pressures. She has developed extensive international collaborations on temporary pond ecosystems with partners across Europe (France, Belgium, Portugal, Spain, Italy, Finland), Africa (Tunisia, Algeria, Mauritania), and North America (USA, Canada). Her work contributes to advancing knowledge on wetland functioning and supports biodiversity conservation strategies in vulnerable ecosystems.

Ahmed Selouani received a technician diploma in surveying, B.Sc. and M.Sc. degrees in geomatics and remote sensing from the University Sultan Moulay Slimane (Béni-Mellal, Morocco). He is the founder and President of Geofly-Morocco, a company specialized in drones and information technologies applications in the fields of geomatics, surveying and civil engineering, precision agriculture, forestry, etc.. Prior to that, he spent more than 20 years in cadastral surveying, geodesy, aerial photogrammetry, and cartography services.

Rajaa Timzioura is a second-year PhD candidate specializing in climate change impacts on forest ecosystems. Her research examines the vulnerability of key Moroccan forest species to environmental stressors and focuses on the development of advanced, science-based adaptation strategies. She employs spatial modeling, species distribution modeling, and selective reforestation approaches to assess and mitigate forest decline. Her work contributes to strengthening the effectiveness and long-term sustainability of national reforestation and climate adaptation programs in Morocco.

Riccardo Valentini is a Full Professor of Forest Ecology at the University of Tuscia. He was lead author of the third and fifth reports of the IPCC (Intergovernmental Panel on Climate Change) on climate impacts and adaptation and author of the IPCC special report "Land and Climate". Member of the European Academy of Sciences and Medal of the Academy of Sciences of the "XL" for Physics and Natural Sciences. He dedicated his research to understand forest role on global biogeochemical cycles with a special focus on Mediterranean forests.

Index

A

B

C

D

E

F

G

U

www.ingramcontent.com/pod-product-compliance
Lightning Source LLC
LaVergne TN
LVHW080906300926
849162LV00003B/569
* 9 7 9 8 3 3 7 3 6 2 3 9 7 *